Replacement, Reduction and Refinement of Animal Experiments in the Development and Control of Biological Products

Developments in Biological Standardization

Vol. 86

This series Dev Biol Stand begins with Vol. 23 and is the continuation of both «PROGRESS in Immunobiological Standardization, Vols 1-5» and «SYMPOSIA SERIES in Immunobiological Standardization, Vols 1-22».

Basel · Freiburg · Paris · London · New York ·
New Delhi · Bangkok · Singapore · Tokyo · Sydney

Replacement, Reduction and Refinement of Animal Experiments in the Development and Control of Biological Products

Langen, Germany
November 2-4, 1994

Volume Editors

Fred Brown
Plum Island Animal Disease Center
Greenport NY, USA

K. Cussler
Paul-Ehrlich-Institute
Langen, Germany

C. Hendriksen
National Institute of Public Health
and Environmental Protection
Bilthoven, The Netherlands

Proceedings of a Symposium organized by
the International Association of Biological Standardization (IABS)

in cooperation with

the European Centre for the Validation of Alternative Methods (ECVAM), Ispra (I),

the Paul-Ehrlich-Intitute (PEI), Langen (D) and

the National Institute of Public Health and Environmental Protection (RIVM), Bilthoven (NL)

70 figures and 78 tables, 1996

Basel · Freiburg · Paris · London · New York ·
New Delhi · Bangkok · Singapore · Tokyo · Sydney

Developments in Biological Standardization

Bibliographic Indices. This publication is listed in bibliographic services, including Current Contents® and Index Medicus.

Drug dosage. The authors and the publisher have exerted every effort to ensure that drug selection and dosage set forth in this text are in accord with current recommendations and practice at the time of publication. However, in view of ongoing research, changes in government regulations, and the constant flow of information relating to drug therapy and drug reactions, the reader is urged to check the package insert for each drug for any change in indications and dosage and for added warnings and precautions. This is particularly important when the recommended agent is a new and/or infrequently employed drug.

Printed in Switzerland by Médecine et Hygiène, Genève
ISBN 3-8055-6260-8

Scientific Committee

M. Balls, Ispra, Italy
P. Castle, Strasbourg, France
R. Dabbah, Rockville, USA
H. v.d. Donk, Bilthoven, The Netherlands
P. Knight, Sevenoaks, UK
J.G. Kreeftenberg, Willowdale, Canada
J. Milstien, Geneva, Switzerland
M. Moos, Langen, Germany
J.M. Spieser, Strasbourg, France
P. Vannier, Ploufragan, France

Programme of the Symposium

Session I: *Laboratory Animals and the Concept of Alternatives*
Chaired by: D. Straughan (Nottingham, UK)
K. Cussler (Langen, Germany)

Session II: *Testing, Regulations and Three Rs Policy*
Chaired by: F. Horaud (Paris, France)
R. Dabbah (Rockville, USA)

Session III: *General Aspects*
Chaired by: R. Winsnes (Oslo, Norway)
C. Folkers (Brussels, Belgium)

Session IV: *Specific Toxicity, Neurovirulence and Other Safety Tests*
Chaired by: E. Fitzgerald (Rockville, USA)
M. Moos (Langen, Germany)

Session V: *Viral Vaccines*
Chaired by: A. Osterhaus (Rotterdam, The Nertherlands)
L. Bruckner (Mittelhäusern, Switzerland)

Session VI: *New Developments in Vaccine Production and Quality Control*
Chaired by: F. Brown (New York, USA)
C. Hendriksen (Bilthoven, The Nertherlands)

Session VII: *Toxoid Vaccines*
Chaired by: J. Milstien (Geneva, Switzerland)
P. Castle (Strasbourg, France)

Workshop I: *Serological Methods and Cell Cultures*
Chaired by: P. Vannier (Ploufragan, France)
R. Gupta (Boston, USA)

Workshop II: *Statistics and Validation*
Chaired by: B. Schneider (Hannover, Germany)
M. Balls (Ispra, Italy)

Session VIII: *Bacterial Vaccines*
Chaired by: S. Houghton (Milton Keynes, UK)
E. Relyveld (Marnes la Coquette, France)

Session IX: *Therapeutic Toxins and Monoclonal Antibodies*
Chaired by: R. Dabbah (Rockville, USA)
H. van de Donk (Bilthoven, The Netherlands)

Poster Session

The Organizing Committee wishes to record gratitude to the following companies, the institutes and organizations who have generously sponsored the symposium.

Companies

Boehringer Ingelheim Vetmedica GmbH, Ingelheim, Germany

Connaught Lab. Ldt., North York, Canada

Hoechst Veterinär GmbH, Wiesbaden, Germany

Impfstoffwerk Dessau-Tornau GmbH, Rossiau, Germany

Intervet International B.V., Boxmeer, The Nertherlands

Sächsisches Serumwerk GmbH, Dresden, Germany

Solvay Duphar BV, Weesp, The Nertherlands

Organizations / Institutes

European Centre for the Validation of Alternative Methods (ECVAM),
Ispra, Italy

European Federation of Animal Health (FEDESA),
Brussels, Belgium

Foundation for the Promotion of Research on Replacement and Complementary
Methods to Reduce Animal Testing,
Mainz, Germany

National Institute of Public Health and Environmental Protection (RIVM),
Centre for Alternatives,
Bilthoven, The Nertherlands

Paul-Ehrlich-Institute,
Langen, Germany

Contents

Session I

Laboratory Animals and the Concept of Alternatives 1
Chairmen: *D. Straughan, K. Cussler*

A Short History of the Use of Animals in Vaccine Development and Quality Control 3
C.F.M. Hendriksen

The Three Rs of Russel & Burch and the Testing of Biological Products 11
M. Balls, D.W. Straughan

Session II

Testing, Regulations and Three Rs Policy 19
Chairmen: *F. Horaud, R. Dabbah*

Alternatives to Animal Testing: Achievements and Recent Developments in the European Pharmacopoeia 21
P. Castle

WHO Activities Towards the Three Rs in the Development and Control of Biological Products 31
J. Milstien, V. Grachev, A. Padilla, E. Griffiths

USDA: Progress Toward in Vitro Tests and Other Trends 41
S.A. Goodman

Alternatives to the Use of Animals in R & D and Quality Control of Veterinary Vaccines: An Industry View 49
C. Verschueren, S. Zänker

The Views and Policy of the Japanese Control Authorities on the three Rs 53
F. Chino

Session III

General Aspects 63
Chairmen: *R. Winsnes, C. Folkers*

Results and Recommendations of an ECVAM Workshop on «Alternatives to Animal Testing in the Quality Control of Immunobiologicals», Utrecht 16-17.4.1994 65
D.W. Straughan

The Five Rs: Refinement, Reduction, Replacement. A Regulatory Revolution 67
R.N. Lucken

Replacement, Reduction or Refinement of Animal Use in the Quality Control of Veterinary Vaccines: Development, Validation and Implementation 73
M.D.O. van der Kamp

Session IV

Specific Toxicity, Neurovirulence and Other Safety Tests 77
Chairmen: *E. Fitzgerald, M. Moos*

Evaluation of New Approaches to Poliovirus Vaccine Neurovirulence Tests 79
D.J. Wood

Testing of Immunomodulatory Properties in Vitro 85
T. Hartung, A. Sauer, A. Wendel

Reducing the Use of the Target Animal Batch Safety Test for Veterinary Vaccines 97
B. Roberts, R.N. Lucken

The Reduction of the Use of Challenge Testing to Provide Evidence of Efficacy in Tests of Immunological Veterinary Medical Products (IVMPs) 103
C.J. Webster

Session V

Viral Vaccines 111
Chairmen: *A. Osterhaus, L. Bruckner*

Use of Animals in the Development and Control of Viral Vaccines 113
P.D. Minor

Rat Immunogenicity Assay of Inactivated Poliovirus 121
J.M. Bevilacqua, L. Young, S.W. Chiu, J.D. Sparkes, J.G. Kreeftenberg

Hepatitis A-Vaccines: a Comparison Between Three Methods of Antigen Determination ... 129
F. Burkhardt, R. Glück, B. Finkel-Jimenez, S. Brantschen

Development and Evaluation of Alternative Testing Methods for the in vivo NIH Potency Test Used for the Quality Control of Inactivated Rabies Vaccines 137
E. Rooijakkers, J.Groen, J. Uittenbogaard, J. van Herwijnen, A. Osterhaus

Serological Responses in Calves to Vaccines against Bovine Respiratory Syncytial, Infectious Bovine Rhinotracheitis, Bovine Viral Diarrhoea and Parainfluenza -3 Viruses 147
M. Tollis, L. Di Trani, P. Cordioli, E. Vignolo, I. Di Pasquale

Replacement of Challenge Procedures in the Evaluation of Poultry Vaccines 157
C. Jungbäck, H. Finkler

Session VI

New Developments in Vaccine Production and Quality Control 165
Chairmen: *F. Brown, C. Hendriksen*

An in Vitro Assay for Acute Pathogenicity of Immunodeficiency Viruses 167
M.T. Dittmar, S. Wagener, P. Fultz, R. Kurth, K. Cichutek

Detection of Extraneous Agents in Vaccines Using the Polymerase Chain Reaction for Newcastle Disease Virus in Poultry Biologicals 175
L. Bruckner, N. Stäuber, K. Brechtbühl, M.A. Hofmann

Session VII

Toxoids Vaccines 183
Chairmen: *J. Milstien, P. Castle*

Opportunities to Reduce the Use of Animals in the Potency and Toxicity Testing of Toxoid Vaccines 185
P.A. Knight

Validation of the Toxin-Binding Inhibition (ToBI) Test for the Estimation of the Potency of the Tetanus Toxoïd Component in Vaccines 199
J. van der Gun, A. Akkermans, C. Hendriksen, H. van de Donk

Use of in Vitro Vero Cell Assay and ELISA in the United States Potency Test of Vaccines Containing Adsorbed Diphtheria and Tetanus Toxoids 207
R.K. Gupta, G.R. Siber

Vero Cell Assay Validation of an Alternative to the Ph. Eur. Diphtheria Potency Tests 217
A.M. Gommer

Passive Haemagglutination Tests Using Purified Antigens Covalently Coupled to Turkey Erythrocytes 225
E.H. Relyveld, M. Huet, L. Lery

Workshop

Serological Methods and Cell Cultures 243
Chairmen: *P. Vannier, R. Gupta*

Comparison of in Vivo and in Vitro Methods for Determining Unitage of Diphtheria Antitoxin in Adsorbed Diphtheria-Tetanus (DT) and Diphtheria-Tetanus-Pertussis (DTP) Vaccines 245
Y.P. Shinde, S.S. Jadhav

Session VIII

Bacterial Vaccines 261
Chairmen: *S. Houghton, E. Relyveld*

Developments in Pertussis Vaccines Leading to Reduction and Replacement of in Vivo Testing 263
K. Redhead

Pertussis Serological Potency Test as an Alternative to the Intracerebral Mouse Protection Test 271
A. van der Ark, I. van Straaten-van de Kappelle, C. Hendriksen, H. van de Donk

Development of a Guinea-Pig Model for Potency/Immunogenicity Evaluation of Diphtheria, Tetanus Acellular Pertussis (DTaP) and *Haemophilus influenzae* Type b Polysaccharide Conjugate Vaccines 283
R.K. Gupta, R. Anderson, D. Cecchini, B. Rost, P. Griffin Jr., K. Benscoter, J. Xu, L. Montanez-Ortiz, G.R. Siber

Potency Tests of Diphtheria, Tetanus and Combined Vaccines Suggestion for a Simplified Potency Assay 297
H. Aggerbeck, I. Heron

Potency Testing of Diphtheria and Tetanus Toxoids. A Reliable, Pragmatic and Economic Approach 303
D.W. Stainer

Session IX

Therapeutic Toxins and Monoclonal Antibodies 309
Chairmen: *R. Dabbah, H. van de Donk*

Requirements for Valid Alternative Assays for Testing of Biological Therapeutic Agents 311
D. Sesardic

Investigation of the Product Equivalence of Monoclonal Antibodies for Therapeutic and in Vivo Diagnostic Human Use in Case of Change from Ascitic to in Vitro Production 319
G. Schäffner, M. Haase, S. Giess

Poster Session 321

Summary and Recommendations 351

Summary and Recommendations 353
H.J.M. van de Donk

Brown F, Cussler K, Hendriksen C (eds): Replacement, Reduction and Refinement of Animal Experiments in the Development and Control of Biological Products.
Dev Biol Stand. Basel, Karger, 1996, vol 86, p 1

SESSION I

Laboratory Animals and the Concept of Alternatives

Chairmen: D. Straughan (Nottingham, U.K.)
K. Cussler (Langen, Germany)

Brown F, Cussler K, Hendriksen C (eds): Replacement, Reduction and Refinement of Animal Experiments in the Development and Control of Biological Products.
Dev Biol Stand. Basel, Karger, 1996, vol 86, pp 3-10

A Short History of the Use of Animals in Vaccine Development and Quality Control

C.F.M. Hendriksen

National Institute of Public Health and Environmental Protection, Bilthoven, The Netherlands

Abstract: Man has been using animals since early times to gain an insight into health, illness and death. The oldest known medical standard work, the Corpus Hippocraticum (circa 350 BC), contains descriptions of experiments on pigs.
Although the first attempt at immunoprophylaxis dates as far back as the 6th century (variolation was practised in China to protect people against smallpox), it was not until the middle of the 19th century that animal experimentation acquired full scientific status in the development and quality control of immunobiological products. It was Louis Pasteur and Robert Koch who, through studies on animals, succeeded in underpinning the causal relationship between infectious diseases and micro-organisms, thus opening the way to the discovery of effective therapeutic and prophylactic agents for a number of these diseases. In several respects, the experimental animal work carried out in the last decade of the 19th century to find an effective and reliable way of treating and preventing diphtheria determined the use of animals. Many common routine animal tests in the quality control of immunobiologicals arose from diphtheria research. Conversely, diphtheria was one of the first diseases where experimental animal research laid the foundation for effectively reducing child mortality. This had a very profound impact on the attitude of society towards animal experiments in those days and almost completely eliminated the growing influence of the antivivisection movement.
The interest in the possibilities of replacement, reduction and refinement (the Three-Rs concept) of the use of laboratory animals is increasing for several reasons, including concern about animal welfare. The root of animal welfare can be traced back to the 18th century with the formulation of utilitarian ethics. One characteristic feature of these ethics was that the interests of any creature which is submitted to any procedure should be taken into consideration.
This presentation sets out some major historical contributions of animal experiments to the development and quality control of immunobiologicals. Attention is also paid to the changing attitude of society towards animal experiments and its impact on the development of alternative methods. It is concluded that, although animal experiments have played an important part, a new area is now beginning in which increasing emphasis will be placed on in vitro methods.

Introduction

Man has been using animals for a very long time to gain an insight into health, illness and death. The oldest known medical standard work, the Corpus Hippocraticum (circa 350 BC), contains descriptions of experiments on pigs. However, until the middle of the last century, animal experimentation gave an impetus mainly to advances in the basic medical sciences such as anatomy and physiology, and not to developments in therapy and immunoprophylaxis.

The first attempts to vaccinate humans were made in China as long ago as the 6th century. An early Chinese medical text, «The Golden Mirror of Medicine», lists several forms of inoculation against smallpox (also called variolation), such as blowing powdered scabs of smallpox lesions into the nose, and stuffing a piece of cotton smeared with the contents of a smallpox vesicle into the nose [1]. Variolation was introduced into Britain in 1721 by Lady Mary Wortley Montague, the wife of the British ambassador in Turkey. Although the treatment was often effective, results were erratic, and 2-3% of those treated died of smallpox obtained from the variolation itself [2]. In the 18th century, variolation was also used to combat cattle plague, a disease which caused major losses in Europe at the time [3].

A Scientific Approach to Vaccine Development

The first scientific attempt to control an infectious disease by means of a deliberate systematic inoculation is attributed to Edward Jenner. After 25 years of study, Jenner published his findings on cowpox and smallpox in 1798 [4]. This publication also contained the results of his successful experiment in 1796 in which an 8-year-old boy, James Phipps, was inoculated with cowpox material and subsequently challenged with smallpox virus (Fig. 1). Like many developments in medicine, the introduction of smallpox vaccination encountered scepticism. Never-

Fig. 1: Edward Jenner inoculating James Phipps with cowpox material.

theless, the method found wide application. With regard to the title of this paper, it is interesting to mention that Jenner's study was entirely based on observation and animal experiments did not contribute to it in any way.

During the 84 years that elapsed between Jenner's treatise on Variolae Vaccinae and the development of the first vaccine based on scientific research – the fowl cholera (Pasteurella multocida) vaccine discovered by Pasteur in 1880 – the field of research was not dormant. Generally speaking, the emphasis was on the development of comparative pathology and the gaining of insight into the aetiology of infectious diseases. The prevailing belief that miasmas were the causal agent was replaced by the realization that diseases might be caused by a living organism (contagium vivum). In 1840 the German pathologist and anatomist Jacob Henle (1809-1885) published his work on contagious diseases, in which he demonstrated that living organisms and not miasmas were the cause of disease. Henle concluded that for establishing the association between causal agent and an infection, it is essential to culture the micro-organism outside a living host. Koch's postulates, published in 1884, have been of historic importance regarding the role of experimental animals in the study of infectious diseases. These postulates stated the criteria to be met in relating a micro-organism to a given infection. The postulates included the isolation of the micro-organism in pure culture, followed by introduction of the pure culture into a suitable experimental animal. This must result in the induction of typical clinical disease in the experimental animal. Koch's postulates gained general acceptance in microbiology and helped to lay the foundation for the intensive use of laboratory animals. The animal assumed a key position within clinical and experimental pathology, and the need for good animal models increased when it was found that not every species is equally sensitive to pathogenic micro-organisms. For example, Koch used mice, guinea pigs and rabbits for his work on anthrax. When it was difficult to find a suitable animal model, as was the case with poliomyelitis, the investigations stagnated. By contrast, rapid results were obtained in research on diphtheria, when it was found that the guinea pig was a relatively simple model [5].

Diphtheria and poliomyelitis are discussed in more detail to illustrate the role of the animal experiment in the development of immunoprophylaxis. It was Pasteur, in particular, who laid the foundation for a rational and experimental approach to vaccine development, in which animal experiments played an important part. For example, he developed vaccines against fowl cholera (1880), anthrax (1881) and rabies (1885).

Developments in the Treatment and Prevention of Diphtheria

In several respects, the experimental animal work carried out in the last decade of the 19th century to find an effective and reliable way of treating and preventing diphtheria determined the use of animals. On the one hand, diphtheria was one of the first diseases where experimental animal work laid the foundation for effectively combatting this disease. On the other, many common routine animal tests in the quality control of immunobiologicals arose from diphtheria (and tetanus) research.

In the 1800s, the mean life expectancy was less than half that of today. With a mortality rate of up to around 40%, diphtheria (also known as «the strangling angel of children») belonged to one of the most feared epidemic diseases of childhood [6]. Diphtheria is now a rare disease in western countries. The sharp decline in both mortality and morbidity has, apart from improved hygiene and diet, been ascribed especially to immunoprophylaxis (antitoxin and toxoid). The experimental animal played an important role, firstly in gaining an insight into the aetiology and course of the disease, secondly in the development of methods of treatment and prevention with antitoxin and toxoid respectively, and finally, in the development of the quality control of these preparations. Important steps in the development of therapy and immunoprophylaxis, and the animal species used, are presented in Table 1. The steps in the development of quality control tests are summarised in Table 2.

Polio Vaccine Development

Although the incidence of paralytic poliomyelitis was not high compared with other diseases, such as measles, it generated much public concern because of its mysterious seasonal incidence, its disfiguring nature and its propensity for paralyzing the respiratory muscles [7].

In contrast to diphtheria, suitable animal models for the study of poliomyelitis were not available at the end of the last century. The lack of methods for studying viruses was an additional complication.

In 1908 Landsteiner and Popper were able to reproduce the disease in monkeys by intracerebral inoculation of a filtrate of central nervous system tissue from a fatal case. Although this discovery was of great importance, the monkey was not a particularly suitable experimental animal for the kind of research necessary to elucidate the characteristics of the poliovirus and its epidemiology or to enhance the development of preventive measures. It was only in 1939 that Armstrong succeeded in adapting the Lansing strain of poliovirus to rodents. However, the

Table 1: Animal experiments in the development of diphtheria treatment and prevention.

Development	Year	Name	Animal species*
• Isolation of the causal micro-organism Corynebacterium diphtheria	1884	Loeffler	pigeon, chicken, rabbit, guinea pig
• Production of the exotoxin	1884	Roux & Yersin	various animal species, GUINEA PIG
• Demonstration of the therapeutic value of antitoxin	1890	Behring	GUINEA PIG, dog, mouse, rat, various other species
• Large-scale production of antitoxin	1894	Roux & Martin	dog, sheep, goat, HORSE, cow
• Toxin-antitoxin mixtures for active immunization	1913	Behring	GUINEA PIG
• Diphtheria toxoid	1923	Ramon	various animal species

* The animal species finally chosen is given in capital letters.

Table 2: Diphtheria antitoxin and toxoid: the development of quality control tests.

Development	Year	Name	Animal species
• Quality control of antitoxin	1892	Behring & Wernicke	GUINEA PIG
• Introduction of a standard preparation in potency testing	1897	Ehrlich	GUINEA PIG
• Potency test using parallel-line bioassay	1937	Prigge	GUINEA PIG
• Multiple intradermal challenge test	1974	Knight	GUINEA PIG
• Mouse/Vero cell test	1985	Kreeftenberg	MOUSE

usefulness of the rodent-adapted virus for experimentation was limited, because it was a single serotype [7].

In the 40-year interval from the isolation of the virus in monkeys to the development of tissue culture techniques, considerable progress was made in understanding the disease, for example, regarding the route of its transmission from studies on chimpanzees. Various efforts were made to prepare vaccines from infected monkey central nervous system tissue. Although some success was reported, the experiments were inconsistent, and the results were disappointing overall. In spite of ambiguous animal data, two investigators, in 1936, independently conducted field trials in humans of virus vaccines prepared from monkey spinal cord. By modern standards, these trials were badly conceived. For example, appropriate tests for safety and efficacy were lacking. Thousands of subjects received these vaccines, some of whom developed paralysis. These findings raised the spectre of persistence of virulent virus in the vaccine, and the trials were terminated [7].

With the availability of tissue culture techniques the development of a vaccine against poliomyelitis became a realistic possibility. Although the first steps towards the culture of cells and tissues were taken in 1885 by Roux, who kept the neural plate of chick embryo alive for a few days, and by Harrison in 1908, who grew neural tube cells from a frog embryo in a drop of coagulated lymph, it was not until the discovery of penicillin by Fleming in 1928 that cell culture conditions could be improved and cell cultures could be adopted for virus vaccine production [8]. In 1936 Sabin and Olitsky reported growth of virus in cultures of human embryonic brain tissue, but they had no success with cells from non-nervous tissues. It was in 1949 that the paper by Enders, Weller and Robbins describing the successful cultivation of the Lansing strain of poliovirus in cultures of non-nervous human tissues, which provided the breakthrough for the development of a vaccine [7].

The Development of Quality Control Tests

With the introduction of immunobiologicals the need for some kind of quality control also arose. This is illustrated by a remark made by Behring (Fig. 2) about diphtheria antiserum: «... irrespective of how the immunization may have been

Fig. 2: Emil von Behring working in his laboratory with guinea pigs.

achieved and what animals were used, be it guinea pigs, rabbits or sheep, the outcome regarding the protective and healing effects of the blood of the immunized animals was qualitatively always the same; it differed only quantitatively» [9].

The work on diphtheria and tetanus also played a key role in the development of quality control tests. Achievements made in this field were applied to the development of other vaccines. Although it is generally very difficult to trace the roots of safety tests, these tests were often introduced after vaccine-related accidents had occurred (Table 3), or accidental findings made some kind of safety testing necessary. For example, after the Cutter incident in the U.S. in 1955, in which five people died as a result of the use of incompletely inactivated vaccine, new requirements for safety testing were introduced [7]. Another example is the isolation from cultures of Rhesus monkey kidney cells of a virus that caused a typical vacuolation in cells from cynomolgus monkeys. This virus, simian virus 40 (SV40), was found to cause a clinically inapparent infection in Rhesus monkeys in the wild. This finding caused a great deal of concern, since SV40 is a DNA virus of the papova virus family and had been shown to cause cancer in several species of animals. Once the presence of the contaminating virus was recognized, a safety test was introduced into quality control to assure SV40 exclusion from the vaccine [7].

Animal Welfare and the Interest in Alternatives to Animal Testing

The ethical implications of animal experimentation were hotly debated almost from the outset. In 1760 the English physicist Ferguson (1710-1776) described the barbaric death-struggle of animals in the «airpump experiments» of Boyle as «... too shocking to every spectator who has the least degree of humanity». The reflection on the man-animal relationship was greatly influenced by the utilitarian ethics formulated by Jeremy Bentham (1748-1832). Bentham counted the capacity

Table 3: Vaccine-related accidents

			Year	Cases	Deaths
Toxin in vaccine	Diphtheria	Dallas	1929	96	10
		Concord	1924	21	?
		Bridgewater	1924	22	?
		Baden	1924	28	7
		Kyoto	1948	600	68
Incomplete inactivation	Polio	Cutter incident	1955	260	5
Contamination with toxin	Tetanus in D-antiserum	St. Louis	1901	20	14
Wrong culture	BCG	Lubeck	1930	135	72

to suffer («the question is not «Can they reason?» nor, «Can they talk?» but, «Can they suffer?») as an essential characteristic which entitles sentient beings to equal benevolence. England was the centre of the opposition to animal experiments. The British Society for the Prevention of Cruelty to Animals (later the RSPCA) was founded in 1842, the first animal protection society in the world. It cannot be denied that the antivivisection movement had success in attacking the scientific community at the end of the 19th century. However, the discoveries in the field of diphtheria treatment also had a major impact. Behring and Kitasato's diphtheria antitoxin completely changed the previously hopeless treament of children suffering from diphtheria. A dangerously ill child could now be cured with a single injection. This result affected many people deeply. The fact that experimental animal research had made the treatment of the feared disease possible convinced many of the value of this type of experimentation and undermined the criticism levelled at it [5]. Nevertheless, with periods of revival and decline, the opposition to the use of experimental animals has continued.

The latest revival is perhaps the most interesting. In 1959 the English researchers Russell and Burch introduced the concept of replacement, reduction and refinement (the Three-Rs concept) as a starting-point for humane treatment of experimental animals. They stated that researchers should continually examine their research for possibilities of replacing the use of higher animals by lower organisms or non-biological material, reducing the number of animals used, or refining the experimental techniques to minimize the level of stress or pain endured by the animal in the experiment [10]. The importance of this approach lies especially in the possibility of combining the maintenance of scientific quality of research and the upholding of ethical principles. The Three-Rs concept is now the basis of European legislation on animal experimentation [11, 12]. Although the interest in alternatives to animal research is often regarded as a feature of our time, the interest in alternative methods can be traced back many years. For example, the recommendations of Marshall Hall (1790-1857), an English physiologist, are still relevant today. He drew up a kind of code of practice for researchers, in which he recommended, among other things, the alleviation of pain and distress within an experiment, the replacement of higher animals by lower organisms, and the prevention of needless repetition of experiments [13]. Another example, relevant to the theme of this presentation, is the introduction of toxoids in 1926. Until then, large numbers of animals (usually horses) were required for the production of antitoxins. The discovery in 1926 by the French veterinary surgeon Ramon that diphtheria and tetanus toxin can be detoxified by formalin ensured that immunotherapy was replaced by immunoprophylaxis, and substantially reduced the number of animals required for antitoxin production.

References

1 Plotkin SL, Plotkin SA: A short history of vaccination, in Plotkin SA, Mortimer EA (eds): Vaccines. Philadelphia,WB Saunders Company, 1988, pp 1-7.

2 Parish HJ: A History of Immunization. London, E & S. Livingstone, 1965.

3 Vries J de: De bestrijding van de runderpest in Friesland gedurende de 18e eeuw. (Combatting cattle plague in Friesland in the 18th century), Argos 10, 1994.

4 Jenner E: An Inquiry into the Causes and Effects of the Variolae Vaccinae. London, Samson Low, 1798.

5 Hendriksen CFM: Laboratory Animals in Vaccine Production and Control, Replacement, Reduction and Refinement. Dordrecht, Kluwer Academic Publishers, 1988.

6 Grundbacher FJ: Behring's discovery of diphtheria and tetanus antitoxins. Immunology Today 1992;13 (5): 188-189.

7 Robbins FC: Polio-Historical, in Plotkin SA, Mortimer EA (eds): Vaccines. Philadelphia, W.B.Saunders Company, 1988, pp 98-114.

8 Hendriksen CFM: Animal experimentation and the concept of alternatives, in Hendriksen CFM, Koeter HBWM (eds). Animals in Biomedical Research, Replacement, Reduction and Refinement: Present Possibilities and Future Prospects. Amsterdam, Elsevier, 1991, pp 10.

9 Behring E von, Wernicke E: Über Immunisierung und Heilung von Versuchstieren bei der Diphtherie. (Immunisation and cure of experimental animals with respect to diphtheria). Zeitschrift für Hygiene 1892;12:10-44.

10 Russell WMS, Burch RL: The Principles of Humane Experimental Technique. London, Methuen, 1959.

11 Council of Europe. European Convention for the protection of vertebrate animals used for experimental and other scientific purposes. Strasbourg, 18 March 1986.

12 European Union. Council Directive 86/609 on the approximation of laws, regulations and administrative provisions of the Member States regarding the protection of animals used for experimental and other scientific purposes. 24 November 1986.

13 Rupke NA (ed): Vivisection in Historical Perspective. London, Croom Helm Ltd., 1987.

Dr. C.F.M. Hendriksen, National Institute of Public Health and Environmental Protection, P.O. Box 1, 3720 BA Bilthoven, The Nertherlands

Brown F, Cussler K, Hendriksen C (eds): Replacement, Reduction and Refinement of Animal Experiments in the Development and Control of Biological Products.
Dev Biol Stand. Basel, Karger, 1996, vol 86, pp 11-18

The Three Rs of Russell & Burch and the Testing of Biological Products

M. Balls[1], *D.W. Straughan*[2]

[1] European Centre for the Validation of Alternative Methods (ECVAM), JRC Environment Institute, Ispra, Italy
[2] Fund for the Replacement of Animals in Medical Experiments (FRAME), Nottingham, UK

Key words: Replacement, reduction, refinement, testing, biologicals, vaccines.

Abstract: The principles of humane animal experimentation proposed by Russell & Burch (1959), namely, *replacement*, *reduction*, and *refinement*, are now commonly known as the Three Rs. These principles are clearly embodied in Article 7 of *Directive 86/609/EEC*. It is instructive, therefore, to consider the priority currently attached to compliance with these principles and to the Directive in the development and control of biological products. Specific comments are made on the need for the application of the Three Rs in relation to the testing of human diphtheria and tetanus vaccines, human pertussis vaccine, inactivated veterinary Gumboro vaccine, veterinary Newcastle disease vaccine, veterinary clostridial vaccines and botulinum toxin. We pose three questions: a. Are the minimum numbers of animals already being used in this area? b. Is any *unnecessary* pain, suffering, distress or lasting harm being inflicted on the animals used? and c. What could and should be done about any shortcomings in current practice? Finally, the role of ECVAM in the promotion of the Three Rs within the European Union will be reviewed.

Introduction

In *The Principles of Humane Experimental Technique* [1], Russell & Burch referred to what had sometimes been seen as an irreconcilable conflict between the claims of science and medicine and those of humanity in our treatment of other animals. They said that «this conflict disappears altogether on closer inspection, and by now it is widely recognised that the most humane possible treatment of experimental animals, far from being an obstacle, is actually a prerequisite for successful animal experiments». This truth applies no less to the testing of biological products than it does to fundamental research or other types of testing.

In considering animal suffering, Russell & Burch distinguished between *direct* suffering, which is an unavoidable consequence of the procedure employed, and *contingent* suffering, which is an inadvertent and incidental by-product. They then went on to discuss ways whereby animal suffering could be diminished or removed, and defined three broad headings, now known as the Three Rs: *replacement* means the substitution for conscious living animals of insentient material; *reduction* means reduction in the numbers of animals used to obtain information of a given amount and precision; *refinement* means any decrease in the incidence or severity of procedures applied to those animals which still need to be used.

They stated that refinement alone is never enough, and that we should always seek reduction and, if possible, replacement. Further, whereas replacement is always a satisfactory answer in terms of animal suffering, reduction and refinement should, whenever possible, be used in combination.

The Three Rs concept of alternatives, as defined by Smyth [2], is now enshrined in a number of national laws on the regulation of animal experimentation, and also in the *Council of Europe Convention for the Protection of Vertebrate Animals Used for Experimental and Other Scientific Purposes* [3] and *Directive 86/609/EEC* of the European Union [4]. For example, Article 7 of the Directive states that:

1. Experiments shall be performed solely by competent authorised persons, or under the direct responsibility of such a person, or if the experimental or other scientific project concerned is authorised in accordance with the provisions of national legislation.
2. An experiment shall not be performed, if another scientifically satisfactory method of obtaining the result sought, not entailing the use of an animal, is reasonably and practicably available.
3. When an experiment has to be performed, the choice of species shall be carefully considered and, where necessary, explained to the authority; in a choice between experiments, those which use the minimum number of animals, involve animals with the lowest degree of neurophysiological sensitivity, cause the least pain, suffering, distress or lasting harm and which are most likely to provide satisfactory results, shall be selected.
4. All experiments shall be designed to avoid distress and unnecessary pain and suffering to the experimental animals.

Partly as a result of such legal requirements, but also as a result of changes in attitudes and practices within the scientific and industrial communities, much effort is now being focused on the development, evaluation and acceptance of methods and strategies which could reduce, refine or replace the use of laboratory animals. Particular attention is being paid to animal tests used to identify hazardous chemicals and to predict the toxic potential of various products, such as medicines and pesticides, as a basis for risk assessment and risk management. At the same time, the natural evolution of the scientific method has led to a trend toward the use of molecular biology and cell biology techniques in our attempts to understand and treat a variety of diseases. Nevertheless, the Three Rs concept is now faced with a new challenge – a likely mushrooming in the development and use of genetically-modified animals.

In comparison with toxicity testing, relatively little attention has been paid to the application of the Three Rs in efficacy and safety tests on biological products, although in their book Russell & Burch devoted quite a lot of attention to vaccine testing and the diagnosis of infectious diseases. They considered virulence testing to be particularly inhumane and saw in vitro tests of virulence as being especially welcome.

Nevertheless, there have been many very good attempts by some of those involved in vaccine development and production to find alternatives [5, 6]. For example, there have been very useful refinements in overall design and in choice of endpoints for potency tests for diphtheria and tetanus toxoid, leading to the acceptance of alternative methods in some pharmacopoeial monographs. However, much more could be done.

Where replacement is not yet possible, reduction and refinement must be our goal. It is the policy of the European Union to seek a reduction in the numbers of animals used in testing in the Member States by the year 2000 [7]. Festing [8] has suggested that better experimental design could contribute a large part of this reduction in animal use.

Meanwhile, the European Commission has set up ECVAM at its Joint Research Centre at Ispra, near Lake Maggiore, in Italy, to coordinate the validation of alternative techniques which could provide the same level of information as that obtained in experiments using animals, but which involve fewer animals or which entail less painful procedures [9].

Concern over Animal Tests for Biological Products

Animal experiments are used in the development of both synthetic pharmaceuticals and biological products. While the total numbers of animals used for the pharmaceuticals is much greater, particular concern with biological products stems from the nature of the efficacy tests conducted in animals, especially when they involve challenge with micro-organisms or toxins and lethal endpoints.

Quality control of biological products usually requires biological tests, whereas for synthetic pharmaceuticals it involves chemical tests. Thus, animal tests are used routinely on batches of production material to ensure that biological products are both safe and efficacious in use. The latter invariably requires potency estimates in animals, often by challenge and lethal or local sign endpoints. For some vaccines, potency tests are able to use serological endpoints and do not require active challenge.

These quality control tests are specified in national product licences and are described in standards set by national and other pharmacopoeias and multinational organisations. For companies or for national agencies licensing animal experiments, there seems to be little or no discretion to vary these test requirements. Our concern stems from the lack or slow rate of harmonisation.

For some products, other validated methods have been accepted by some authorities, but their use may be limited, because they are not universally accepted.

The consequence of such slow and inadequate harmonisation must be that, in testing some products, animals continue to be used in larger numbers and/or with more suffering than the minimum necessary to ensure that products are safe for human or animal use. This is not compatible with *Directive 86/609/EEC*. It is therefore certainly not acceptable to ECVAM, to the public or to us personally. Where there are already validated and accepted alternatives which allow obsolete animal tests to be replaced, refined or reduced, then such alternatives should be accepted and used without unnecessary delay. Notwithstanding any administrative difficulties, we suggest that regulatory agencies and scientists should take immediate action to ensure that outdated tests are replaced as soon as possible.

It is our hope that the following questions will be addressed at this Symposium and on every appropriate occasion in the future, until satisfactory progress has been made toward the effective application of the Three Rs principles in biological product development, production and testing:

a. Are the minimum numbers of animals being used in every scientifically justifiable procedure?

b. Is any unnecessary pain, suffering, distress or lasting harm being inflicted on the animals that are necessarily used?

c. What is being done to eliminate any deficiencies?

Specific Issues

We believe that the real commitment to other methods in any field can be best evaluated with respect to their response to specific issues and actions. We will therefore highlight and comment on a few specific organisational and scientific problems. Hopefully, these few examples of potency tests will complement other presentations at this Symposium, and particularly the detailed report of the recent ECVAM Workshop held in Utrecht in April 1994 [10].

In considering the main objectives of this Symposium, we hope that the experts present will make real progress with some of these specific issues. In particular, we hope that they will identify and agree problems and priorities: i) for research into the Three Rs and particularly into replacement alternatives to the present animal tests applied to a range of specific biological products; ii) for their validation; iii) for their incorporation into regulatory test requirements; and iv) for changes to scientific and regulatory organisational attitudes and practices.

1. Animal use committees within bodies requiring animal tests

National pharmacopoeial committees and product licencing agencies and the European Pharmacopoeial (EP) committees do not at present have a formal obligation to consider animal welfare issues, experimental design, or the minimal introduction of alternative methods when requirements for animal test data on biological products are initially agreed or reviewed.

Similar issues have been considered recently within the UK, in a report of the Animal Procedures Committee (APC) to the Home Office, which involved the present authors [11]. This report noted *inter alia* that contacts between the UK regulatory bodies for human and animal medicines which *require animal tests* and those *actually licensing the tests* were ad hoc and appeared less satisfactory than those for chemicals and pesticides. This situation seems likely to pertain elsewhere.

We do not dispute that the primary duty of regulatory bodies requiring animal tests on biological products (i.e. those for human and animal medicines) is to ensure safe and efficacious products. However, we consider that they also have an ancillary duty toward animal welfare. We suggest that the relevant national and European regulatory bodies should now set up formal animal use committees on review bodies in-house. This would provide reassurance that human and animal medicine regulatory bodies are not setting test requirements which are difficult to justify in terms of *Directive 86/609/EEC*. In addition, there should be effective arrangements for formal liaison between those requiring animal tests and those authorising them. This should clarify misunderstandings, improve public accountability, and might even expedite the introduction of alternatives in line with the Three Rs.

2. Required accuracy with flexible test design

In most cases, pharmacopoeial and product licence requirements allow some flexibility by specifying the accuracy required in a potency assay with a minimum or agreed optimal test design. However, on occasion, the EP or certain national licencing agencies have stipulated inflexible test designs that are excessive and insupportable statistically. As a matter of general policy, we would suggest that required tests should always allow the use of the minimum numbers of animals required to ensure accuracy.

3. Improved co-ordination of research effort in alternatives

Within the UK, the APC report [11] has urged the establishment of an interdepartmental group involving appropriate UK Government departments and non-governmental organisations to discuss priorities for work on alternatives, including validation studies and their funding. We suggest that this concept be extended to include national and European product licencing agencies and pharmacopoeial committees.

With your support, ECVAM will undertake this role within Europe. In setting priorities for action to achieve alternatives *for individual biological products*, we suggest that this symposium considers the level of suffering and numbers of animal used per year and the feasibility and level of effort/expense anticipated for achieving a satisfactory outcome.

There may be particular problems in giving priority to research into alternative methods and their validation for biological products which have a limited production, limited numbers of producing companies and/or low profit margins. These will increase, particularly when public funds for research are limited. We have no easy solutions to these problems and would welcome discussion on these issues.

4. Cost implications of alternatives

In some specific cases, it is argued that the application of alternatives is inconvenient and/or would make products uncompetitive commercially. This argument is sometimes used to justify the production of monoclonal antibodies by ascites tumours in vivo instead of by using in vitro methods. We suggest that the difficulties here are more imagined than real. Thus, large-scale production in industry and small-scale production in academia by in vitro methods are now commonplace.

Also, the price of hollow fibre/dialysis systems to concentrate antibody is now reasonably low. Since some European nations have already restricted antibody production by ascites tumours, we see no compelling reason why this should not become general policy within Europe.

It is also argued sometimes that it is uncompetitive to produce horse and rabbit sera except by bleeding schedules and routes which may be considered excessive. This does not seem a compelling argument for not agreeing to the least severe protocols. Again, some European nations have taken a lead here. We hope that we will either be able to agree now on a general policy to apply within Europe, or that we can set up task forces with early target dates to report and advise on the best ways forward.

5. **Proposed code of practice**

The above suggestions will be complemented and augmented if and when all concerned accept and implement the code of practice proposed at the Utrecht Workshop [10].

6. **Human diphtheria and tetanus vaccines**

At present, some national monographs, e.g. the British Pharmacopoeia, only specify non-lethal endpoints (dermal toxicity and local paralysis). However, the EP still allows obsolete lethal endpoint tests to be used. We would like to see a clear and urgent commitment by the EP to resolve this matter, together with their proposed timetable for its resolution. As a matter of principle, when further improvements in potency tests in animals are accepted into the EP (e.g. the single dilution assays discussed at the Utrecht Workshop), a clear commitment and timetable should be made for removal of the unimproved/obsolete methods.

7. **Human pertussis vaccine**

For the time being, there appears to be a continuing need for the Kendrick test for assaying the potency of whole-cell vaccines. It is common experience that, because of variability, substantial numbers of animals are needed to get statistically acceptable results. If the sources of variation could be identified, improvements in test design could result, with the benefit of considerable reductions in animal use and also in costs. The Utrecht Workshop discussed possible refinements in challenge tests, as well as replacements – which are the ideal, as Russell & Burch pointed out in 1959.

8. **Inactivated veterinary Gumboro vaccine**

Infectious bursal disease (IBD) in fowl is undoubtedly a serious problem, for which several types of vaccines are available. Although a live attenuated vaccine is obtainable there is still a demand for the inactivated product. We assume that the latter provides genuinely superior protection, which is not merely a reflection of the dosing schedule. In one country, virus for inactivation is produced by infecting several hundred thousand fowl per year with IBD virus and harvesting the virus-rich bursal gland. Further enquiries are required on the extent to which an equivalent product could be produced by using in vitro methods (probably with a genetically-engineered expression system) to grow the virus or express appropriate marker proteins. If this approach is viable, then decisions will be needed on how to encourage and expedite such work and who should pay for it.

9. **Newcastle disease vaccine**

In recent years, alternative, minimal-severity, serological methods have been accepted into pharmacopoeial monographs and have come into general use. However, the EP still retains an alternative lethal endpoint challenge assay, which now requires a minimum of 20 birds at each of three dose levels for test and reference preparations. For some potent vaccines, there is published and unpublished evidence that efficacy could still be assured while using simpler assay designs or still smaller numbers of fowl per dose group. This is a classic example of the kind of inflexible and excessive test requirement noted in section 2 above. The justification for the present EP test design is not known, and we are not aware that the expert and experienced views of Orthel and his collaborators [12] have been rebutted.

10. **Veterinary clostridial vaccines**

For determining the potency of clostridial vaccines, the Utrecht Workshop gave priority to the need to replace conventional challenge tests with serological tests. More will be said on the matter later in this Symposium. In the meantime, we would only comment that *Clostridium chauovei* (Blackleg) must have a special place in our considerations, since, although it involves relatively few animals, it is considered to be particularly painful, as well as being ultimately lethal.

11. **Botulinum toxin**

With only one manufacturer in the EU, this remains a matter of local interest. From newspaper accounts, it appears that there has been concern over the design of the lethal endpoint potency assay. Also, it can be inferred from the published statistics that potency testing under the product licence involves the use of fairly substantial numbers of mice each year. We suggest that the development of an in vitro replacement test for the assay of this toxin should be a high priority, at least in the UK.

The Role of ECVAM

ECVAM should be able to help to promote the Three Rs in the context of the testing of biological products in a number of ways. For example, ECVAM could support future workshops, such as that held in Utrecht [10], and symposia such as this one [13], or organise task forces charged with achieving specific, defined objectives. Secondly, ECVAM could help to finance promising replacement alternative methods as they approached the validation stage, and could manage and/or finance international validation studies designed to evaluate the relevance and reliability of particular tests or testing strategies. Thirdly, ECVAM could provide a route for communication with other services of the European Commission, with the appropriate authorities in the Member States of the European Union, and with other appropriate national or international bodies.

References

1 Russell WMS, Burch RL: The Principles of Humane Experimental Technique. London, Methuen, 1959, p 238.

2 Smyth D: Alternatives to Animal Experiments. London, Scolar Press, 1978, p 218.

3 European Convention for the Protection of Animals Used for Experimental and Other Scientific Purposes. Strasbourg, Council of Europe, 1986, p 51.

4 Council Directive of 24 November 1986 on the Approximation of Laws, Regulations and Administrative Provisions in the Member States for the Protection of Animals Used for Experimental and Other Scientific Purposes. Off J Eur Comm 1986, L262, pp 1-29.

5 Hendriksen CFM: Laboratory Animals in Vaccine Production and Control: Replacement, Reduction and Refinement. Dordrecht, Kluwer Academic Publishers, 1988, p 175.

6 van der Kamp MDO: Ways of Replacing the Use of Animals in the Quality Control of Veterinary Vaccines. Lelystad, Institute for Animal Science and Health (ID-DLO), 1994, p 197.

7 Resolution of the Council and the Representatives of the Governments of the Member States of 1 February 1993 on a Community programme of policy and action in relation to the environment and sustainable developments. Off J Eur Comm 17/5/1993, C138, pp 1-98.

8 Festing MFW: Reduction of animal use: experimental design and quality of experiments. Lab Animals 1994, 28;212-221.

9 Establishment of a European Centre for the Validation of Alternative Methods (ECVAM). Communication from the Commission to the Council and the European Parliament. Brussels, Commission of the European Communities, 1994, p 6.

10 Hendriksen CFM, Garthoff B, Aggerbeck H, Bruckner L, Castle P, Cussler K, Dobbelaer B, van der Donk M, van der Gun J, Lefrançois S, Milstien J, Minor PD, Mougeot M, Rombaut B, Ronneberger PD, Spieser JM, Stolp R, Straughan DW, Tollis M, Zigtermans G: Alternatives to Animal Testing in the Quality Control of Immunobiologicals: Current Status and Future Prospects. The report and recommendations of ECVAM Workshop 4. ATLA 1994;22:420-434.

11 Animal Procedures Committee: Report to the Home Secretary on Regulatory Toxicity. London, HMSO, 1994, p 20.

12 Orthel FW, Lütticken D, Jacobs J: An economic potency test for inactivated Newcastle Disease vaccine, in Hennessen W, Moreau Y (eds): Proceedings of the International Symposium on the Immunization of Adult Birds with Inactivated Oil Adjuvant Vaccines. Dev Biol Stand, Basel, Karger 1982;51:55-58.

13 Brown F, Cussler K, Hendriksen C (eds): Replacement, Reduction and Refinement of Animal Experiments in the Development and Control of Biological Products. Dev Biol Stand. Basel, Karger, 1995, vol 86.

Prof. M. Balls, European Centre for the Validation of Alternative Methods (ECVAM), JRC Environment Institute, 21020 Ispra (VA), Italy

Brown F, Cussler K, Hendriksen C (eds): Replacement, Reduction and Refinement of Animal Experiments in the Development and Control of Biological Products.
Dev Biol Stand. Basel, Karger, 1996, vol 86, p 19

SESSION II

Testing, Regulations and Three Rs Policy

Chairmen: F. Horaud (Paris, France)
R. Dabbah (Rockville, USA)

Brown F, Cussler K, Hendriksen C (eds): Replacement, Reduction and Refinement of Animal Experiments in the Development and Control of Biological Products.
Dev Biol Stand. Basel, Karger, 1996, vol 86, pp 21-29

Alternatives to Animal Testing: Achievements and Recent Developments in the European Pharmacopoeia

P. Castle

European Pharmacopoeia Secretariat, Strasbourg, France

The European Pharmacopoeia has declared its full support for thorough application of the three Rs in the tests that comprise the official standards. Significant progress has been achieved in the last few years in the fields of pyrogen testing, hormones, blood products and, to a lesser extent, vaccines. Many projects are at present under way. Until recently, the approach was more or less passive, awaiting a consensus on new methods before incorporation into the official standard. However, a more active approach has now been adopted, notably by the organisation of validation exercises both to establish alternative methods and to encourage their wide use. All users of the Pharmacopoeia are concerned by this problem and they should be aware that all proposals and initiatives will be followed up within the means at our disposal.

Introduction

The quality standards of the European Pharmacopoeia can have important implications for animal welfare in a positive or a negative manner: when progress in the three Rs can be incorporated into the official text of the European Pharmacopoeia, this will ensure that it is rapidly taken up by manufacturers and control authorities in the testing of medicines; the EP now covers more than twenty European countries so that decisions that imply progress in animal welfare have a very wide effect; the EP also acts as an effective forum for discussion of new ideas among the control authorities. The effects can also be negative: if the official standards are static, or tardy in taking account of newly developed alternatives, this will perpetuate the unnecessary use of animals; the application of the EP still varies from country to country, especially on the use of validated alternative methods instead of the official method so that as long as an animal test is present in a monograph it will be carried out at least in some countries and also in cases of dispute. Since the monographs are also used as models for newly developed products, any positive or negative effects for animal welfare will be spread further than the products covered by the monographs.

European Pharmacopoeia Policy

The year 1986 marked a turning point for the policy of the EP. Our parent organisation, the Council of Europe, opened for signature in that year the European Convention for the Protection of Vertebrate Animals used for Experimental and other Scientific Purposes; later in the year the provisions of the convention were incorporated in Directive 86/609/EEC, thus giving rapid effect for EC member countries.

Some of the provisions of the convention are very encouraging for animal welfare:

Article 6

1. A procedure shall not be performed for any of the purposes referred to in Article 2 if another scientifically satisfactory method not entailing the use of an animal is reasonable and practicably available.
2. Each Party should encourage scientific research into the development of methods which could provide the same information as that provided in procedures.

Article 7

When a procedure has to be performed, the choice of species shall be carefully considered and, where required, be explained to the responsible authority; in a choice between procedures, those should be selected which use the minimum number of animals, cause the least pain, suffering, distress or lasting harm and which are most likely to provide satisfactory results.

These provisions are legally binding on all states that ratify the convention and in any case form part of EC law. It is clear that despite this the convention and directive are not being applied fully; we all know of cases where established practices carry on despite the existence of alternatives and sufficient encouragement is not given in all cases for finding alternatives.

The European Pharmacopoeia Commission, which was represented by a Secretariat observer during negotiation of this convention, fully endorsed the aims of the convention and encouraged all the groups of experts working on monographs and general methods that implied the use of animals to take account of this in new monographs and also to review all existing animal tests.

Applying the Three Rs in the European Pharmacopoeia

Progress in application of the three Rs seems to follow a cyclical pattern conditioned by scientific and technical developments. After an initial period when activity was mainly directed at replacing animal tests, it became clear that there was a fairly large core of monographs on biological products, especially vaccines, where no alternative to the use of animals was available or in prospect. The value of the other two Rs, reduction and refinement, seems to be generally undervalued but in the monographs, after the first cycle of replacement, application of reduction and refinement to the remaining tests was the only possibility at the time.

When the animal tests in monographs are reviewed, various typical situations are met as listed below; in some of these situations, the EP can take effective action, in others the action taken depends for its effectiveness on action taken by licensing authorities, control authorities and manufacturers; in some situations the EP can do little but wait for further developments in techniques or basic scientific knowledge.

Replacement of an EP Animal Test: Typical Situations

1. replacement is possible immediately in the present state of knowledge;
2. replacement seems possible but collaborative testing of the new method is needed;
3. replacement seems possible but basic research and development are needed;
4. replacement is possible but the new method is product-specific;
5. omission of an animal test is possible for routine testing once consistent results on consecutive batches have been established.

Reduction in the Number of Animals used in an EP Test: Typical Situations

1. reduction possible while remaining with the present test system;
2. reduction possible by moving to a new test system after collaborative testing;
3. reduction seems possible but basic research and development are needed;
4. reduction is possible but the new method is product-specific;
5. reduction possible by simplifying or omitting a test for routine use once consistent results have been demonstrated;
6. reduction possible by moving the test 'upstream' in the production process;
7. reduction possible by combining test systems.

Situation 1 in the list is obviously the easiest and the EP can act immediately. Situation 2 is also a case where action by the EP is appropriate to organise collaborative testing. In situation 3, the EP cannot act since there is no direct involvement in basic research. Situation 4 is problematic for the EP since the monographs are intended not to be product-specific. At best, the monograph can mention the possibility of using an alternative method after approval by the licensing authority; this may seem a rather weak response but if convention no. 123 is applied properly, the possibility of using a method is an obligation to try. In situation 5, the EP can open up the possibility and encourage the application of this approach but since it depends on a direct knowledge of the situation of each manufacturer, it can only be applied with the consent of the licensing authority. Situation 6 again depends on the licencing authority which will have to authorise the application of this possibility. Situation 7 is potentially applicable to combined vaccines and as new methods are developed it will be borne in mind to derive full benefit from the point of view of animal welfare.

Typical Examples from EP Monographs

Situation 1: assays of insulin, somatropin, etc.; rabbit pyrogen tests replaced by LAL; potency determination of rabies immunoglobulin.

Situation 2: for diphtheria and tetanus vaccines, reduction and refinement by use of immunogenicity tests; further reduction possible since situation 5 applies if one-dilution assay is used; veterinary clostridial vaccines.

Situation 3: inactivated poliomyelitis vaccine.

Situation 4: rabies vaccine (glycoprotein and ribonucleoprotein determination instead of mouse challenge test).

Situation 5: delayed hypersensitivity test for BCG vaccine (carried out on first five batches from each working seed lot); one-dilution assays for diphtheria, tetanus and rabies vaccines.

Situation 6: abnormal toxicity test carried out on final bulk for vaccines; potency determination of tetanus immunoglobulin; potency determination of DTP vaccines.

Situation 7: diphtheria and tetanus vaccines with new test systems; veterinary clostridial vaccines.

The Abnormal Toxicity Test

This is undoubtedly the most controversial test in the Pharmacopoeia, which has been accused of prolonging its application by inclusion in monographs. The test is now being abandoned for many monographs although the situation is not yet resolved for vaccines.

There is no need to go into the difficulties that are encountered when discussing the basis for a decision on the removal of this test. Early in the review of all EP texts, it was stated that the first step in replacing an animal test should be the definition of its aims; for the abnormal toxicity test, this definition is not possible so that the algorithm for replacement would not get beyond the first step. In fact, the basis for removing the test from EP monographs has always been a historical review of results. Where the abnormal toxicity test had been carried out for decades with no rejection of a batch, this was taken as an indication that the test was not useful; where an occasional positive test had been encountered, the decision is more difficult if the reason for the positive result was not identified. However, such cases have rarely been encountered in the review.

Pyrogens Test in Rabbits

Wherever possible, the pyrogens test in rabbits is being replaced by the test for bacterial endotoxins. For large-volume parenterals, the situation is complicated by the fact that the EP does not, in most cases, contain monographs on final preparations but on the ingredients; since the pyrogens test is carried out on the final product, the question cannot be dealt with fully in the EP. The general monograph on preparations for parenteral use states the conditions in which a pyrogens test is required and also the conditions in which a bacterial endotoxins test can be used instead. Thus the EP only opens up a possibility in this case but clearly, under convention no. 123, it is the responsibility of each licensing authority to ensure that reasonable efforts are made by manufacturers to develop bacterial endotoxins tests to replace the rabbit test.

For most vaccines there is no pyrogens test. For those vaccines where a test is prescribed the general provision for replacement of the test by a validated bacterial endotoxins test applies and manufacturers should be encouraged to do this by the licencing authorities.

Antibiotics

Many monographs on antibiotics used to contain : (1) the pyrogens test in rabbits; (2) the abnormal toxicity test; (3) in a few cases, a test in the cat for depressor substances. For many antibiotics, the pyrogens test has already been replaced by the bacterial endotoxins test and replacement is programmed for all the others once satisfactory data are available on validation of the test for each product.

The abnormal toxicity test will shortly be removed from all monographs on antibiotics on the basis of the criterion described above.

The test for hypotensive substances applied, for example, to kanamycin is now being reconsidered since it has been demonstrated that proper control of the manufacturing process will ensure absence of these substances.

Hormones

There has been considerable progress for hormones especially in the replacement of animal bio-assays by physico-chemical methods, usually liquid chromatography: insulin, desmopressin, somatropin, gonadorelin and somatostatin are examples. The monographs on oxytocin and calcitonin are at present under review for replacement of the bio-assay. Some hormones extracted from tissues are not amenable to this kind of assay because of the complex nature of the mixture of components: the monographs on gonadotrophins still have a bio-assay with a large number of animals. Reduction and refinement have been applied to the monograph on corticotrophin and the synthetic derivative tetracosactide by inclusion of an assay in isolated rat adrenal cells; animals are still needed for the assay but in far smaller numbers.

Progress for Blood Products

In the first monographs on blood products, animal tests were included for pyrogens and abnormal toxicity. For tetanus immunoglobulin the assay prescribed used a toxin neutralisation test in mice. For rabies immunoglobulin, situation 1 applied: the standard method of assay was by virus neutralisation in mice but an in vitro test was available (rapid fluorescent-focus-inhibition test); publication was withheld until collaborative testing had been carried out and the test could be included in the monograph.

Using the criterion described above, the test for abnormal toxicity was removed from the monographs on albumin, immunoglobulins and antithrombin III; in view of the complex nature of these products there was a certain reluctance to remove the test despite the absence of any indication of its usefulness. It seems that the fact that the pyrogens test in rabbits is still carried out was influential in overcoming this reluctance. For fibrin sealant, the test for abnormal toxicity was considered irrelevant and was not included in the recently published monograph.

For blood coagulation factors VIII and IX, in view of reports of occasional rejection of batches through failure of the test for abnormal toxicity, the test has been maintained but a further review is being carried out to see whether the reasons for failure can be established and more specific tests not using animals devised.

The test for pyrogens using rabbits has been maintained in monographs on blood products because of the occurrence of false negative results when the bacterial endotoxins test is used, i. e. batches may pass the bacterial endotoxins test yet fail the pyrogens test.

For tetanus immunoglobulin the state of knowledge is such that neither situation 1 nor situation 2 can be said to apply and an intermediate approach has been proposed in the current revision: the toxin-neutralisation test in mice is prescribed as a validation requirement but for routine assay of batches an in vitro immunochemical method can be used. This should ensure that for any given product the test in animals is carried out on only a few occasions.

Vaccines: a Special Problem

The EP contains monographs on vaccines for both human and veterinary use. After the first cycle of application of the three Rs, it is clear that vaccines represent the most difficult area. Animal tests are prescribed throughout the manufacturing process starting with the seed lot and cell bank through to the final product. The abnormal toxicity test is at present a standard feature of vaccine monographs.

Outline of Animal Testing for Vaccines in the EP

Virus seed lots	Extraneous agents Neurovirulence (oral polio vaccine)
Cell banks	Extraneous agents (suckling mice, adult mice, guinea-pigs, rabbits) Tumorigenicity
Harvests	Neurovirulence (oral polio vaccine)
Bulk purified toxoids	Absence of toxin, irreversibility of toxoid
Final lot	Safety (veterinary vaccines, target animal test) Specific toxicity (DTP and veterinary clostridial vaccines)

Delayed hypersensitivity (BCG vaccine)
Extraneous agents (poultry vaccines)
Abnormal toxicity
Pyrogens (polysaccharide vaccines)
Potency/Assay (D, T, P, hepatitis B, hepatitis A, inactivated poliomyelitis, rabies , haemophilus + many veterinary vaccines)

In some monographs, standardisation is based on a combination of physicochemical and immunochemical tests:

1. inactivated influenza vaccines
2. polysaccharide vaccines (pneumococcal polysaccharide vaccine and meningococcal polysaccharide vaccine).

The recently adopted monograph on typhoid polysaccharide vaccine requires no use of animals in any test.

For live virus vaccines, standardisation is on the basis of the virus titre and in the monographs on veterinary vaccines an animal potency test is given to be carried out once to establish the acceptable virus titre.

Why can some vaccines be accepted on the basis of physicochemical and immunochemical tests whereas for others an animal assay is judged necessary? Influenza vaccines seem to be a special case. For the two polysaccharide vaccines, immunochemical determination of the polysaccharide together with its molecular size gives sufficient assurance on the immunogenicity of the products; haemophilus vaccine, also a polysaccharide vaccine, is a more complex product because of the conjugated protein and the present draft requires an animal assay.

For many animal tests for vaccines, hardly any are in situation 1; a number are in situation 2. For diphtheria and tetanus vaccines, reduction and refinement of the potency test seem possible after collaborative testing; this has been done for diphtheria vaccine and will be started for tetanus vaccine for human use; for the veterinary vaccine, collaborative testing has already been carried out and it will be possible to amend the monograph on this basis. For other veterinary clostridial vaccines, reduction and refinement is an important issue because of the degree of suffering that is implied by the present potency tests. The process of revision has been initiated and we are counting on the co-operation of many of the participants at this symposium to be able to carry out this work.

A lot of tests not in situation 1 or 2 still remain. This means that solutions further down the line have to be explored. Some of those that have been applied or proposed for the EP are:

One-dilution assays. Several monographs on vaccines contain three-point parallel-line assays: diphtheria vaccine, tetanus vaccine, rabies vaccine etc. The use of a one-dilution assay has been described by several authors including the WHO compendium on the assay of DTP vaccines. Until recently, the view taken for the EP was that the official assay using three dilutions had to be kept since it constituted the reference method; use of one-dilution assays was seen as a matter for the manufacturer and licencing authority. Clearly, Convention no. 123 requires that one-dilution assays be carried out wherever feasible but despite this their use is not widespread; the EP cannot do a lot about such a situation but recently it has been decided that encouragement should be given by mentioning in a footnote to the monographs the one-dilution assay and the circumstances in which it can be used. A revision proposal on these lines for veterinary rabies vaccine has recently been published in Pharmeuropa and depending on the reception this receives the notion can be extended to many other monographs.

Testing of final bulks instead of final lots. In some instances a final bulk is filled on separate occasions and until recently the EP required the manufacturer to carry out all tests on each final lot. In recent monographs, performance of some tests on final bulk rather than final lot is allowed; such tests include that for abnormal toxicity, potency determinations of DTP vaccines and testing for virulent mycobacteria in BCG vaccine; this will bring about a moderate but worthwhile reduction in the use of animals.

Validation requirements and consistency testing. Until a few years ago EP monographs all had the same structure: definition, identification, tests, assay. Now monographs on vaccines have an additional Production section that gives mandatory requirements for manufacturers. The presence of this section has made it possible to reduce the use of animals by classifying some tests as validation tests or tests to be carried out on a number of consecutive batches, for example the first ones from a new working seed lot. The test for excessive dermal reactivity of BCG vaccine, for which the monograph has just been revised, is such a case.

The Biological Standardisation Programme

Recently, the EP has moved from a more or less passive approach, incorporating results of research and development work carried out independently, and has adopted a more active approach via the Biological Standardisation Programme. The programme is jointly financed by the Council of Europe and the European Union and is run by the EP Secretariat. Most of the projects are aimed at establishing reference preparations for biological products but in each phase at least one project is directed at replacement of animal tests by the organisation of validation trials for alternative methods. A trial of the Vero-cell assay for diphtheria vaccine has already been carried out; this will lead to reduction and refinement for many assays of diphtheria vaccines. A similar exercise for tetanus vaccine is planned. A current project for hepatitis B vaccine is intended to establish the feasibility of a working standard for all vaccines on the European market and of an in vitro assay to replace, at least partly, the animal assay.

It is hoped that such projects will become a regular feature of EP work. The projects proposed for inclusion in the programme should concern methods in the final stages of development where collaborative testing is needed before adoption as the official test. It is not intended that the programme be used to support basic research.

Prospects

Projects for the immediate future, some of which have been mentioned above, include refinement of the potency tests for inactivated swine erysipelas vaccine and for several veterinary clostridial vaccines where toxin-neutralisation tests in mice for antibody determination can be replaced by in vitro methods; collaborative testing of these new methods will probably be necessary so that this will be a major project.

The EP will continue to follow up all proposals received for improvements in the existing monographs. At the Utrecht workshop there was a free discussion of proposals for improvement. Many of these will be taken into the EP, some in the near future, some after a longer period because of the need to organise collaborative testing of new methods, others over a much longer period because fundamental research is still needed.

At Utrecht, the test for extraneous agents in feline and canine vaccines using intracerebral injection in mice was criticised; this question has been taken up with the group of experts concerned and a revision proposal will be published shortly to remove the test which is considered redundant in view of the present methods of production and the other tests carried out during manufacture on seed lots etc.

The abnormal toxicity test in mice for swine erysipelas vaccine was also criticised at Utrecht since it has not been found necessary in other monographs on veterinary vaccines; a revision proposal will be published shortly to remove the test.

We often find ourselves in a situation where individual manufacturers validate for their product a new method which is preferable from the point of view of animal welfare; their know-how for such a method is of course a matter of industrial property and there may be reluctance to divulge the information but in the interests of animals a very strong appeal must be made to all manufacturers to try to generalise any method which has benefits in terms of the three Rs. Undoubtedly the participants will leave this symposium convinced of this point.

The process of application of the three Rs has been said to be cyclic: as far as we are concerned; the cycling must of course continue as long as animal tests are prescribed in the EP.

Dr. P. Castle, European Pharmacopoeia Secretariat, B.P. 907, F-67029 Strasbourg Cedex 1, France

Brown F, Cussler K, Hendriksen C (eds): Replacement, Reduction and Refinement of Animal Experiments in the Development and Control of Biological Products.
Dev Biol Stand. Basel, Karger, 1996, vol 86, pp 31-39

WHO Activities Towards the Three Rs in the Development and Control of Biological Products

J. Milstien, V. Grachev, A. Padilla, E. Griffiths

Biologicals Unit, World Health Organization, Geneva, Switzerland

Key words: Collaborative studies, in vitro tests, quality control, neurovirulence test, cell substrates, international requirements

Abstract: WHO supports the concept of replacement, reduction and refinement of the use of in vivo methods for biologicals production and control, and regularly conducts reviews of its recommended procedures to allow reduction in the use of animals. The coordination of collaborative studies, publication of standardized methods, and holding of workshops on the use of these methods contributes to their use. The neurovirulence test for oral poliovaccine is probably the single most visible animal test for which alternative methods are sought. Collaborative studies on alternative methods for screening products are currently being sponsored by WHO. The use of Vero cells rather than primary monkey kidney cells for poliovaccine production can avoid the use of many monkeys. Cell banks of Vero and HEp-2 cells have been developed by WHO, tested for virus sensitivity and freedom from adventitious agents, and are available for vaccine production and control, replacing primary animal cells. For the future, final product testing will increasingly be directed towards establishment of consistency of production rather than potency. By supporting the validation and use of this approach, WHO can effectively influence more rational animal use in biologicals production and control.

Introduction

The production of a biological requires strict observance of certain procedures to ensure the safety and efficacy of the product. International requirements established by WHO are available to manufacturers and national control authorities. Because of the nature of biological medicines and their mode of production, their development and control often involve in vivo procedures, and these have served well for many years. However, it is estimated that 10 million laboratory animals are used annually for production and quality control of vaccines [1], of which only nine of over 1.2 million species account for 97% of the use [2].

Recognizing that the use of animals for biologicals production and control and for other biomedical uses is necessary, WHO staff members actively collaborated, with the encouragement of the WHO Advisory Committee on Medical

Research, as part of the Council for International Organizations of Medical Sciences (CIOMS) to publish, early in 1985, International Guiding Principles for Biomedical Research Involving Animals [2, 3]. This document was used as a basis for «Guidelines for the humane use and care of laboratory animals» developed by the European Centre for the Validation of Alternative Methods [1]. This will be published in a laboratory manual being prepared by WHO for vaccine potency testing (Manuel of laboratory methods for potency testing of vaccines used in the WHO Expanded Programme on Immunization, WHO/BLG/95.1).

General Activities to Replace, Reduce, and Refine the Use of Animals in Manufacturing and Control of Biologicals

WHO supports the concept of replacement of in vivo by in vitro methods taking into consideration new scientific and research data. Major initiatives in this direction are in three general areas: (1) regular review of production requirements and testing methods; (2) facilitating the use of in vitro tests where possible; and (3) promoting the use of continuous cell lines rather than primary cells as substrates for vaccine production and control.

Review of requirements

The Expert Committee on Biological Standardization, the expert group which approves the so-called WHO Requirements and establishes international standards and reference reagents, regularly reviews, through its secretariat in WHO, all of its recommended procedures in the light of current developments and information. In 1980 [4], all tests on virus vaccines were reviewed. This process led to many recommendations for revisions in the WHO Requirements, some of which reduce the use of animals. It was recommended that the monkey neurovirulence test for oral polio vaccine (OPV) be done either by the manufacturer or by the control authority, with tissue sections made available for inspection, rather than having both parties do the test. Another change recommended omitting tests in dog and guinea-pig kidney and for testing for neurotropic agents for measles vaccine pools, as these tests were not justified.

Other recent changes included the recommendation to replace, whenever possible, the potency test for yellow fever vaccine in mice with a cell culture test [5]. A revision of the diphtheria-tetanus-pertussis (DTP) vaccine requirements [6] suggested use of single-dilution assays in place of multiple-dilution assays for testing the potency of components of DTP vaccines and the possibility of using properly validated in vitro methods for potency testing of the toxoids. Similar changes in methods are being considered for the testing of tetanus immunoglobulins [7].

Facilitating use of in vitro tests

The support of WHO for newer methods does not stop with changing the WHO Requirements. The coordination of collaborative studies, publication of standardized methods, and holding of workshops on the use of these methods contributes to their use. WHO regularly commissions or monitors collaborative studies, the results of which are presented to the Expert Committee on Biological Standardization for the purpose of assigning international units to standards and reference materials. This activity, besides assigning units to standards, defines the testing methods which will be used.

In recent years, a series of workshops had been held on in vitro test methods for the childhood vaccines. These workshops have resulted in published documents of laboratory methods (Laboratory methods for the testing of potency of diphtheria (D), tetanus (T), pertussis (P) and combined vaccines, BLG/92.1; Laboratory methods for the titration of live virus vaccines using cell culture techniques, BLG/EPI/89.1) which will soon be published as a WHO Laboratory Manual (Laboratory Manual for the Testing of EPI Vaccines, WHO/BLG, in preparation) to replace that currently in existence [8].

Promoting non-primary cell substrates

In the light of developments in the biological sciences, WHO convened a Study Group in 1985 to consider the use of continuous cell lines as substrates for the production of biologicals. It was felt that continuous cell lines might possess distinct advantages over primary and diploid cell cultures [9] because of the possibility to guarantee consistency of production through the cell seed lot system; their less demanding growth characteristics; the fact that they could be used in microcarrier or suspension culture; their high sensitivity to a wide spectrum of viruses; and their low cost.

The Study Group concluded that continuous cells lines are acceptable as substrates for the production of biologicals [10], but that differences in the nature of the products and in the characteristics of the manufacturing process must be taken into account when making a decision on the use of a given product. They recommended the establishment of characterized cell lines of value to national control authorities and manufacturers of biologicals.

Thus, WHO has developed a WHO Cell Seed Bank for Vero cells, as these cells appeared to offer the best immediate prospect for improving the quantity and quality of vaccines now being produced in other cell systems. The Master Cell Bank of Vero cells was donated to WHO by a manufacturer at the 134th passage. The maximum passage level permitted for production, with a large safety margin from studies of tumorigenicity in newborn rats, is 150. Collaborative studies in ten laboratories established safety and freedom from adventitious agents. Producers of biologicals and national control laboratories can obtain cultures of these cells free of charge along with background information on request from the Chief, Biologicals, WHO, Geneva, Switzerland [11].

The use of Vero cells rather than primary monkey kidney cells for poliovaccine production can avoid the use of many monkeys. Since the establishment of the Vero cell bank, specific production requirements have been written for vaccines produced in continuous cell lines [12, 13]. Similarly, use of other continuous cell lines rather than primary cells for testing of vaccines and other biologicals will reduce the use of animals. Cell banks of Vero and HEp-2 cells for potency testing have been developed, tested for virus sensitivity and freedom from adventitious agents, and are available for laboratories to use in vaccine control, on request from Chief, Biologicals, WHO.

Product-Specific Examples

Measles, mumps and rubella vaccines and combined vaccines

Previous requirements for measles vaccine [14] required extensive testing on the virus seed lots for extraneous agents and neurovirulence, including testing in at least 10 adult mice, at least 20 suckling mice, at least five guinea-pigs, and in at least ten measles-susceptible monkeys. In the revision of these requirements [15] the small animal tests have been omitted. Noting that some kind of neurovirulence test on the seed lot was important and in the absence of a more satisfactory test, the Expert Committee urged the development of a better testing method and the preparation of virus seed lots in large quantities to avoid the unnecessary use of monkeys.

Polio neurovirulence test

The neurovirulence test for oral poliovaccine [13] is probably the single most visible animal test for which alternative methods are sought. Thousands of monkeys are used by producers for this test each year. One producer in 1993 used 601 monkeys for testing 56 lots of OPV plus one lot of working seed.

It has been shown by using sensitive mutant analysis by the polymerase chain reaction and restriction enzyme cleavage (MAPREC) that revertants of attenuated type 3 polio with C at position 472 rapidly accumulate during virus propagation in vitro. Abundance of revertants in this position quantitatively correlates with neurovirulence in monkeys, and is predictive of the monkey neurovirulence test results [16, 17]. The presence of revertants in vaccine lots suggests that mutants in all three types of oral poliovaccine (OPV) should be routinely evaluated by MAPREC to ensure vaccine consistency [18]. WHO thus initiated a Collaborative Study on the use of MAPREC to evaluate the use of an alternative method to the monkey neurovirulence test for screening type 3 OPV for routine production and quality control purposes. The study aimed to establish the procedure in individual laboratories, harmonize the performance of the test, and identify potential problems in its implementation.

Two of the three phases have been completed. Phase A was performed with synthetic DNA samples and concentrated on technical aspects to allow all participants to become competent. Phase B included complete MAPREC tests of virus samples with varying amounts of 472-C and adequate controls to test discrimination among samples and consistency of results. In Phase 3, vaccine samples will be analysed to determine whether participants can discriminate between vaccine batches which passed or failed the monkey neurovirulence test.

D and T potency tests

DTP vaccine is probably the single largest consumer of animals for testing purposes. Recognizing that the potency test for diphtheria toxoid was one of the most animal-intensive tests, the Expert Committee on Biological Standardization noted [6]:

> *Means of reducing the number of animals required, without prejudice to the principle of expressing potency in terms of International Units, have therefore been sought, the emphasis being on the use of the minimum number of animals necessary to provide assurance that the potency of the vaccine is indeed greater than the minimum required. A further method of reducing the number of animals used in three-dilution assay systems is to determine the individual titres of antitoxin of laboratory animals such as mice or guinea-pigs by toxin neutralization tests in cell culture. Further development of a variety of methods specific for the assay of diphtheria antitoxin should also result in a reduction of the number of laboratory animals used.*
>
> *The number of animals used in tests based on challenge can also be reduced by assaying both the test and reference vaccine at a single dilution, provided that the test is performed by laboratories with extensive experience of vaccines on which three-dilution assays have been regularly and successfully performed.*

These requirements for the first time advised that potency tests, ***once consistency of production and testing had been established***, be done by animal-sparing methods. Three of these methods will be discussed in this paper: the single-dilution method for toxoid potency testing, the Vero-cell method for testing of diphtheria toxoid, and the toxin binding inhibition (ToBI) test for potency testing of tetanus toxoid.

The single-dilution assay

Standard three-dilution assays for potency testing give upper and lower fiducial limits and allow an actual check of the validity of the potency estimates by testing linearity, slope, and parallelism of the dose-response curves. The single-dilution method is based on reducing the number of vaccine dilutions to one, and demonstrating that the response is significantly higher than a minimum level. No actual checks on linearity, slope significance, or parallelism are thus possible. Therefore, this kind of test cannot be used until adequate validation of the test procedure on a specific product has been done. The test cannot be used for comparison of different products prepared by different methods unless the specific test procedure in use has been validated for all these products. Moreover, use of the test will reduce the precision of the results.

Before adopting the single vaccine dilution assay system, a control laboratory should have acquired adequate experience with multiple-dilution assays on the type of vaccines it wishes to test. The results of such assays should provide:

- proof of consistency in production and testing;
- evidence of a highly significant regression of response on dose;
- justification of the assumptions of linearity and parallelism of the dose-response relationship for the vaccines under study.

Table 1: Impact of test method on animal use in diphtheria potency test.

Test method	Multi-dilution	Single-dilution	Vero cell*	Vero cell* + single dilution
# Animals/lot	101 guinea-pigs	32 guinea-pigs	64 mice	20 mice
Response parameter	death	death	bleed	bleed

* Can be used to test tetanus toxoid potency in same mice.

Vero-cell assay for diphtheria toxoid

The test aims to replace the lethal challenge test in guinea-pigs. As the mouse is not sensitive to diphtheria toxin, serum samples are collected after immunization of mice with various dilutions of the reference as well as the test vaccine and are examined for antibodies against diphtheria toxoid [19]. Non-neutralized toxin is detected by addition of Vero cells. The cells will die due to the cytotoxic activity of non-neutralized toxin in the incubation mixture. If toxin is neutralized by antibodies present in the serum, cells will remain alive, causing a colour change from red to yellow. The more antibodies present in the serum the more dilutions will show a positive reaction. The potency of the test vaccine can be calculated by comparing the dose response curves by parallel line analysis.

The test replaces guinea-pigs with mice and allows the assay of both diphtheria and tetanus components in the same group of animals. Table 1 shows the potential impact of this test on animal usage.

ToBI test for tetanus toxoid

The ToBI test resulted from investigations to develop an in vitro tetanus toxoid neutralization test analogous to the diphtheria potency test in Vero cell cultures [20]. It is a modification of an ELISA test based on detection of an unbound toxin in a toxin-antitoxin mixture. It is therefore similar to a toxin neutralization test, differing in the way the free toxin is detected. The procedure can be used for the estimation of the potency of tetanus vaccine, and by using diphtheria antitoxin/toxin, the potency of diphtheria toxoid as well. Table 2 shows the potential impact of this test.

Table 2: Impact of test method on animal use in tetanus potency test.

Test method	Multi-dilution	Single-dilution	ToBI*	ToBI* + single dilution
# Animals/lot	108 mice or 102 guinea-pigs	32 mice or guinea-pigs	64 mice	20 mice
Response parameter	death	death	bleed	bleed

* Can be used to test diphtheria toxoid potency in same mice.

Somatropin

The potency of therapeutic somatropin (recombinant DNA-derived human growth hormone) is determined by in vivo bioassay in hypophysectomized rats. A collaborative study was organized in 1988 to investigate whether the in vivo bioassay could be eliminated for routine batch control [21]. Physicochemical assay methods, such as reverse phase and size exclusion high performance liquid chromatography (HPLC), showed ability to discriminate between degraded and intact samples of somatropin. Study participants concluded that the bioassay could be removed from routine batch control, and that an International Standard, calibrated primarily by physicochemical assays of somatropin content, should be established.

A second collaborative study was then organized to characterize a preparation as a candidate WHO International Reference Preparation, which would provide a means of relating mass units to biological units of activity of the current International Standard of human growth hormone (pituitary origin). A meeting to discuss the interim report of the collaborative study [22] indicated that such a means of conversion could be established, and an activity could be thus assigned to the candidate International Reference Preparation.

As a consequence of these international collaborative studies, and by the establishment of the first WHO International Standard for Somatropin, a set of physicochemical methods can sufficiently guarantee the batch to batch consistency of somatropin and are appropriate to define the content of somatropin monomer.

Future Directions

The ECVAM Workshop [1] made several recommendations to WHO to encourage the use of alternatives to animal testing, as follows:

- consider and hopefully implement the «Guidelines for the humane use and care of laboratory animals»;
- set guidelines for interlaboratory validation procedures and initiate and coordinate these studies. Initiate production of reference preparations;
- review and modify monographs in light of the three Rs concept;
- omit the present animal test for abnormal toxicity or replace the test on final lot by test on final bulk, and evaluate the use of a second species;
- identify clinical endpoints to replace lethal endpoints in protection tests.

The first three are in progress, as noted in this document. The last one will need further information, and researchers are encouraged to communicate experience with nonlethal endpoints to the Chief, Biologicals Unit, WHO.

The abnormal toxicity test has been under continuous review for several years. The impact of changing the recommendations to omit it, at least for vaccines which are already tested extensively at the final bulk stage in small animals, could be significant. Table 3 summarizes some of this information. For a manufacturer who produces four final lots from each final bulk of DTP vaccine, for example, performing the test only in mice would reduce the number of animals used per final lot for this one test from 7 to 1.25.

Table 3: Impact on animal use of the abnormal toxicity test.

Test	Abnormal toxicity	Abnormal toxicity bulks only	In mice on bulks only
# Animals/lot	2 guinea-pigs + 5 mice	2 guinea-pigs/(# final lots per bulk) + 5 mice/(# final lots per bulk)	5 mice/(# final lots per bulk)
Response parameter	death	death	death or non-lethal end-point

For the future, final product testing will increasingly be directed towards establishment of consistency of production rather than potency. Two examples of this concept, one old and one new, are BCG and acellular pertussis vaccine. For these products, where no laboratory test which correlates with protection in humans has been defined, the strategy is to establish consistency of production through a test which measures an in vitro characteristic of a lot for which field efficacy has been established. Good Manufacturing Practice will become even more important in quality control of biologicals. By supporting the validation and use of this approach, WHO can effectively influence more rational animal use in biologicals production and control.

References

1 European Centre for the Validation of Alternative Methods: Guidelines for the humane use and care of animals in the production and quality control of vaccines, in Hendriksen C, Garthoff B (eds): Alternatives to animal testing in the quality control of immunobiologicals; state of the art and future prospects. Report of an ECVAM Workshop, Annex 1. Utrecht, 1994.

2 Howard-Jones N: A CIOMS ethical code for animal experimentation. WHO Chronicle 1985;39:51-56.

3 Council for International Organizations of Medical Sciences: International guiding principles for biomedical research involving animals. Geneva, 1985.

4 Expert Committee on Biological Standardization: A review of tests on virus vaccines. World Health Organization Technical Report Series 1982; 673: Annex 3.

5 Expert Committee on Biological Standardization: Requirements for yellow fever vaccine (revised 1975). World Health Organization Technical Report Series 1976; 594: Annex 1.

6 Expert Committee on Biological Standardization: Requirements for diphtheria, tetanus, pertussis and combined vaccines (revised 1989). World Health Organization Technical Report Series 1990; 800: Annex 2.

7 Sesardic D, Wong MY, Gaines Das RE, Corbel MJ: The First International Standard for Antitetanus Immunoglobulin, Human; pharmaceutical evaluation and international collaborative study. Biologicals 1993;21:67-75.

8 World Health Organization. Manual of details of tests required on final vaccines used in the WHO Expanded Programme of Immunization. BLG/UNDP/82.1 Rev.1, Rev.Corr.1)

9 Grachev VP: World Health Organization attitude concerning the use of continuous cell lines as substrates for production of human virus vaccines, in Mizrahi A (ed): Viral Vaccines Advances in Biotechnological Processes, Wiley-Liss, Inc, 1990, vol 14, pp 37-67.

10 Expert Committee on Biological Standardization: Requirements for continuous cell lines used for biologicals production (adopted 1985). World Health Organization Technical Report Series 1987;745.

11 Expert Committee on Biological Standardization. World Health Organization Technical Report Series 1990;800:11.

12 Expert Committee on Biological Standardization: Requirements for rabies vaccine (inactivated) for human use produced in continuous cell lines (adopted 1986). World Health Organization Technical Report Series 1987; 760: Annex 9. Amendment 1992. World Health Organization Technical Report Series 1994;840: Annex 5.

13 Expert Committee on Biological Standardization: Requirements for poliomyelitis vaccine (oral) (revised 1989). World Health Organization Technical Report Series 1990;800:Annex 1.

14 Expert Committee on Biological Standardization: Requirements for measles vaccine (live) (revised 1987). World Health Organization Technical Report Series 1988;771:Annex 5.

15 Expert Committee on Biological Standardization: Requirements for measles, mumps and rubella vaccines and combined vaccine (live). World Health Organization Technical Report Series 1994;840:Annex 3.

16 Chumakov KM, Powers LB, Noonan KE, Robinson IB, Levenbook IS: Correlation between amount of virus with altered nucleotide sequence and the monkey test for acceptability of oral poliovirus vaccine. Proc Natl Acad Sci 1991;88:199-203.

17 Chumakov KM, Norwood LP, Parker ML, Dragunsky EM, Ran Y, Levenbook IS: RNA sequence variants in live poliovirus vaccine and their relation to neurovirulence. J Virol 1992;66:966-970.

18 Chumakov KM, Dragunsky EM, Norwood LP, Douthitt MP, Ran Y, Taffs RE, Ridg J, Levenbook IS: Consistent selection of mutants in the 5'-untranslated region of oral poliovirus vaccine upon passaging in vitro. J Med Virol 1994, in press.

19 Kreeftenberg JG, van der Gun JW, Marsman FR, van Asten JAAM, Sekhuis VM, Hendriksen CFM: A mouse model to estimate the potency of the diphtheria component in combined vaccines. Dev Biol Standard 1986;64:21-27.

20 Hendriksen CFM, van der Gun JW, Kreeftenberg JG: Combined estimation of tetanus and diphtheria antitoxin in human sera by the in vitro Toxin-Binding Inhibition (ToBI) test. J Biol Standard 1989;17:191-200.

21 Bristow AF, Jeffcoate SL: Analysis of therapeutic growth hormone preparations: Report of an interlaboratory collaborative study on growth hormone assay methodologies. Biologicals 1992;20:221-231.

22 Bristow AF, Schulster D, Jeffcoate SL: Report of an International Workshop on assays, standardization and labelling requirements of somatropin. Pharmeuropa 1994;6:60-67.

Dr. J. Milstien, Biologicals Unit, WHO, Av. Appia, CH-1211 Geneva 27, Switzerland

Brown F, Cussler K, Hendriksen C (eds): Replacement, Reduction and Refinement of Animal Experiments in the Development and Control of Biological Products.
Dev Biol Stand. Basel, Karger, 1996, vol 86, pp 41-47

USDA: Progress Toward In Vitro Tests and Other Trends

S.A. Goodman

USDA, APHIS, BBEP, Veterinary Biologics, Hyattsville, MD, USA

Abstract: The Animal and Plant Health Inspection Service of the United States Department of Agriculture has demonstrated a commitment toward replacement, reduction and refinement of animal use in the development and control of biological products. This presentation describes some specific approaches with which APHIS has reduced the number of animals used in testing by replacing host or laboratory animal potency tests with validated in vitro tests, reduced the number of animals required for tests by allowing sequential use of animals for tests of immunologically distinct entities, and replaced host or laboratory animal challenge studies with serological tests. It also describes APHIS' plans to reduce pain and suffering of animals by allowing euthanasia when death from causes unrelated to the test is expected. Finally, it reports on refinements in extraneous agent testing, which began when host animal tests were replaced with an in vitro test method and continued when the in vitro test was replaced with a more sensitive in vitro test. The status of these approaches is discussed in the context of APHIS' current regulatory framework.

The statute governing the licensure of veterinary biological products in the United States is the Virus-Serum-Toxin Act (VSTA). This act requires USDA to establish that veterinary biological products are pure, safe, potent and efficacious before they may be marketed. The implementing regulations for the VSTA are published in Title 9 of the Code of Federal Regulations (9 CFR).

Control is exercised through licencing by the Veterinary Biologics staff (VB), inspection by the Veterinary Biologics Field Operations (VBFO) and confirmatory testing of 5-10% of the serials by the National Veterinary Services Laboratories (NVSL). For the purposes of this presentation, the focus is on testing by the manufacturers and the NVSL.

Acceptable demonstrations of safety and efficacy are integral parts of the licencing process. In addition, the USDA relies on testing of the source material (master seed testing) and the final product testing by the manufacturer and by the NVSL. Once licenced, each lot or serial of product must be tested by the firm and meet criteria for purity, safety and potency specified during the licencing process before it is released to market. The potency test which the product must pass is designed as a marker for efficacy and these tests routinely involve use of animals. It is in replacement, refinement and reduction in the use of animals in these potency tests where major progress has been made by USDA.

Reliance on final product testing makes the USDA system distinct from the FDA and the European systems for registration of veterinary drugs. Although the overall system embraces many of the principles of GMP, this is not the major control mechanism.

Final product testing by both manufacturers and NVSL is performed according to proscribed procedures. These procedures are documented and defined in two basic ways.

Standard Requirements, published in Title 9, Code of Federal Regulations (9 CFR) define tests intended to establish purity, potency, safety, and efficacy (immunogenicity) test requirements for products. Some tests were developed by industry and adopted as standard requirements after evaluation and validation by NVSL. Some tests were developed by NVSL and made available as a service to the industry.

Outlines of Production on file with USDA define tests intended to establish purity, potency, safety and efficacy (immunogenicity) test requirements for products. Outlines of Production may refer to 9 CFR Standard Requirement tests or to tests developed by the licencee when an applicable test is not addressed in the 9 CFR standard requirements. Once at least three licences have been issued for a product fraction, a Standard Requirement test may be codified into the 9 CFR.

Identified below are specific areas in which progress has been made by APHIS toward refinement, reduction or replacement of animal experiments in the development and control of biological products. For example, in vitro potency tests are under development by NVSL. Once validated, the test procedures and reagents will be made available to the industry. A list of the tests is shown in Figure 1. In vitro immunoassay tests are being developed for Escherichia Coli Bacterin, Erysipelothrix Rhusiopathiae Bacterin, Leptospira Bacterins, Clostridium Chauvoei Bacterins, Tetanus Antitoxin, and Bovine Virus Diarrhea Vaccine.

In Vitro Tests

Escherichia Coli Bacterin

In October, 1994 a proposed standard requirement was published in the *Federal Register* that would provide for in vitro immuno-assay tests for potency of Escherichia Coli Bacterin pilus antigens. Currently, there is no standardized test for these products, each manufacturer specifying its own procedure in the Outline of Production for the product. Such tests have included mouse tests, guinea pig tests, and enzyme-linked immunosorbent assays (ELISAs).

To standardize test procedures and reduce animal testing, an in vitro test has been developed to assess levels of pilus antigens in Escherichia Coli Bacterins. This provides a sensitive and specific potency test based on the widely accepted protective nature of the pilus antigens. Test procedures and reagents (including monoclonal antibodies) have been developed for K99, K88, 987P and F410 pilus types. NVSL has supplied the monoclonal antibodies and the test procedures to manufacturing firms for use in the test. Firms will be required to qualify the efficacy of their own references in vivo. An ELISA test can then be used to compare the antigen level in the reference that has been shown to elicit protection in the host, with the antigen level in each serial of product using a parallel line assay.

- Escherichia coli
- Erysipelothrix rhusiopathiae
- Leptospira serovars
- Clostridium chauvoei
- Tatanus antitoxin (and toxoid?)
- Bovine Virus Diarrhea

Fig. 1: In vitro potency tests being developed by NVSL, USDA.

It is APHIS' intention to extend the use of the in vitro parallel line assay for determining relative potency to several killed products in the future. The proposed rule also recommends that the firms identify serological titres which are correlated with protection during the initial qualification of the reference. If the correlation is sufficient, further host animal challenge studies to requalify references might be unnecessary.

Erysipelothrix Rhusiopathiae Bacterin

Vaccines containing Erysipelothrix Rhusiopathiae Bacterins are some of the most widely used veterinary biological products in the United States. Initially, the potency/immunogenicity test for this bacterin was a challenge test in pigs. The endpoint was based upon clinical signs, including high body temperature. A number of years ago, a correlation of host animal efficacy with a mouse test was established and the host animal test was superseded with a relative potency challenge test performed in mice.

Recently, however, NVSL initiated studies to replace the mouse test with an in vitro parallel line assay for determining relative potency. As a result of these studies, a protein of mol.wt. 65'000 has been characterized which is responsible for protection in both mice and swine. A monoclonal antibody has been produced which specifically recognizes this antigen, and an ELISA has been developed using this monoclonal antibody in conjunction with an NVSL reference. NVSL has supplied the monoclonal antibody and test procedures to firms for use in their own ELISA.

Each manufacturer must qualify its own reference in host animals. The potency of this reference will be measured in the ELISA developed by NVSL. Release of serials will be based on an in vitro parallel line assay for determining the relative potency of individual serials compared to the firm's validated reference preparation.

APHIS has announced its intention to approve in vitro immuno-assay tests in Outlines of Production for firms which have adequately validated a reference. This should provide field experience with the procedure and allow manufacturers to evaluate the test. In addition, APHIS is drafting a proposed rule intended for incorporation into 9 CFR requiring use of the in vitro immuno-assay test for potency testing of erysipelas bacterins. Once this rule is made available for public comment and these comments are considered, a final rule will be published, establishing this test as a Standard Requirement in the regulations. This example illustrates the move from testing in host animals to testing in laboratory animals, followed by a potential replacement of all animal testing for serial release by an in vitro test. Further information on this test may be obtained from two poster sessions [1, 2] to be presented at this symposium.

Leptospira Bacterins

In vitro tests are also being developed for leptospira bacterins such as those made from L. interrogans, serovars pomona and canicola. Other serovars will be addressed in the future. The test has been a vaccination/challenge test in hamsters. The endpoint is death of the animals. Over the past few years, however, NVSL has developed monoclonal antibodies which recognize several serovars present on NVSL challenge material. These antibodies have protected hamsters against challenge with virulent micro-organisms. The antibodies have been supplied to firms to use in an ELISA. Firms must validate a reference for use in the ELISA based on an acceptable protective level of the bacterin in hamsters. Potency testing for serial release will be based on determining relative antigen content in comparison with the validated reference, using a parallel line assay.

NVSL has supplied the reagents to several firms for evaluation. The results of the ELISA relative potency test have compared favourably with the results in the hamster test. It is intention of APHIS to codify this test into law, thereby eliminating the hamster test for use in serial release. Two poster presentations provide additional information on the technical details of this test system [3, 4].

Clostridium Chauvoei Bacterin

Developmental work is underway at NVSL to establish an ELISA test for Clostridium Chauvoei Bacterin. Classically, the potency test has been a challenge test in vaccinated guinea pigs. The goal of test under development is two-fold: the first is to develop a serological test which can be the subject of an ELISA; the second is to use the same rabbit sera in the assay that is available from currently required potency tests for bacterins of other clostridial species generally contained in the same products.

A flagellar antigen has been identified which is recognized by antibodies in sera from vaccinated rabbits. Cooperative studies between NVSL and manufacturers of veterinary biological products have generated data correlating response in rabbits with response in host animals for bacterins of *C. novyi* and *C. sordellii*. Additional cooperative studies are underway to provide the basis for establishing an appropriate correlation between the response in host animals and the ELISA for this species. Preliminary work suggests that the ELISA results compare well with results of the guinea pig test. If validation of the test is successful, the test will be incorporated into Standard Requirements in the regulations [5, 6].

Tetanus antitoxin

Levels of tetanus antitoxin have been measured classically by toxin neutralization in guinea pigs. The endpoint of the assay is death of the animals. An ELISA is currently under development at NVSL as an alternative to this test. This is possible without further validation of references by firms, since antibody levels displaying both therapeutic and prophylactic properties have been accepted. The only comparison necessary is the correlation between results of the in vitro test and the guinea pig test. The results of these tests have correlated well. Details of the developmental status of this test are available at a poster session [7].

Bovine Virus Diarrhea vaccine, killed virus

An ELISA is being developed to detect levels of Bovine Virus Diarrhea (BVD) antigens in killed virus vaccines. The current 9 CFR test for killed BVD vaccines requires the use of susceptible (seronegative) calves. The test measures seroconversion in response to vaccination as increases in virus neutralizing antibody titres. If the serological results do not meet the criteria for a successful test, the animals are challenged with virulent virus and observed for clinical signs. Once the in vitro test is fully developed and validated, it could be used to replace the host animal tests currently defined in the Standard Requirements for these vaccines. Additional details of this development are described in the poster [8].

Other Approaches

Euthanasia of test animals

Recently a proposed revision of the 9 CFR was published which would allow euthanasia of test animals which would be expected to die anyway. This revision should reduce the degree and duration of pain and suffering of animals used in tests, including those subjected to challenge with virulent micro-organisms as a reflection of a natural disease process.

Specifically, the revision proposes that animals be euthanised if they are expected to die after challenge with virulent micro-organisms as a part of the test or after accidental illness or injury. The revision allows acceptable criteria for morbidity or expected mortality to be specified in the Outline of Production for the vaccine before initiation of the test. Under these circumstances, euthanised test animals could be included in the analysis of results.

The proposed revision has been published for public comment. The comment period is usually 60 days. Once public comment has been received and considered, the proposed rule will be published as a final rule. At that point, it will become a part of the regulations.

Polymerase Chain Reaction (PCR) technology

PCR technology is being used to assess chicken anaemia virus contamination of virus master seeds intended for use in poultry. This technology will provide an alternative to the detection of such contamination by observation of seroconversion in birds. This technology would eliminate the use of host animals as test subjects.

Serology as an alternative to challenge

Where serology can be validated as an appropriate alternative test for efficacy, APHIS has allowed this in lieu of host animal challenge. This can eliminate the pain and suffering resulting from challenge with virulent micro-organisms. Use of serology has proven an acceptable surrogate marker for potency for a variety of vaccines, including avian reovirus, infectious bronchitis virus, tetanus, and clostridial antigens. Other serological responses have been accepted on a case-by-case basis when adequate validation is available.

Sequential or concurrent use of the same animals and reduction in the number of animals

Changes have been made in in vitro tests which have resulted in a reduction in the number of animals used. Classically, potency tests for *C. perfringens* C&D toxins have been toxin neutralization tests using both rabbits and mice. In this test, the ability of immune rabbit serum to neutralize the effects of toxin in mice is measured. Specifically, serum is obtained from rabbits vaccinated with a specified fraction of the host animal dose. A standardized amount of toxin is mixed with the rabbit serum, and the complexes are injected into mice, which serve as indicators for the ability of the rabbit serum to neutralize the effects of the toxin.

Currently, amendments to the standard requirements for these toxins are under consideration which will 1) reduce the number of mice required for the test by allowing the use of pooled sera, rather than requiring testing of individual sera, and 2) reduce the number of rabbits required for potency testing of these products used in combination products with *C. novyi* and *C. sordellii.* The latter would be accomplished by modifications to the test which provide consistency with standard requirements for *C. novyi* and *C. sordellii*, to permit use of the same group of rabbits.

There are other situations in which sequential or concurrent use of immunologically distinct entities have been allowed in studies to establish efficacy. These are allowed on a case-by-case basis where adequate justification is provided by the manufacturer.

Limitations of Alternatives for In Vitro Testing

This topic should not be addressed without mentioning the limitations of some of the in vitro approaches which have been described in this presentation. The major limitation is the lack of ability of an in vitro test to reflect the total protective response of the host. Usually, only one parameter is being measured with any sort of surrogate or in vitro test, e.g., humoral or cell-mediated but not both; antibody to an epitope but not the entire range of specificities made by the host. Actually, the host response is a composite of a variety of interacting factors. At best, in vitro tests can be expected to define a response which can be measured and which has a meaningful relationship to the host response.

A second limitation relates to the ability of the test to identify a minimal antigen level which can elicit a protective host response. This relationship is often expensive to define and may require significant levels of host animal testing. In addition, other questions must be addressed relative to surrogate testing, including variability of the assay, stability of the reagents, expense of test validation, and requirements for re-qualification of the test and reagents.

A third limitation relates to the re-qualification of references which constitute a critical component of a valid in vitro test. The best approach to re-qualification has not yet been identified. It will probably be necessary to evaluate each situation on an individual basis. Despite these limitations, APHIS has accepted the concept and is making progress toward development of surrogates to animal testing.

References

1 Henderson LM, Jenkins PS, Scheevel KF, Walden DM: Characterization of a monoclonal antibody for in vitro potency testing of erysipelas bacterins. Poster presented at this symposium.

2 Henderson LM, Scheeval KF, Walden DM: Development of an enzyme-linked immunosorbent assay for potency testing of erysipelas bacterins. Poster presented at this symposium.

3 Kolbe DR: Use of ELISA to quantitate the antitoxin content of commercial equine tetanus antitoxin. Poster presented at this symposium.

4 Hauer PJ, Whitaker MS, Henry LA: A serological potency assay for *Clostridium chauvoei*. Poster presented at this symposium.

5 Hauer PJ, Rosenbusch RF: Identifying and purifying protective immunogens from cultures of *Clostridium chauvoei*. Poster presented at this symposium.

6 Ruby KW: Evaluation of the *Leptospira pomona* ELISA and its correlation to the hamster potency assay. Poster presented at this symposium.

7 Ruby KW, Walden DM, Wannemuehler MJ: Development of an in vitro assay for measuring the relative potency of leptospiral bacterins containing serovar *canicola* and its correlation to the hamster potency assay. Poster presented at this symposium.

8 Ludemann LR, Katz JB: Enzyme-linked immunosorbent assay assessment of Bovine Viral Diarrhea virus antigen in inactivated vaccines using polyclonal or monoclonal antibodies. Poster presented at this symposium.

Dr. S.A. Goodman, USDA, APHIS, BBEP, Veterinary Biologics, 4700 River Road, Unit 148, Riverdale, MD 20737-1237, USA

Brown F, Cussler K, Hendriksen C (eds): Replacement, Reduction and Refinement of Animal Experiments in the Development and Control of Biological Products.
Dev Biol Stand. Basel, Karger, 1996, vol 86, pp 49-51

Alternatives to the Use of Animals in R & D and Quality Control of Veterinary Vaccines: An Industry View

C. Verschueren, S. Zänker

European Federation of Animal Health (FEDESA), Brussels, Belgium

INTRODUCTION

Animals are often used in the routine quality control of veterinary vaccines. This is a reason why, given the growing recognition of animal welfare and ethics in the use of animals in research, it is in the area of veterinary vaccines that the need for alternative methods to animal testing has been identified. Several meetings have been dedicated to this issue and a report has been prepared: detailed. FEDESA supports all these endeavours and is pleased to participate in this debate. This paper reviews the importance of quality control, analyses the weaknesses in the EC legislation with respect to animal testing and presents the role of FEDESA in this discussion.

THE IMPORTANCE OF QUALITY CONTROL

Quality control is a key component in the production of medicinal products in general, and of immunological veterinary medicinal products in particular. It is one of the facets of integrated quality assurance necessary to ensure the quality of the finished product and the consistency of this quality from one batch to another. The question, however, is the relative importance of this component.

While quality control (QC) remains important, it is recognised that QC on its own has some weaknesses in infallibly detecting defects. Nowadays, the emphasis in modern production is placed much more on concepts such as total quality management and integrated quality assurance. These principles govern Good Laboratory Practice (GLP) and Good Manufacturing Practice (GMP) that have been EC law since 1987 and 1993 respectively, and are wholeheartedly supported by the industry. It is expected that when these principles are uniformly applied throughout the industry and an efficient GMP monitoring system is operational, some of the finished product testing will become redundant, certainly the one that is imposed as a repetition by state control laboratories. As pointed out by the regulatory authorities of a Member State at a workshop organised by the Dutch Ministry of Agriculture and of Health on June 2, 1994 «The authorities should have the courage to allow finished product testing to be abandoned».

EC Legislation and Finished Product Testing

The core text of the EC Directive 81/851 and its extension to veterinary vaccines (EC Directive 90/677) govern the legal requirements for finished product testing and batch release.

Article 30 of the main Directive states that the Qualified Person, i.e. the person responsible on behalf of the company, is the primary one responsible for releasing a particular product batch. This person is therefore also responsible for carrying out the necessary tests for ensuring batch quality and consistency. The onus is therefore placed on the manufacturer. These tests are normally carried out as part of the manufacturing process.

Article 3 of 90/677 recognises that Immunological Veterinary Medicinal Products (IVMPs) are of a special nature and that Member Sates may require batches of vaccines for carrying out (and therefore in most cases repeating) specific tests. In FEDESA's view, some Member States are abusing this legal provision and require batches of each production lot and of all vaccines. Furthermore, when a batch has been examined by one Member State and is being exported to another Member State, re-control by the importing Member State is not permitted, unless the veterinary situation justifies it and the EC Commission has been notified. This provision is also subject to potential over-interpretation. Indeed, some Member States consider that only a physical examination (i.e. testing) of some samples of a certain batch is what counts in this context. In this regard, Member States that no longer test each batch but carry out a control of the manufacturer's quality assurance system would not be considered as having done a physical examination of each batch. This is against the principles of mutual recognition of the equivalence of methods and needs to be adapted. The EC Commission is stimulating mutual recognition by having developed an EC procedure and form for the batch release. A Member State releases a vaccine batch by using this form at the request of the manufacturer, to enable the release of the same batch in other Member States without further retesting.

Finished Product Testing and Import from non-EC Countries

In accordance with the EC legislation, a vaccine batch manufactured outside the EC has to be re-tested on entry into the EC by a company's Qualified Person. Since this testing is normally carried out at the manufacturing site, it is only repeated in the EC. This results in redundant animal tests, a situation which is untenable. Proper implementation and monitoring of GMP and bilateral agreements on the mutual recognition of GMP inspections between the EC and key countries such as the USA, Japan, Canada, Australia, New Zealand and Switzerland should remedy this situation.

EC Legislation and Alternatives to Animal Testing

All situations described above stem from poorly written or over-interpretation of the EC legislation. They result in undue repetition of animal testing. Modification of this legislation would lead to considerable reduction in animal testing.

Another aspect that the legislation should take into account is the need for flexibility in the technical requirements. Flexibility expressed in the wording of EC Guidelines or European Pharmacopoeia monographs should enable the use of in-vitro methods provided these are properly validated. Unfortunately, some of the testing requirements are written in such a way that alternative methods to the ones described are not authorised. This disconcerts the research community which is adamant in wishing to develop alternative methods.

The Role of FEDESA

FEDESA (the European Federation of Animal Health), as the representative organisation of the animal health industry in Europe, is the key partner to European regulatory authorities. The Federation adheres to scientific principles and supports the moves towards the development of alternatives to animal testing. It proposes to use the excellent report by M.D.O. van der Kamp [1] which provides a good basis for discussion and future work. It also suggests that the ECVAM could play a stimulating and co-ordinating role in this area.

Reference

1 Van der Kamp MDO: Ways of replacing, reducing or refining the use of animals in the quality control of veterinary vaccines. Study carried out at the Institute for Animal Science and Health (ID-DLO) in Lelystad, The Netherlands, 1994.

Dr. C. Verschueren, European Federation of Animal Health (FEDESA), rue Defacqz 1, Bte 8, B-1050 Brussels, Belgium

Brown F, Cussler K, Hendriksen C (eds): Replacement, Reduction and Refinement of Animal Experiments in the Development and Control of Biological Products.
Dev Biol Stand. Basel, Karger, 1996, vol 86, pp 53-62

The Views and Policy of the Japanese Control Authorities on the Three Rs

F. Chino

Department of Safety Research on Biologics, National Institute of Health, Tokyo, Japan

Key words: JNIH, testing and regulations, neurovirulence test of OPV, endotoxin test, abnormal toxicity test.

Abstract: NIH Japan has tested and regulated the three Rs of animal experiments in the development and control of biological products in a stepwise manner. (1) The number of monkeys was reduced from 108 to 72 for the neurovirulence test of OPV in each type, since paralysed monkeys inoculated intraspinally revealed a linear relationship between average scores of the lumbar lesion and cumulated paralysis occurrence ratio (%). (2) Rabbits for the pyrogen test were replaced by the endotoxin test for PPF, albumin and interferon products. The endotoxin is measured by the parallel line assay method using both turbidimetric kinetic and colorimetric methods. (3) Histopathological examination was introduced to the abnormal toxicity test as a refinement. Mean body weight loss of two guinea pigs inoculated with five ml. of an albumin product in each was far below the mean weight of pooled guinea pigs used ($p \leq 0.01$) and appeared repeatedly. The histopathological examination showed focal necrosis in the liver. This finding was suggestive of the presence of endotoxin in the product. The product contained 0.1 EU/ml of endotoxin. The same amount of the reference endotoxin produced a similar change in guinea pigs. The mechanism of the liver cell damage by endotoxin has been investigated by an in vitro method.

INTRODUCTION

I present the views and policy of NIH JAPAN on three Rs, Replacement, Reduction and Refinement of animal experiments, in the development and control of biological products Some biological products have been highly purified and their components crystallized and their structures elucidated at the molecular level. Their properties could not be evaluated by chemical or physical tests alone, but required biological procedures [1, 2]. Animal experiments are contained in the biological procedures as an essential control element (Table 1). Among testing and regulations, three examples are underlined in the Table; (1) reduction of the number of monkeys used for the neurovirulence test of OPV, (2) replacement of the rabbits in the pyrogen test by endotoxin test, and (3) refinement of the use of guinea pigs into the abnormal toxicity test by the application of a histopathological method.

Table 1: Views and policy of NIH Japan on Three Rs of animal experiments. Biological products testing and regulations.

Development	Pest vaccine, Interferon, and HA vaccine
Control	Potency tests Safety tests
General tests	1) Pyrogen test-*endotoxin test* *2) Abnormal toxicity test* 3) Sterility test 4) Physico-chemical test
Specific toxicity tests	*1) Neurovirulence test of OPV* 2) Others

Materials and Methods

Reduction in number of monkeys at neurovirulence test of OPV

Poliovirus Sabin type 1 was used in this experiment. Bulk lot numbers 111-1 were produced by Connaught and 113 by the Japan Poliomyelitis Research Institute. Both lots were shown not to be statistically different in their neurovirulence ($p < 0.5$). These two lots were adjusted to $10^{7.5}$ CCID50/ml by monkey kidney cell-tube titration. Intraspinally, 0.1 ml doses in each lot were inoculated in the lumbar region of 36 cynomolgus monkeys. Paralysis was found in 49 of 72 monkeys inoculated with either lot. All the monkeys were autopsied 17 to 19 days after inoculation. After removing the central nervous system, they were processed in the usual manner. The lumbar spinal cord was dissected into at least 12 sections. The lesions observed microscopically were scored according to criteria given by WHO [3]. The average score shows either 0, 1, 2, 3, or 4 in the individual lumber spinal cord.

Replacement of rabbit pyrogen test by endotoxin test

Two kinds of the limulus amebocyte lysate (LAL) test reagents [4] (ES-Test Wako and Endospecy) were purchased. The first reagent was used for the turbidimetric kinetic method and the latter for the colorimetric method. The reference endotoxin was prepared from Escherichia coli UKT-B strain. Samples and the reference endotoxin were diluted with distilled water. The endotoxin contained in the samples and the reference was measured according to the parallel line assay method. The amount of endotoxin was expressed as endotoxin units (EU) which corresponds to the reference endotoxin in which 1 ng = 8 EUs. The amounts of endotoxin estimated by the LAL test in the samples were correlated with the rise of rabbit body temperature.

Refinement of the abnormal toxicity test by a histopathological method

When mean body weight loss of two guinea pigs inoculated with 5 ml of product into the abdominal cavity was still far below the mean and standard deviation of those of pooled guinea pigs ($p < 0.01$) at 1, 2, 3, or 7 days after inoculation the two guinea pigs were autopsied, after the fifth consecutive test. After fixation, the removed organs were dissected and processed in the usual manner to make histological sections. The presence of lesions would require specific examination to determine the cause.

Results

Reduction in the number of monkeys in the neurovirulence test of OPV

As shown in Table 2, average lesion scores in the lumbar region were demonstrated in all the monkeys with paralysis. Figure 1 shows a linear relationship in normal distribution paper between the average lesion score in the lumbar region and the cumulative paralysis occurrence ratio (%). The lesions caused by lumbar inoculation therefore reflect the occurrence of paralysis in monkeys. The lumbar inoculation is adequate to predict the occurrence of paralysis. Another inoculation such as the intracranial route is no longer necessary. We could therefore reduce the number of monkeys which are used for the intracranial inoculation.

Replacement of rabbit pyrogen test by the endotoxin test

To establish a parallel line assay, parallelism was examined between the reference endotoxin and the endotoxin contained in human plasma protein fraction (PPF). Figure 2 shows the parallelism between the two lines in the endotoxin dose (EU/ml) response measured by the rabbit body temperature. After we established parallelism between the dose response lines for the reference endotoxin and the sample-containing endotoxin, the amount of endotoxin in PPF samples were measured by the parallel line assay method. The correlation coefficient was 0.547 in a total of 52 samples by the colorimetric method and 0.642 in a total of 48 samples by the turbidimetric kinetic method. Figure 3 shows the correlation between the rise of body temperature and the endotoxin dose in a total of 19 samples to which were added various dose of endotoxin using the colorimetric method assay. The correlation coefficient was 0.934. Figure 4 shows the same correlation as indicated in Figure 3 by the turbidimetric kinetic method assay. The correlation coefficient was 0.911. Good correlation was obtained by both methods. These results suggest the replacement of the rabbit pyrogen test by the endotoxin test.

Refinement of abnormal toxicity test by histopathological method

Two guinea pigs inoculated with 5 ml. of albumin into the abdominal cavity were autopsied at the fifth consecutive test, since their mean body weight loss was still far below the mean and standard deviation of that of approximately 2.000 pooled guinea pigs ($p < 0.01$). The liver shows focal necrosis in both the guinea

Table 2: Relationship between average lesion score and paralysis occurrence.

Average lesion score in lumbar region of each monkey	**0**	**1**	**2**	**3**	**4**
Rate of cumulative number of monkeys with paralysis	0/49	1/49	24/49	46/49	49/49
Cumulative paralysis ratio (%)	0	2.0	49.0	93.9	100

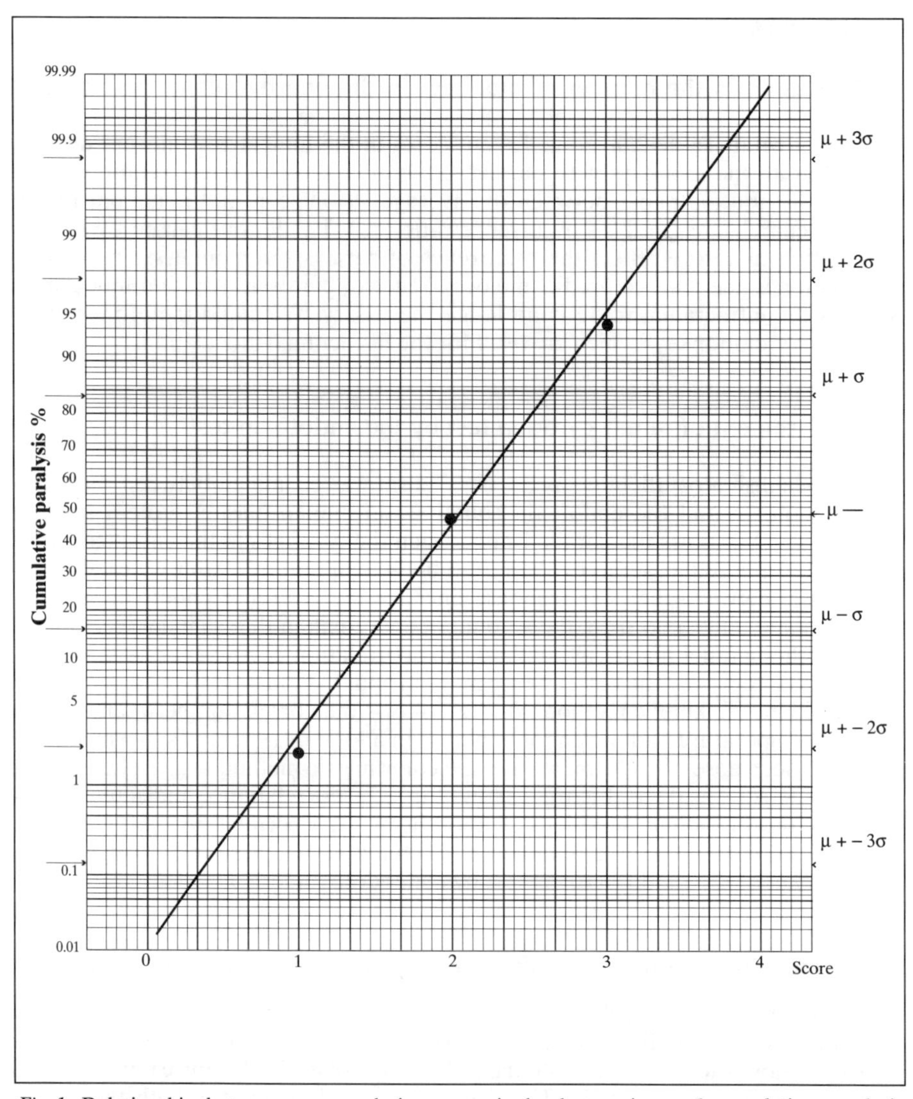

Fig. 1: Relationship between average lesion score in lumbar region and cumulative paralysis occurrence ratio.

pigs. Figure 5 reveals coagulative necrosis of liver cells associated with round cell infiltration. The infiltrate consisted of mononuclear cells and polymorphonuclear leukocytes. The mixture of polymorphonuclear leukocytes was suggestive of the presence of exogeneous endotoxin in an albumin product. The product contained 0.1 EU/ml of endotoxin.

The same amount of reference endotoxin produced a similar change in the liver as shown in Table 3. Further, the liver injury with the endotoxin was confirmed by

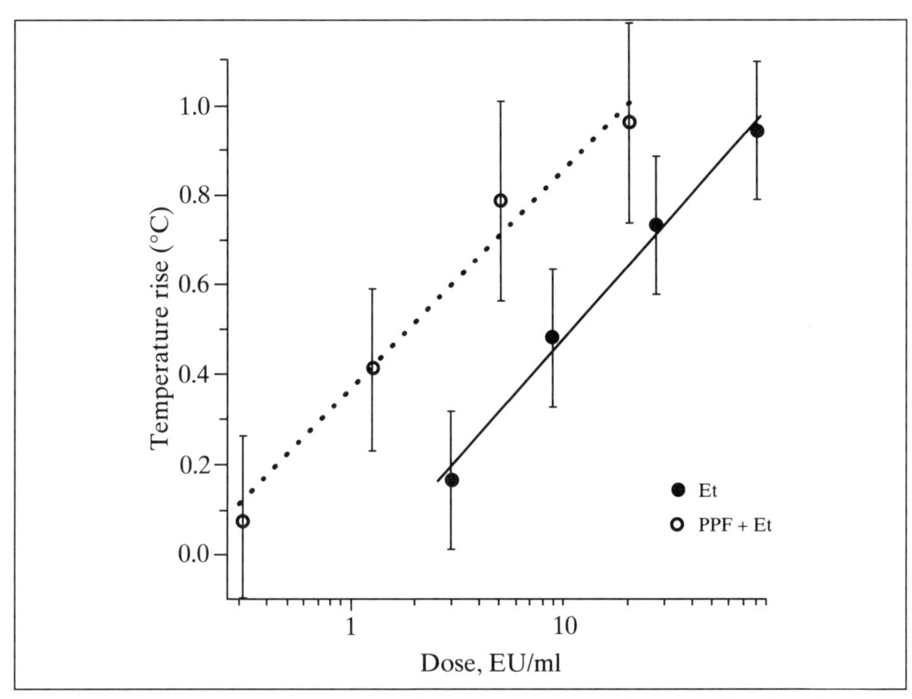

Fig. 2: Enhancing effect of PPF injection on to the pyrogenicity of *E. coli* endotoxin.

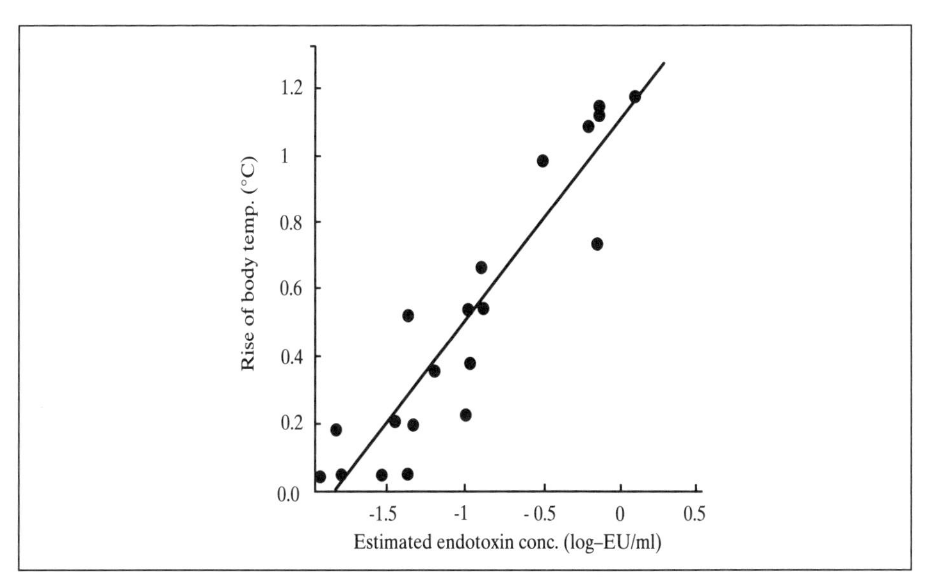

Fig. 3: Correlation between LAL-Test (Colorimetric Method) and Rabbit Pyrogen Test. Total of 19 samples to which were added various doses of endotoxin were assayed. Correlation coefficient is 0.934.

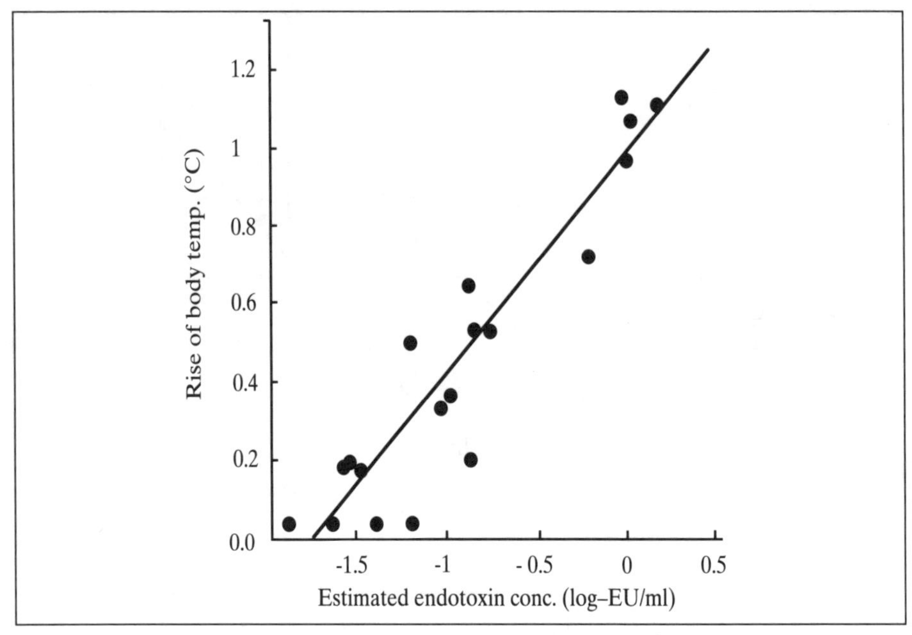

Fig. 4: Correlation between LAL-Test (Turbidimetric Kinetic Method) and Rabbit Pyrogen Test. Total of 19 samples to which were added various doses of endotoxin were assayed. Correlation coefficient is 0.911.

GPT-dose response release in the serum of mice injected with serially graded doses of endotoxin (LPS) (Fig. 6). The mechanism of the liver cell damage by LPS has been investigated in vitro. Figure 7 suggests that LPS first stimulated Kupffer's cells or macrophages which might release several cytokine-like substances which would affect the liver cells.

Discussion

Reduction in the number of monkeys in the neurovirulence test of OPV

Based on the relationship of paralysis occurrence with lesion score in the lumbar region, we consider that the lumbar inoculation is adequate to assure the safety of OPV. Seventy-two monkeys have therefore been used, with two or three dilutions in both the test and reference vaccines (Table 4). Both the mean lesion scores are compared statistically by the t-test in each dilution. From the number of monkeys with paralysis in each dilution, paralysis-inducing virus dose 50 (PID_{50}) can be obtained by the Reed-Muench method. The PID_{50}s of both the vaccines were compared.

Replacement of rabbit pyrogen test by endotoxin test

The same type of examination used for the PPF product has also been performed for albumin and several types of interferon. The upper limit was deter-

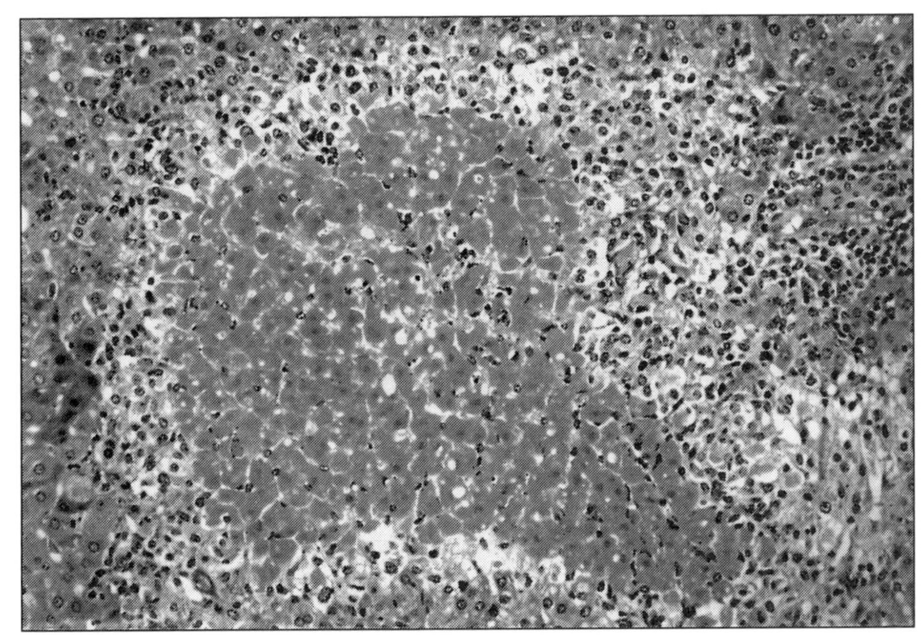

Fig. 5: A focal necrosis lesion, in which coagulative necroses of liver cells are associated with round cell infiltration. The infiltrate consists of mononuclear cells and poly-morphonuclear leukocytes. The mixture of polymorphonuclear leukocytes is suggestive of the presence of endotoxin. H.E. staining. X50.

mined as follows: PPF < 0.2 EU/ml, albumin < 0.6 EU/ml and interferon < 0.25 EU/ 6×10^5 IU (Table 5). We are examining other products such as several types of globulins and vaccines.

Refinement of the abnormal toxicity test by a histopathological method

In general, when the mean body weight loss of two guinea pigs or mice is larger than the mean and standard deviation of pooled animals ($p \leq 0.01$), the test would

Table 3: Number of sections with focal necrosis in the livers of two guinea pigs inoculated with endotoxin or saline.

Inoculation	**Focal necrosis**		
	(+)	**(–)**	
Endotoxin 0.1 EU/ml	21	24	45
Saline	3	41	44
	24	65	89
		χ^2 – Test : P < 0.001	

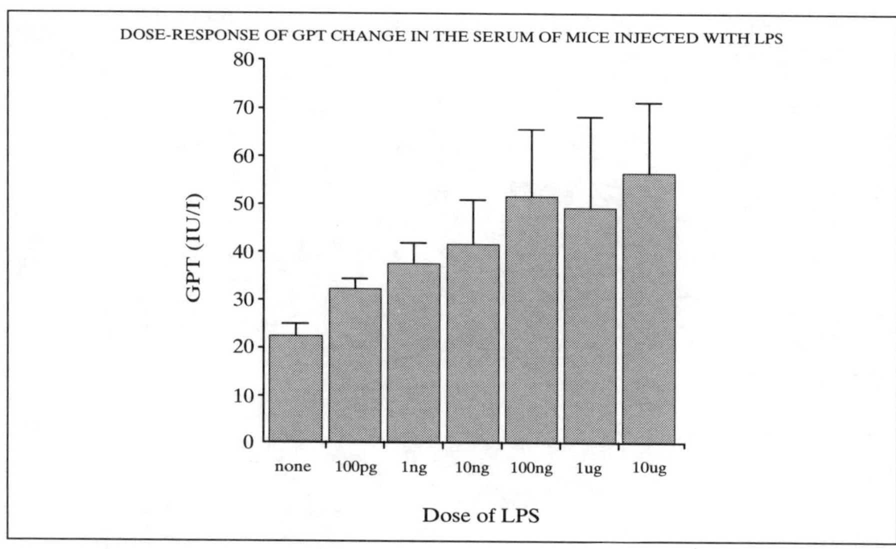

Fig. 6: DBA/2 mice were bled four hours after LPS injection for the assay of serum GPT. The change of serum GPT level was observed in a dose dependent manner and a significant change was observed at a dose of endotoxin as low as 100 pg.

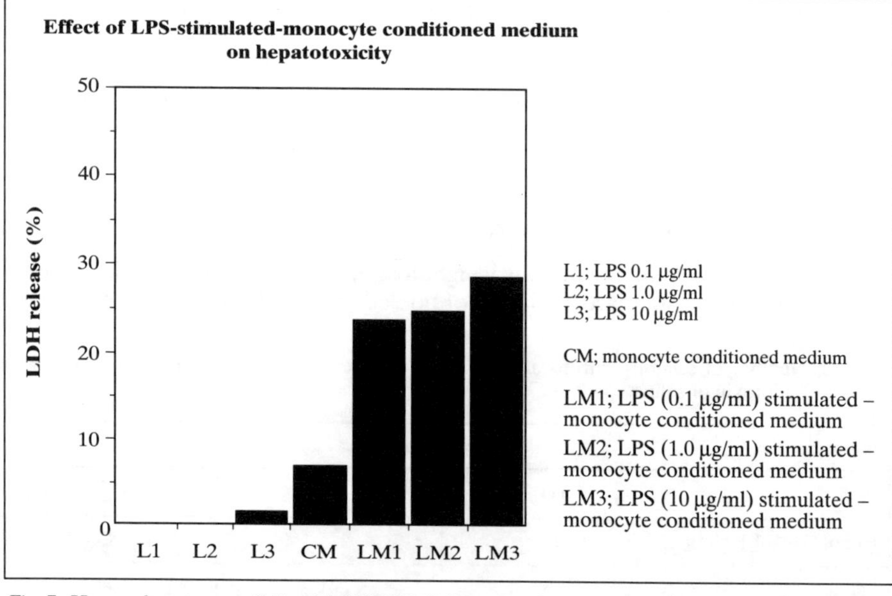

Fig. 7: Human hepatoma cell line PLC/PRF/5 (1-3) was maintained in RPMI 1640 medium supplemented with 10% foetal calf serum, penicillin (100 U/ml) and streptomycin (100 g/ml) at 37°C in 5% CO_2, 95% air. When the cells were 90-100% confluent they were exposed to LPS (*E. coli*, 0111: B4, Shigma) or LPS-stimulated-monocyte conditioned medium (LM). Then the microtitre plate was incubated for 48 h; LDH release assay was performed for culture medium.

Table 4: Neurovirulence test of OPV in monkeys: Comparaison of paralysis-inducing virus $dose_{50}$ between test and reference by Reed-Muench method.

	Test	Reference
Inoculation/Dilution	Monkeys	Monkeys
I.S. (0,1 ml) 10^0	12 (18)	12 (18)
10^{-2}	12 (18)	12 (18)
10^{-4}	12 (0)	12 (0)

() Number of monkeys in type3. I.S., Intra-spinal inoculation.
Comparison of mean lesion scores by t-test in each dilution is added.

be repeated up to three times. If, after the last test, the mean body weight loss is still far below the level $p \leq 0.01$, histopathological examination is introduced into the abnormal toxicity test. When the histopathological examination results are positive, a specific toxicity examination would be prepared (Table 6). In the case of an albumin product, GPT-dose response in the serum of mice injected with a serially graded dose of endotoxin was measured to confirm the liver injury induced by the contaminating endotoxin. Furthermore, the mechanism of liver cell damage by endotoxin has been studied using a human hepatoma cell line. The results are in agreement with those reported by Hartung and Wendel [5, 6]. The released substances from the endotoxin-stimulated Kupffer's cells or macrophages were analysed as TNF or IL-1 [7, 8]. On the other hand, the cytokine-release system from the endotoxin-stimulated macrophages might be available for the pyrogen test [9]. The macrophage, moreover, could be responsible for pyrogen such as muramyl dipeptide other than the endotoxin.

Conclusion

NIH Japan has tested and regulated the three Rs of animal experiments in the development and control of biological products in a stepwise manner: (1) the number of monkeys was reduced from 108 to 72 for the neurovirulence test of OPV of

Table 5: Endotoxin test.

Parallel line assay method **Turbidimetric kinetic method and Colorimetric method**	
Reference entodoxin	*E. coli* UKT-B strain
Upper limit of endotoxin	PPF <0.2 EU/ml
	Albumin <0.6 EU/ml
	Interferon <0.25 EU/6×10^5 IU

Table 6: Abnormal toxicity test

Mean body weight loss	Larger than mean and S.D. of that of pooled animals ($P \leq 0.01$)
	Repeat test three times
	In the last test, mean body weight loss still $P \leq 0.01$, histopathological examination
Specific toxicity examination after the above examination is positive	

each type; (2) rabbits for the pyrogen test were replaced by the endotoxin test in PPF, albumin and interferon products; (3) histopathological examination was introduced for the abnormal toxicity test or general safety test as a refinement.

Acknowledgements

The author would like to express many thanks to colleagues for their cooperation, Drs. S. Ohkawa (the former Chief of Laboratory of Toxicological Pathology), S. Ishida (the former Chief of Laboratory of Biological Statistics) and K. Eto (the former Chief of Laboratory of General Toxicology), Drs Y. Horiuchi (the Chief of Laboratory of Biological Statistics), N. Goto, S. Asakawa, T. Uchida and A. Masumi-Fukazawa, and others.

References

1 WHO: Development of a national control laboratory for biological substances (A guide to the provision of technical facilities). WHO Tech Rep Ser, 1970, No. 444 Annex 3, pp 71-86.

2 Biological standards Act of UK 1975.

3 WHO: Requirements for poliomyelitis vaccine (oral) (Requirements for biological substances N° 7). WHO Tech Rep Ser, 1990, No. 800 Annex 1, pp 46-49.

4 Levin J: Bacterial endotoxins: structure, biomedical significance, and detection with the limulus amebocyte lysate test, history of the limulus lysate test, in Liss AR (ed): New York, 1985, p 3.

5 Hartung T, Wendel A: Endotoxin-inducible cytotoxicity in liver cell cultures I. Biochem Pharmacol 1991; 42:1129-1135.

6 Hartung T, Wendel A: Endotoxin-inducible cytotoxicity in liver cell cultures II. Biochem Pharmacol 1992; 43:191-196.

7 Nagakawa J, Hishinuma I, Hirota K, Miyamoto K, Yamanaka T, Tsukidate K, Katayama K, Yamatsu I: Involvement of tumor necrosis factor-α in the pathogenesis of activated macrophage-mediated hepatitis in mice. Gastroenterology 1990; 99: 758-765.

8 Chensue SW, Terebuh PD, Remick DG, Scales WE, Kunkelt SL: In vivo biologic and immunohistochemical analysis of interleukin-1 alpha, beta and tumor necrosis factor during experimental endotoxemia. Am J Pathol 1991; 138:395-402.

9 Taktak YS, Selkirk S, Bristow AF, Carpenter A, Ball C, Rafferty B, Poole S: Assay of pyrogens by interleukin-6 release from monocytic cell lines. J Pharm Pharmacol 1991; 43: 578-582.

Dr. F. Chino, Dept. of Safety Research on Biologics, National Institute of Health, Gakuen 4-7-1, Musashimurayama, Tokyo 208, Japan

Brown F, Cussler K, Hendriksen C (eds): Replacement, Reduction and Refinement of Animal Experiments in the Development and Control of Biological Products.
Dev Biol Stand. Basel, Karger, 1996, vol 86, p 63

SESSION III

General Aspects

Chairmen: R. Winsnes (Oslo, Norway)
C. Folkers (Brussels, Belgium)

Brown F, Cussler K, Hendriksen C (eds): Replacement, Reduction and Refinement of Animal Experiments in the Development and Control of Biological Products.
Dev Biol Stand. Basel, Karger, 1996, vol 86, p 65

Results and Recommendations of an ECVAM Workshop on «Alternatives to Animal Testing in the Quality Control of Immunobiologicals», Utrecht 16-17.4.1994

D.W. Straughan

Fund for The Replacement of Animals in Medical Experiments (FRAME), Nottingham, UK

A workshop on alternatives to animal testing in the quality control of vaccines was organized by the Centre for the Validation of Alternative Methods (ECVAM) of the European Union, April 16-17, 1994. The aim of the workshop was to evaluate potential alternatives, to identify obstacles in development, validation and implementation and to discuss possible solutions. Participants were affiliated to national control laboratories, regulatory bodies, vaccine manufacturers or academia.

A wide range of topics was discussed, including the use of humane end-points, in vitro methods in the quality control, for example, of clostridial vaccines and rabies vaccine, abnormal toxicity tests and safety tests. The workshop identified a number of areas where further studies were needed and made specific recommendations to ECVAM and to the European Parmacopoeia, the WHO, national control authorities, manufacturers and the scientific community. The participants also felt that criteria for validation procedures were urgently needed. Finally, to express the concern of the scientific community for the welfare of experimental animals, the workshop drafted guidelines for the humane use and care of laboratory animals. Guidelines were set for humane treatment of animals and for the implementation of the Three Rs concept.

The full paper has been published in ATLA 1994;22:420-434.

Brown F, Cussler K, Hendriksen C (eds): Replacement, Reduction and Refinement of Animal Experiments in the Development and Control of Biological Products.
Dev Biol Stand. Basel, Karger, 1996, vol 86, pp 67-72

The Five Rs: Refinement, Reduction, Replacement. A Regulatory Revolution

R.N. Lucken

Veterinary Medicines Directorate, Addlestone, Surrey, England

Abstract: Regulatory requirements have frequently been cited as a major reason for the continued performance of batch control tests using animals. Although such claims have not always been justified regulatory authorities and associated organizations such as the pharmacopoeial committees have sometimes been slow to respond to emerging knowledge and the potential of novel techniques. The regulatory approach to quality control tests in the wider context of quality assurance and total quality management is discussed together with the role which the regulatory authorities can play in promoting reduction in the use of animal based tests and the opportunities which exist for them to encourage manufacturers to explore other approaches.
The need for caution on the part of regulatory authorities in requesting additional data which would necessitate further animal testing is stressed and the concept is proposed of a formal review of the benefits of animal testing applied on a batch basis whenever a marketing authorization is issued or renewed. The types of animal test most amenable to deletion or modification are considered together with those tests which present particular problems. Specific proposals are made for an immediate reducation in the numbers of animals used.

Introduction

In examining the role of animal experiments in the development and control of Biological Products the justification for the use of such tests is, in principle, the outcome of a risk versus benefit analysis. In the very early days of vaccine use, before adequate quality assurance and quality control measures were in place, safety and efficacy problems with both human and veterinary products occurred. The existence of these problems led directly to increased quality control testing and particularly to the use of animal-based tests. The regulatory authorities were, at least in part, responsible for the introduction of some of these tests and regulatory requirements are frequently cited by manufacturers as one reason for the continued performance of batch control tests using animals. Although the regulatory authorities cannot shoulder the entire responsibility it is true that they and associated bodies such as pharmacopoeial committees have sometimes been slow to respond to emerging knowledge and the opportunities for change arising from the development of novel techniques.

The time is right for a revolution in regulatory affairs in which the authorities, working with the manufacturers, should re-appraise, and continue to re-appraise, the level of quality control applied, and question its appropriateness. In fact, this re-appraisal needs to encompass all aspects of the testing of products including that performed at the development stages and that conducted by the regulatory authorities themselves.

This re-appraisal needs to be performed in a structured way. The first question that must be asked is what is the risk that the test seeks to measure or minimize? For example, in the case of a safety test or an abnormal toxicity test, irrespective of the species of animals used, what we are attempting to provide is some evidence that the product will be safe when administered to the target species. Similarly, in the case of a potency test, irrespective of the species, dose or route used in the test, the risk that is being assessed is that the product will not be efficaceous when used as recommended in the target species.

Having established the risk to be assessed the next step is to question whether the test is capable of fulfilling the required function. In the case of a veterinary vaccine, can a safety test, which might use only a single vial from a batch of several thousand vials, in just two animals, really provide satisfactory assurance of the safety of the whole batch for the whole population? Clearly it cannot but the test is not designed to provide incontrovertible proof, it is designed to provide evidence. Equally, in a potency test for a vaccine we would expect to observe variation between the responses of individual animals and, even if the test is carried out in the target species, the conditions of the test will often not fully reflect the conditions in the field. In all cases, therefore, reliance is placed upon a correlation between the test and the field situation which will be more or less good but certainly not absolute. The extent to which the test fulfils the function will therefore vary from test to test. Presumably, in the case of tests applied routinely, someone, at some time, determined that the test fulfilled the function sufficiently well so that the benefits of the test justified its performance.

In re-appraising such tests one of the most important points that has to be considered is whether the risk that we are trying to assess has changed. Because any test will provide only evidence, not proof, a reduction of the risk of failure may influence the balance of the judgement about the benefit of the test. In other words each test contributes to the overall assurance of quality. If the general level of quality assurance has improved or if standards of GMP have increased it is likely that the benefit derived from the test will be reduced, while at the same time the risk of the fault which the test seeks to detect is also reduced.

In recent years there has been a great deal of emphasis placed on Total Quality Management (TQM) in all areas of business. However, TQM, as applied in, for example, the engineering sector, is not entirely appropriate to the biologicals business because of the inherent variability associated with biological systems. Nevertheless, the essential principle of building in quality rather than testing for quality can, and should, be applied. The greater the assurance that each batch of product will be similar to the first, the less benefit is derived from testing to show that this is the case. To achieve this level of confidence the principles of quality assurance must extend throughout the entire organization and it is particularly important that QA principles are applied within the R&D areas to minimize the testing burden for new products. Clearly, if these initiatives are to result in regulatory agreement that tests be discontinued, there is a need for the regulatory authorities to be involved in and to understand the steps which the manufacturer has taken to ensure product quality. This can only be achieved by the regulatory authorities working much more closely with the companies than has generally been the case in the past.

Safety Tests

The regulatory authorities have a responsibility to promote and encourage the manufacturers to make the changes necessary to permit replacement, reduction, refinement or even removal of a test and to bring proposals for such test amendments forward. The National Office for Animal Health, together with the Veterinary Medicines Directorate, have been active in encouraging companies to perform risk/benefit analyses for their products, particularly in relation to the safety test and to apply to the regulatory authority to discontinue testing in those cases where this is considered inappropriate [1]. Where it is agreed that the benefit derived from the test does not make any additional, useful contribution to the overall assurance of product quality the regulatory authorities should have the courage to allow, or even advocate, that the test be discontinued.

Similarly, the abnormal toxicity test is another area where the regulatory authorities should be actively encouraging re-appraisal if not actually taking the initiative. There is seldom any evidence that the abnormal toxicity test reflects target species safety [2] and it is therefore questionable whether the test can usefully contribute to the overall assurance of product quality.

Potency Tests

Both safety tests and abnormal toxicity tests are attractive targets for reduction of animal use because the case can be argued without the need to commit resources to development and validation of alternatives. The same is not true of potency tests. In the veterinary field in particular, there has been a natural emphasis on the replacement of challenge models [3]. Unfortunately, in many cases, insufficient knowledge is available about the nature of the protective antigens or the type of response which is required to be able to recommend meaningful alternatives. Nevertheless, there are areas where progress can be made and where the regulatory authorities are ideally positioned to promote change. For example, Leptospira challenges for use in the Ph. Eur. test for potency of vaccines for dogs are often maintained by continual passage through hamsters. In some companies more animals are used in maintaining the cultures than in testing the product. Not only is this not particularly good science because the organism is likely to change but it is also not necessary. The regulatory authorities are in a position to encourage the manufacturers to pool their experiences in order to provide advice which would result in an immediate decrease in animal usage.

In considering whether it is possible to replace a vaccination and challenge model for potency with an alternative test it is necessary to assess whether the basic principle of equivalent or better assurance of product quality can be achieved. While it may be possible to show a correlation between protection and, say, a serological response, it is quite likely that, because the serological response between individuals is variable, more animals are needed to obtain the same precision of the estimate of potency [4]. A judgement must then be made as to whether the elimination of the challenge and its replacement with a blood sample justifies using the extra animals. Here again, the regulatory authorities need to adopt a pragmatic approach and should not fall into the trap of blindly requiring statistical equivalence between the methods.

The ideal, of course, is to move to a totally in vitro alternative but it is important that the alternative proposed should actually fulfil the same function as the test which it replaces. In the veterinary field in particular there has been a recent trend towards attempts to replace potency tests for inactivated vaccines with «alternative» techniques which measure the amount of antigen present. For live vaccines this basic approach is often acceptable. For example, the validity of viral titre as a measure of the potency of a live viral vaccine is usually not questioned and, within limits, the two parameters normally show reasonable correlation. However, some of the antigen quantification tests proposed as alternatives for inactivated vaccines show a very worrying failure to grasp the basic problems. In these cases, the regulatory authorities are undoubtedly right to stand firm and insist that the existing in vivo tests be continued but, at the same time, they also have a responsibility to explain why such tests are unacceptable and to provide advice to enable the company to propose more effective alternatives.

The most obvious difficulty with using antigen quantification as a measure of potency is that this approach takes no account of the effect of any adjuvant in the host or the interaction of the antigen with the adjuvant. Although some manufacturers routinely perform a chemical analysis of adjuvant as part of their raw material control there is no assessment made of the adjuvant activity of batches of adjuvant on a regular basis even for materials where the activity, and often the composition, is known to vary.

A second common problem with such antigen quantification tests is that the techniques used are often excessively specific. For example, monoclonal antibodies are commonly used as the detection system. Unfortunately, these techniques do not usually provide any assurance that the antigen being measured is either complete or in an appropriate conformation to elicit an immune response. Furthermore, in many cases, there is no evidence that the particular antigen is of direct importance in the efficacy of the vaccine and often it is known that a single antigen does not provide complete protection.

It is important that enthusiasm to develop an in vitro alternative is not allowed to cloud the evaluation of that alternative. However, the regulatory authorities must not be dogmatic. Proposals for such replacement tests must be considered in the wider context of the whole of the manufacturing and testing procedure. A high level of proven batch consistency and an effective quality assurance system can go some way towards compensating for risks associated with reliance upon a test which is not a true potency test. It is, therefore, entirely possible that for some vaccines a measure of antigen content, perhaps in conjunction with effective quality control of adjuvants, may provide satisfactory assurance of potency. Indeed, proposals for such approaches have already been made in respect of some toxoid vaccines [5].

Opportunities for the Regulator

The regulatory authorities together with the statutory National and European Medicinal Products Committees and potentially the European Medicines Evaluation Agency (EMEA) all have opportunities to promote the reduction of animal testing. For example the review of Immunological Veterinary Medicinal Products (IVMPs) which is currently underway could be used to question the need for animal-based tests. In particular, the need for safety and abnormal toxicity tests should be carefully considered and should be co-ordinated throughout Europe so that tests regarded as unnecessary by some authorities are not still performed because others have either not considered the issue or reached a different decision. In some cases, a valid alternative to an animal test may be employed by one manufacturer who, for commercial reasons may be unwilling to share its knowledge, or more often their reagents, with competitors. The European agencies should consider ways in which they might actively encourage better communication between companies working in similar fields to ensure that maximum use is made of all available developments.

It has, in the past, been quite common in the veterinary field, to request additional data during the registration process which necessitates the use of animals. Such requests must be very carefully considered and the rationale for requiring additional information must always be clear and fully justified. This is particularly so when the proposal is for the introduction of additional batch testing but the principle applies equally to requests for specific experimental data to be provided on a one-off basis. It is not acceptable for regulatory authorities to require companies to perform additional testing of any sort, but especially involving the use of animals, for reasons of scientific curiosity.

In developing regulations and procedures the European agencies can contribute to the ongoing reduction in animal tests by requiring that the competent authorities specifically and formally review the need for all animal tests whenever a marketing authorization is granted or renewed. This would ensure that maximum benefit was derived from emerging knowledge and novel techniques in a timely fashion. Guidance notes for assessors might usefully require conclusions of such a review to be included in the assessment report.

The European agencies could also contribute by taking a more central role in the co-ordination of field trials, especially for veterinary products. Currently, the mechanisms for control of field trials vary widely between member states and there is inconsistency in the type of information required for authorization of trials and the level of control and influence which the regulatory authorities exercise. One consequence of this is that the data emerging from trials do not always fulfil the requirements for subsequent marketing authorization.

Both in connection with trials and in the case of batch tests which use animals, the regulatory authorities could take a more active role in ensuring that those tests which are performed make the best possible use of the animals involved and the data generated. The agencies involved at both national and European level should question more closely some of the experimental designs proposed by companies. This suggests that these agencies should adopt a more consultative role than has previously been the case. In addition, the close links that the regulatory authorities and European agencies already have with the European Pharmacopoeia could be exploited more fully by taking a more pro-active role in proposing changes to monographs in cases where the inclusion of animal-based tests is considered to be of limited value or where alternative techniques have been developed. More important, the European agencies might be the most appropriate organizations to co-ordinate liaison on these issues with other regulatory authorities, especially those in the U.S., the Far East and Australasia.

In addition to the various initiatives discussed, there are two actions which could be taken at a European level which would have an immediate and very significant impact on the numbers of animals used in the quality control of veterinary vaccines. First, the requirement imposed by Article 30 of Directive 81/851 for retesting of IVMPs on first importation into the EU should be removed. In principle the standards applied within a manufacturing facility outside the EU should not be any different from those within the Union and hence this retesting fails to contribute to assurance of product quality and cannot be justified. Secondly, and following the same logic, repeat testing by authorities as part of an official batch release system is difficult to defend if the manufacturer is shown to be operating within the principles of GMP and with an effective QA system. In such cases the regulatory authorities should be positively discouraged from repeating any tests which involve the use of animals.

References

1 Roberts B, Lucken RN: Reducing the use of the target animal batch safety test for veterinary vaccines. Dev Biol Stan 1995;86:97-102.

2 Kamp M van der: Ways of replacing, reducing or refining the use of animals in the quality control of veterinary vaccines. Lelystad, ID-DLO 1994.

3 Tilleray J, Knight PA: Prospects for the refinement and reduction of animal usage in the potency testing of *Cl. chauvoei* vaccines. Dev Biol Stan 1986;64:111-119.

4 Hanley JA: If nothing goes wrong is everything all right? JAMA 1983;249(13):1743-1745.

5 Knight PA: Are potency tests for tetanus vaccine really necessary? Dev Biol Stan 1986;64:39-47.

Dr. R.N. Lucken, Fron Fawr, Boncath, Pembrokeshire, SA37 0HS, England

Brown F, Cussler K, Hendriksen C (eds): Replacement, Reduction and Refinement of Animal Experiments in the Development and Control of Biological Products.
Dev Biol Stand. Basel, Karger, 1996, vol 86, pp 73-76

Replacement, Reduction or Refinement of Animal Use in the Quality Control of Veterinary Vaccines: Development, Validation and Implementation

M.D.O. van der Kamp

Netherlands Centre Alternatives to Animal Use, Utrecht, The Netherlands

Keywords: Replacement, reduction, refinement, vaccines, validation.

Abstract: Various different methods have been developed to replace, reduce or refine the use of animals in batch quality control tests of veterinary vaccines. The development, validation and implementation of those alternatives which lead to a significant reduction in the numbers or the suffering of the animals used must receive priority. Validation is one of the hurdles to the implementation of alternative methods into relevant legislation, so it deserves special attention. For proper validation studies, co-operation between various organisations of both government and industry is necessary. International and national organisations such as ECVAM, FRAME, SIATT, ZEBET and the recently established Netherlands Centre Alternatives to Animal Use (NCA) could contribute significantly to the development, validation and implementation of alternative methods by stimulating and co-ordinating research projects, by providing information through databases, and by co-ordinating validation studies.

With regard to the three Rs concept in the quality control of immunobiologicals, three stages can be distinguished before an alternative method is actually used on a routine basis. The first stage is that of the scientific development of an alternative method. When the method seems to be satisfactory, this is followed by an extensive validation study. Only after an alternative method has been validated properly, can it be implemented into relevant regulations and requirements.

At the Dutch Institute for Animal Science and Health (ID-DLO, formerly the Central Veterinary Institute) in Lelystad, the Netherlands, an inventory study has been carried out to investigate the feasibility for developing and implementing alternative methods in the batch quality control of veterinary vaccines [1].

It is difficult to obtain the precise numbers of animals which are actually used in the batch quality control of veterinary vaccines. In the Netherlands, there is an accurate system for registering the use made of vertebrate animals. Every vertebrate used in the Netherlands for experiments must be reported to the Veterinary Public Health Inspectorate.

In 1993, approximately 150,000 animals were used for purposes related to human and veterinary immunobiologicals. It is estimated that approximately 40,000 of these animals were used by manufacturers and official control institutions in the Netherlands solely for the routine batch quality control of veterinary vaccines. About 15% of these were subjected to severe suffering. Looked on European or world-wide basis, a manifold of this number of animals is used for these purposes.

Animals are routinely used for several types of tests. For live vaccines, animals are mainly used in safety tests (usually 10-100 doses of vaccine administered to two to 10 target animals) and tests for the detection of extraneous agents. The latter tests are to some extent also carried out in cell cultures or eggs, but especially for poultry vaccines, many animals are used for the detection of extraneous agents. For inactivated vaccines, animals are used for safety tests (usually a double dose administered to two or three target animals) and potency tests. Potency tests can be performed in either the target species or in laboratory animals. They can be vaccination-serology tests, whereby the potency of a vaccine is estimated from serological parameters, or vaccination-challenge tests, whereby the potency of the vaccine is measured by actual protection of animals after vaccination and subsequent challenge. For the vaccination-challenge tests, usually one or more groups of animals are needed, and in total many animals are used per test. These tests cause severe suffering in the test animals, especially in the unvaccinated controls, but also in animals which are not fully protected. The end-point of vaccination-challenge tests is often death. The concept of a humane end-point should be considered, and clear clinical symptoms of disease should be used as criteria in such tests, euthanizing animals showing these symptoms as soon as possible. Vaccination-challenge tests are still routinely performed for a wide range of veterinary vaccines. Replacement, reduction or refinement of these tests should get the highest priority. It is good to realise that for quite a number of vaccines, alternative methods have been developed or are being developed at the moment. Among others, projects on the potency tests of, for instance, rabies, tetanus and erysipelas vaccines are currently carried out.

Some general recommendations which were made in the author's report at the Lelystad symposium [1] are intended to lead to a debate about some fundamental principles in the quality control of vaccines. A great impact concerning the reduction of animal use can be expected from the following:

- abolishing the use of abnormal toxicity tests as safety tests: a safety test in target animals is preferable and is considered to be more meaningful;
- the possibility of performing sequential tests in the same animals (already done for safety tests with horse and dog vaccines);
- the possible potency testing of more than one vaccine component in one animal test if it has been demonstrated that the vaccine components concerned do not interact;
- the possible combination of safety tests and the tests for the detection of extraneous agents for live vaccines (to some extent already introduced for poultry vaccines);

- abolishing tests for the detection of extraneous agents in animals for live vaccines if sensitive in vitro methods are available for all relevant pathogens;
- a reassessment of the need to perform safety tests for every batch of vaccine and the necessity of the use of a double dose for the safety tests of inactivated vaccines;
- the possible combination of safety tests and potency tests for inactivated vaccines, using a specific dose of vaccine;
- the possibility of performing more than one safety test in the same animals at the same time (already done to some extent with dog vaccines).

The rationale underlying current testing practice should be critically reviewed and discussed, in the light of the aim to reduce animal use. At the Lelystad symposium in June 1994, it was generally agreed that a discussion on several items should be held with scientists, industry and regulatory bodies such as the European Pharmacopoeia and the Committee for Veterinary Medicinal Products (CVMP).

In general, it has been accepted that after an alternative method has been developed, its acceptability must be tested in a validation study. Because the extrapolation of test results from in vitro methods or tests in an animal species different from the target animals may cause difficulties, and responses to vaccines are specific for each vaccine, alternative methods have to be developed for each test and each vaccine. After development, the reliability of the method, its relevance to the specific goal for which it is intended and its acceptability must be examined in a validation study. The validation of alternative methods may cause scientific problems, which may or may not be resolved. Currently one of the biggest problems with validation studies is caused by the lack of guidelines or general descriptions with which a validation study has to comply in order to have a method validated and generally accepted. Regulatory bodies such as the CVMP and the European Pharmacopoeia ought to try to specify definitions and conditions for the validation of alternatives and should stimulate validation projects. This should preferably be done in co-operation with scientists and industry. After an alternative method has been properly validated, it should be implemented into relevant regulations and monographs.

It appears that often, after an alternative method has been scientifically developed, it is not taken up for validation studies. This is because scientists, who developed the alternative method, have completed their initial task. Initiatives have to be taken to stimulate further validation and eventual implementation of the alternative method. To achieve this, co-operation between regulatory bodies, scientists and pharmaceutical companies is necessary. International and national organisations can contribute significantly to this co-operation. There are several national organisations, such as FRAME (Fund for the Replacement of Animals in Medical Testing, UK), SIATT (Swiss Institute for Alternatives to Animal Testing, Switzerland), ZEBET (Centre for Documentation and Evaluation of Alternative Methods to Animal Experiments, Germany), and the recently established NCA (Netherlands Centre Alternatives to animal use). Ideally, each national centre should have a network of animal welfare officers, national working groups, regional coordination points for animal welfare, scientists, regulatory bodies and industry. Being the centre of such a network, national centres can contribute significantly to co-operation on a national level. National organisations are also part of an international network, preferably co-ordinated by organisations as ECVAM (European Centre for the Validation of Alternative Methods, Ispra, Italy), thereby enabling co-operation at both national and international level.

In conclusion it can be said that some goals have already been achieved, but that more research and validation studies remain to be done. Apart from research, it is important that definitions and conditions for validation studies are specified by the relevant regulatory authorities, in co-operation with scientists and industry. Finally, several national and international organisations may contribute significantly to the necessary co-ordination and co-operation for the development, validation and implementation of alternative methods. Both in the above-mentioned report and at the international symposium in Lelystad, the Netherlands, it was acknowledged that the development, validation and implementation of alternative methods is a responsibility for regulatory bodies, industry and scientists.

Reference

1 Van der Kamp MDO: Ways of replacing, reducing or refining the use of animals in the quality control of veterinary vaccines. Institute for Animal Science and Health, Lelystad, The Netherlands, 1994.

Dr. M. D.O van der Kamp, Netherlands Centre Alternatives to Animal Use, Yalelaan 17, De Uithof, NL-3584 CL Utrecht, The Netherlands

Brown F, Cussler K, Hendriksen C (eds): Replacement, Reduction and Refinement of Animal Experiments in the Development and Control of Biological Products.
Dev Biol Stand. Basel, Karger, 1996, vol 86, p 77

SESSION IV

Specific Toxicity, Neurovirulence and Other Safety Tests

Chairmen: E. Fitzgerald (Rockville, USA)
M. Moos (Langen, Germany)

Brown F, Cussler K, Hendriksen C (eds): Replacement, Reduction and Refinement of Animal Experiments in the Development and Control of Biological Products.
Dev Biol Stand. Basel, Karger, 1996, vol 86, pp 79-83

Evaluation of New Approaches to Poliovirus Vaccine Neurovirulence Tests

D.J. Wood

National Institute for Biological Standards and Control, Potters Bar, Herts, UK

Key words: Poliovirus vaccine, quality control, neurovirulence tests, transgenic mice, polymerase chain reaction, restriction enzyme analysis.

Abstract: Transgenic mice that express the human receptor for poliovirus and mutant analysis by PCR followed by restriction enzyme cleavage (MAPREC) are two new approaches to poliovirus vaccine neurovirulence testing. This review assesses the validity and current status of these tests. Both approaches require further rigorous validation and development. They are most likely to be used as corollary rather than replacement tests for the current neurovirulence test.

INTRODUCTION

The recent certification of poliomyelitis eradication in the Americas [1] is a highly impressive testament to the efficacy of the live attenuated poliovirus vaccines developed by Sabin. The vaccines also have a consistent record of safety that is maintained by well controlled production methods and by rigorous testing of all batches for virulence in primates. Recent advances in our understanding of the molecular biology of poliovirus have resulted in the development of alternative approaches to neurovirulence tests for live attenuated polioviruses [2-4]. The aim of this review is to develop a realistic understanding of the validity and status of these alternative test methods.

THE NEUROVIRULENCE TEST FOR LIVE ATTENUATED POLIOVIRUS VACCINES

Neurovirulence tests in primates were key tests used by Dr Albert Sabin in the 1950's to select the now widely used Sabin vaccine strains and derivative pools of uniformly low monkey neurovirulence [5]. Extensive field trials and subsequent large scale use of vaccines derived by limited passage from the Sabin original (SO) vaccines showed that these strains are well attenuated for human recipients when given orally. There is a high degree of assurance that newly manufactured vaccine batches which show the same low level of primate neurovirulence as previous batches will be satisfactory when used for human vaccination.

Originally a variety of methods for monkey neurovirulence testing of this vaccine was used in different laboratories but a standardised method was introduced in 1982 [6] which enabled a reduction in the use of animals. The aim of the test is to show consistency of monkey neurovirulence of vaccine batches. Various manufacturing parameters, such as the number of passages from the SO virus, can be shown to increase primate neurovirulence [7] and therefore be undesirable. Vaccine batches that show an adequately low level of neurovirulence in primates are associated with a low level of vaccine-associated poliomyelitis in human recipients, estimated at one per 500,000 primary vaccinees [8] or more than one in 2.5-3.6 $\times 10^6$ recipients overall [9].

New Approaches

Developments in the molecular biology of poliovirus have resulted in several new approaches to neurovirulence testing. For example identification and isolation of the gene for the human cellular receptor for poliovirus led to the development of transgenic mice which express this receptor and become susceptible to poliovirus infection [2, 3]. These animals offer an alternative in vivo model. On the other hand an understanding of the genetic basis of attenuation of the Sabin live attenuated poliovirus vaccines [10] has led to the development of in vitro assays. These include an in vitro translation test [11], plaque hybridisation and quantification of RNA sequence variants by the MAPREC test [4]. The latter test is the most developed of the in vitro tests and will be the only one discussed here.

Evaluation of Transgenic Mice

Several transgenic mice lines have been developed which are sensitive to poliovirus infection by a variety of routes [2, 3]. Good progress has been made in designing the format of suitable neurovirulence tests in transgenic mice. Vaccine batches will be inoculated concurrently with a reference preparation and rejected if they show significantly increased activity. The validity of this approach depends on the assumption that factors which increase neurovirulence in transgenic mice are the same as those that increase neurovirulence in primates. This may or may not be the case since there are likely to be differences in the physiology of transgenic mouse and primate central nervous system (CNS) tissue, such as temperature and cellular factors, which may influence selective pressures applied by the different environments. Thus transgenic mouse CNS tissue may assay different properties of the virus population in a vaccine batch than simian CNS tissue. This question has been addressed only for Sabin type 2 poliovirus so far and, reassuringly, the same determinants of attenuation are important in both primates and transgenic mice [12].

Further work is required to standardise the model. Preliminary study showed that attenuated vaccine strains could be distinguished from virulent wild type isolates by intracerebral inoculation [13]. There is general consensus however that intraspinal inoculation is required to achieve the sensitivity required to discriminate between suitable and unsuitable vaccine batches, at least for poliovirus type 3. One transgenic mouse line, PVR Tg21 [3], has been selected for use by a WHO international collaborative study of type 3 poliovirus and a standard intraspinal inoculation method is being employed in several laboratories. This study will also evaluate the merits of clinical scoring schemes compared to a histopathological scoring scheme modeled on the standard test. The reference preparation used in the standard test is being evaluated in the transgenic mouse collaborative study which will also provide information on the reproducibility and robustness of this model. Early results from the study are encouraging and suggest that a batch of vaccine that failed the standard neurovirulence test also has increased activity in transgenic mice. However, very many more vaccine batches will have to be tested before firm conclusions can be drawn about the correlation between primate and mouse neurovirulence.

Evaluation of MAPREC

Mutant analysis by PCR and restriction enzyme cleavage (MAPREC) [4] is totally different conceptually from the current neurovirulence test. This test quantifies the amounts of specific RNA sequence variants in vaccine batches and has been most extensively developed for poliovirus type 3. The genomic RNA of poliovirus is approximately 7500 nucleotides in length yet the virulent progenitor strain differs from the Sabin type 3 vaccine strain by only 10 nucleotides. Two nucleotide changes were found to affect virulence [10], namely: a C-U change at nucleotide (nt) 472 in the 5' non-coding region of the genome; and a C-U change at nt 2034 which causes an amino acid substitution at residue 91 of the capsid protein VP3. The possible effect of a third mutation, U-C at nt 2493 [14], on attenuation is not proven.

The MAPREC test is a polymerase chain reaction-based assay to measure the proportion of genomes which have reverted to the virulent base at specific nucleotide positions. The assay for nt 472 involves amplification with a mismatched primer introducing a *MboI* restriction site into the revertant sequence and a *HinfI* site into the vaccine sequence. Fragments are resolved after digestion and the ratio of cut to uncut fragments gives the proportion of reverted genomes present. Different primers and restriction enzymes are used to quantify other nt positions.

The MAPREC test therefore provides for a quantitative molecular evaluation of a vaccine batch. The validity of this approach depends on the assumption that direct back mutation at the selected base is the major genetic change resulting in reversion to virulence. The correlation between the proportion of genomes which reverted at nt 472 and the histological MLS of a large group of type 3 vaccines grown in primary monkey kidney or human diploid cells is very good [15], which suggests that for poliovirus type 3 the standard neurovirulence test assays variation in the proportion of reverted 472 genomes only. However, it is known that at least one other mutation, at residue 91 of VP3, also attenuates the type 3 virus. Conditions of production seem to make this mutation stable [15] but virulent revertants from vaccine-associated cases frequently do not show direct back-mutation. Instead, other mutations are found in the capsid protein at a variety of positions which seem to have the effect of suppressing the attenuated phenotype conferred by the phe-91 of VP3 change [16]. While these changes may not occur in vaccine batches they emphasise that mechanisms by which mutations are suppressed may be subtle and unexpected. This has implications for assays of types 1 and 2 poliovirus vaccines, or for type 3 vaccines produced on substrates other than primary monkey kidney or human diploid cells.

Technical details of the type 3 MAPREC test for nt 472 have recently been established in a WHO Collaborative Study. As the proportion of 472 revertants that correlates with failure in the monkey neurovirulence test is less than 1%, a high level of precision is required from the test. The Collaborative Study showed that because of between-laboratory variation in the absolute values of 472 revertants, a ratiometric approach will be required. Therefore a standard preparation will be essential but it is not yet determined whether such a standard should be a virus or DNA preparation. Moreover, statistical parameters for evaluation of the test are yet to be finalised. The laboratories participating in the collaborative study successfully ranked artificial mixtures with very large differences in mutant ratios but it is not yet known whether a large number of laboratories could successfully discriminate pass/fail vaccine samples.

Conclusions

The primate neurovirulence test has been widely used for many years and Sabin vaccines which show an acceptably low level of neurovirulence have a good safety record. New approaches to neurovirulence testing are now possible and two of these, transgenic mice and the MAPREC test, are being extensively developed. Both these tests assess vaccine bulks in ways that are different from the current test. Further validation and development of both methods is now required. It is also essential to develop criteria to support the eventual application of these new approaches in vaccine testing. Parallel testing of products from each manufacturer with the new and the standard technique must be mandatory for several years. For established manufacturers as many archived samples as possible should be tested in addition to newly produced batches. For new manufacturers, or for established manufacturers who change production methods, a substantial number of parallel tests would be required.

The new approaches are likely to be used as corollary rather than replacement tests for the standard neurovirulence tests because of the major implications of either an increase in the rate of neurovirulence in children or the rate of rejection of vaccine batches.

References

1 Expanded Programme on Immunisation: Certification of poliomyelitis eradication – the Americas 1994. Weekly Epidemiological Record 1994;69: 293-295.

2 Ren R, Costantini F, Gorgacz EJ, Lee JJ, Racaniello VR: Transgenic mice expressing a human poliovirus receptor: a new model for poliomyelitis. Cell 1990; 63:353-362.

3 Koike S, Taya C, Kurata T, Abe S, Ise I, Yonekawa H, Nomoto A: Transgenic mice susceptible to poliovirus. Proc Natl Acad Sci USA 1991;88:951-955.

4 Chumakov KM, Powers LB, Noonan KE, Roninson IB, Levenbook IS: Correlation between amount of virus with altered nucleotide sequence and the monkey test for acceptability of oral poliovirus vaccine. Proc Natl Acad Sci USA 1991;88:199-203.

5 Minor PD: Use of animals in the development and control of viral vaccines. Dev Biol Stand 1995;86:113-120.

6 WHO Technical Report Series: Requirements for poliomyelitis vaccine (oral), 1983;687.

7 Boulger LR, Marsden SA, Magrath DI, Taffs LF, Schild GC: Comparative monkey neurovirulence of Sabin type III poliovirus vaccines. J Biol Stand 1979;7:97-111.

8 Assaad F, Cockburn WC: The relationship between acute persisting paralysis and poliomyelitis vaccine – result of a ten year enquiry. Bull WHO 1982;60:231-242.

9 Furesz J, Contreras G: Some aspects of the monkey neurovirulence test used for the assessment of oral poliovirus vaccines. Dev Biol Stand 1993;78:61-70.

10 Minor PD, Macadam AJ, Stone DM, Almond JW: Genetic basis of attenuation of the Sabin oral poliovirus vaccines. Biologicals 1993;21:357-363.

11 Svitkin YV, Alpatova GA, Lipskaya GA, Maslova SV, Agol VI, Kew O, Meerovitch K, Sonenberg N: Towards development of an in vitro translation test for poliovirus neurovirulence. Dev Biol Stand 1993;78:27-32.

12 Macadam AJ, Pollard SR, Ferguson G, Skuce R, Wood DJ, Almond JW, Minor PD: Genetic basis of attenuation of the Sabin type 2 vaccine strain of poliovirus in primates. Virology 1993;192:18-26.

13 Dragunsky E, Gardner D, Taffs R, Levenbook I: Transgenic PVR Tg-1 mice for testing of poliovirus type 3 neurovirulence: comparison with monkey test. Biologicals 1993;21:233-237.

14 Tatem JM, Weeks-Levy C, Georgiou A, Dimichele SJ, Gorgacz EJ, Racaniello VR, Cano FR, Mento SJ: A mutation present in the amino terminus of Sabin 3 poliovirus VP1 protein is attenuating. J Virol 1992;66:3194-3197.

15 Chumakov KM, Norwood LP, Parker ML, Dragunsky EM, Ran Y, Levenbook IS: RNA sequence variants in live poliovirus vaccine and their relation to neurovirulence. J Virol 1992;66:966-970.

16 Macadam AJ, Arnold C, Howlett J, John A, Marsden S, Taffs F, Reeve P, Hamada N, Wareham K, Almond JW, Cammack N, Minor PD: Reversion of attenuated and temperature sensitive phenotypes of the Sabin type 3 strain of poliovirus in vaccinees. Virology 1989;174:408-414.

Dr. D.J. Wood, National Institute for Biological Standards and Control, Blanche Lane, South Mimms, Potters Bar, Herts, EN6 3QG, UK

Brown F, Cussler K, Hendriksen C (eds): Replacement, Reduction and Refinement of Animal Experiments in the Development and Control of Biological Products.
Dev Biol Stand. Basel, Karger, 1996, vol 86, pp 85-96

Testing of Immunomodulatory Properties in Vitro

T. Hartung, A. Sauer, A. Wendel

Biochemical Pharmacology, University of Konstanz, Konstanz, Germany

Keywords: Leukocytes, blood cytokine response, endotoxin, liver cells, immunopharmacology.

Abstract: The immune response of different species to a given stimulus varies considerably. The in vitro evaluation of immunomodulatory properties of test compounds therefore prompts the use of human cells. We have conducted experiments on human whole blood incubations which offer the advantages of few preparation artefacts, natural cell environment and easy performance. Ten different immune stimuli were used to initiate leukocyte mediator release. Out of > 20 factors as readout, each and every stimulus released a unique set of factors with different kinetics and concentration dependences. We also used liver macrophages as an alternative cellular model. In this model, over-activation of the macrophages by endotoxin released a toxic combination of factors which killed co-cultured hepatocytes. Co-culture experiments were carried out with primary rat as well as with human liver cells to check for common mechanisms. Furthermore, we added human neutrophil granulocytes to these co-cultures which synergized with the macrophages in killing hepatocytes. Since a similar cellular interaction exists in vivo, this extended cell system bears additional characteristics of the in vivo situation. Therefore, these in vitro models of basic mechanisms of inflammation might be suitable for the evaluation of pro- and anti-inflammatory properties of test compounds.

INTRODUCTION

Immunomodulatory properties of compounds are of increasing interest in pharmacology and toxicology. Due to considerable species differences, extrapolation from animal models to the human situation is difficult. Thus, models assessing immunomodulatory properties of substances based on human cells are required. Cytokines have been identified as primary messengers of the immune system e.g. in inflammation, anti-infectious defence, allergy and transplant rejection. The release of cytokines by immunocompetent cells appears to be a valuable measure of the immunocompetence of leukocytes and their donors. The induction of a cytokine response by a test compound might be a measure of immunostimulatory properties.

A number of human leukocyte cell lines are available. However, the predictive value of experiments making use of cell lines is limited due to lack of differentiation and to artificial culture conditions. Within the non-specific immune system where granulocytes and macrophages are primarily responsible for the immediate defence against micro-organisms, only a few cell lines with restricted metabolic competence are available. Therefore, primary human leukocytes are the tool of choice for the assessment of basic mechanisms of the immune system. Although isolation of different leukocyte populations from human blood is well established, we preferred a whole blood incubation model of cytokine release (BCR, i.e. blood cytokine response) to minimize preparation artefacts and to simplify the assay procedure.

Endotoxins (LPS) from Gram-negative bacteria are well known inducers of inflammatory responses. But a variety of other compounds was also shown to initiate the release of cytokines in human whole blood. In our system, each stimulus induced a unique pattern within the twenty cytokines and related factors that we measured.

Two types of information on a test compound can be obtained in our whole blood assay: firstly, test compounds can be checked for their potential to initiate a cytokine response. Comparing the amounts of cytokines formed as well as the pattern of pro- and anti-inflammatory cytokines permits the determination of immunostimulatory properties of an unknown agent. Secondly, the effect of the test compound on BCR induced by established stimuli can be measured. This information on the modulation of a standard response characterizes the immunotoxicological and immunomodulatory properties of a test compound.

More complex models allow a more integrative measure of the effect on inflammatory processes to be obtained. We have therefore established an in vitro model of septic organ failure [1]. LPS-induced liver injury in laboratory animals is known to be mediated by macrophages via cytokines. An in vitro correlate of early steps of septic liver failure was built up by co-culturing rat hepatocytes and Kupffer cells. Addition of LPS induces the release of cytokines such as tumour necrosis factor (TNF) and eicosanoids such as thromboxane and leukotriene D_4, which contribute to liver injury [2]. This in vitro system of LPS-induced macropahge-mediated hepatocytotoxicity will be termed the macrophage in vitro model. In this assay, a number of pharmacological test compounds showed a similar pattern of protection as demonstrated by the animal experiment. As a further step towards complexity, by addition of human neutrophilic granulocytes (PMN) to these cocultures additional features of the in vivo situation could be modelled [3]. This LPS-induced PMN-amplified hepatocytotoxicity will be termed the macrophage/PMN in vitro model.

To overcome the limitations of the use of rodent cells, we extended our work to the use of human liver cells. We currently characterize the relative contribution of different mediators in rodents compared to man.

Materials and Methods

Materials

All plastic materials for tissue culture were obtained from Greiner (Frickenhausen, Germany), RPMI 1640, Culture media and supplements were purchased from Biochrom (Berlin, Germany). LPS from Salmonella abortus equi was obtained from Sigma (Deisenhofen, Germany). Mono-Poly Resolving Medium was purchased from ICN (Meckenheim, Germany). The polyclonal sheep anti-murine TNF antibody was prepared in our own laboratory. Pharmacological inhibitors were from Sigma except BW 755C (Wellcome), FPL 55712 (Fisons) and l-660-177 (Merck Frosst).

Blood cytokine response (BCR)

Citrate blood was withdrawn from healthy volunteers between 8 and 10 a.m. to account for diurnal variations in the BCR. Blood was diluted 5-fold using RPMI 1640 culture medium supplemented with penicillin/streptomycin and heparin (2.5 IE/ml). One ml of this diluted blood was incubated in the presence of the different stimuli. Standard incubations lasted 4 or 24 h at 37°C in the presence of 5% CO_2. When kinetics were assessed other time points were also included (0.5, 1, 2, 6, 8, 12 and 18 h). After the incubation period blood cells were sedimented by centrifugation (2000 g, 3 min.). Cell-free supernatants were frozen at –80°C until cytokines were measured. The following maximal concentrations as well as 1/10 and 1/100 of these agents were used: LPS [10 µg/ml], MDP [10 µg/ml], SEA [1 µg/ml], SEB [1 µg/ml], LTA [100 µg/ml], SLO [2.5 haemolysing U.], heat killed S.aur. [0.01% cells v/v], C5a [100 ng/ml], PHA-M [15 µg/ml] and PMA [100 nM]. All stimuli were purchased from Sigma (Deisenhofen, Germany) except MDP which was from Bachem (Heidelberg, Germany).

Cytokine determination

Commercial ELISA kits were used throughout: TNF-α, TNF-β, sTNF-R I, sTNF-R II, IL-1, IL-1ra, IL-6, IL-8, G-CSF, GM-CSF (Quantikine, H. Biermann, Bad Nauheim, Germany); IL-2, IFN-γ (Endogen, Biomar, Marburg, Germany); IL-10 (Titerzyme, H. Biermann), MIP1α (Cytokit, Genzyme, Germany).

The macrophage in vitro model

Co-cultures of rat liver cells were performed as described [1]. Briefly, liver cells were prepared from male Fischer rats (F344) according to the collagenase perfusion technique. The fraction of Kupffer cells (KC) was increased by modification of the differential centrifugation scheme used to isolate hepatocytes. After 4 h of adherence to plastic (1.25 million hepatocytes in 1ml RPMI 1640 plus 10% newborn calf serum per well of a 6-well culture dish, 37°C, 40% oxygen tension) 3-5% Kupffer cells were identified by peroxidase staining. Cell medium was renewed and putative effectors were added. Thirty minutes later the co-cultures were challenged with 10 µg/ml LPS from Salmonella abortus equi for 16 h of further incubation. LPS-inducible hepatocytotoxicity was measured by lactate dehydrogenase release.

Bioassays for TNF, IL-1 and G-CSF

TNF bioactivity was measured in cell supernatants by the WEHI 164 clone 13 assay [4]. IL-1 bioactivity was determined by the D10G4 assay [5]. G-CSF bioactivity was assessed by the NFS-60 assay [6]. In all cases recombinant murine cytokines served as standard (provided by Dr. Adolf, Bender & Co., Vienna, Dr. Vosbeck, Ciba Geigy, Basel and Dr. Stevens, Amgen, Thousand Oaks, U.S.A., respectively).

The macrophage/PMN in vitro model

PMN separation was performed using a one-step centrifugation method: 25ml heparinized blood from human volunteers were layered over 21ml of Mono-Poly Resolving Medium. After centrifugation (300 g; room temperature, 30 min) the PMN layer was collected and PMN were washed twice with PBS containing 1mM EDTA at 250 g for 10 min (4°C). Erythrocyte contamination was removed by water lysis. PMN were resuspended in RPMI 1640 and kept on ice until further use. PMN viability was greater than 98% as determined by trypan blue exclusion. Purity of PMN was greater than 99% as judged by myeloperoxidase staining.

Liver cells were plated on 96-well cell culture plates, at a cell number of 3×10^4 hepatocytes per 100 µl RPMI 1640 plus 10% newborn calf serum per well. After an adherence phase of four hours, medium was replaced by 100 µl of serum-free medium. 25 µl of inhibitor stock solution, solvent control or mere medium were added. Inhibitors were dissolved in DMSO or RPMI. DMSO concentration at no time exceeded 1% (v/v). After 5 min 50 µl of PMN suspension or RPMI were added. After 30 min of pre-incubation, cell cultures were stimulated with 25 µl of LPS stock solution or an equivalent volume of RPMI as control. Final LPS concentration was 10 ng/ml (in contrast to 10 µg/ml in the macrophage in vitro model). Sixteen hours after stimulation hepatocytotoxicity was determined by measuring MTT reduction of living cells. Briefly, to remove PMN, cells were washed once with PBS (37°C). Cells were then incubated in the presence of 100 µl/well of RPMI containing 0.8 mg/ml MTT for 30 minutes. Parallel incubations of liver cells alone (without PMN or any other treatment) were taken as control.

The macrophage and the macrophage/PMN in vitro model with human liver cells

Human liver cells were prepared from surgical excisions obtained from donors without signs of liver disease or acute infections for histopathological evaluation. The preparation protocol for rat liver cells was essentially followed; since blood vessels could not be cannulated, a cannula was inserted into the cubic piece of liver (about 1 g) which was thereby perfused from the centre to the periphery due to a hydrostatic pressure of 30 cm H_2O of the perfusion fluid. Cells were plated in 96-well microtitre plates coated with collagen. LPS-inducible cytotoxicity was determined after 16 h of incubation by the MTT reduction assay.

Results

Cytokine release from human whole blood (BCR)

The original approach was based on the method of Wilson [7] established for the measurement of LPS-inducible TNF from human whole blood. In this study, blood was diluted five-fold using RPMI 1640 tissue culture medium. We checked the effect of dilution on the LPS-inducible release of TNF and G-CSF, i.e. an early and a late formed cytokine: incubations with blood alone, 80%, 60%, 40% as well as 20% (= standard incubation) were performed and stimulated with 10 µg/ml for 24 h. When calculated per ml. blood there was no significant difference for TNF or G-CSF production between these different dilutions. Thus, dilution by RPMI 1640 did not alter TNF and G-CSF release by blood. 20% blood, i.e. a five-fold dilution, was used subsequently.

Data in Figure 1 show the kinetics of LPS-inducible cytokines representing factors released early (TNF) or in an intermediate time (IL-1β, G-CSF). The kinetics of release depended on the stimulus employed. For example, TNF release was maximal at 4 h of incubation for LPS, PHA or SEB stimulation. TNF formation was maximal after 10 h of incubation in response to LTA, SEA, MDP or S.aur. In the case of stimulation with PMA the highest amounts of TNF were detected after 24 h of incubation. These differences in kinetics were independent of the amount of cytokine released, e.g. PMA induced more TNF than SEA, SEB, PHA or MDP. The concentration of TNF-α was the only parameter which decreased during the ongoing incubation after having passed a maximum at about 12 h of incubation in the presence of LPS. Other factors either increased steadily until 24 h or reached a plateau concentration. The duration of incubation to determine the maximum of a given cytokine in the blood supernatant depended both on the stimulus and the cytokine measured. Thus, the standard incubation time was arbitrarily fixed at 24 h.

Data in Figure 2 show the monokines TNFα, IL-1β and G-CSF as examples of LPS concentration/response-curves. Thus, for a given stimulus the concentrations for optimum factor release from monocytes differ among the various cytokines released. To obtain a maximal stimulation of cytokine production as a measure of release capacity, the standard concentration for the major stimulus LPS was set at 10 µg/ml.

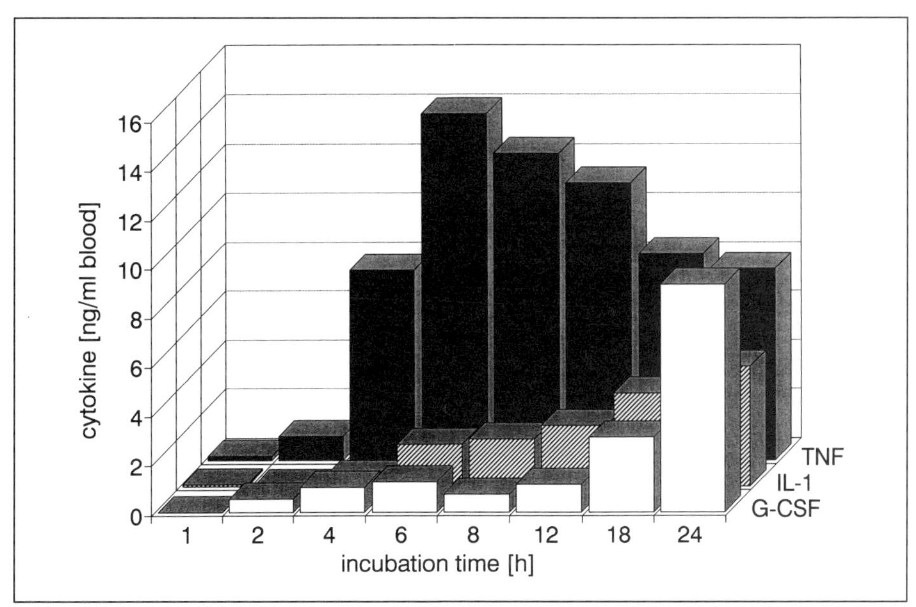

Fig. 1: **Kinetics of LPS-inducible cytokine release in human whole blood.** Whole blood from seven healthy donors was diluted five-fold in RPMI 1640 and incubated in the presence of 10 μg/ml LPS. The cell supernatant was taken at the time points indicated and cytokine release was measured by ELISA.

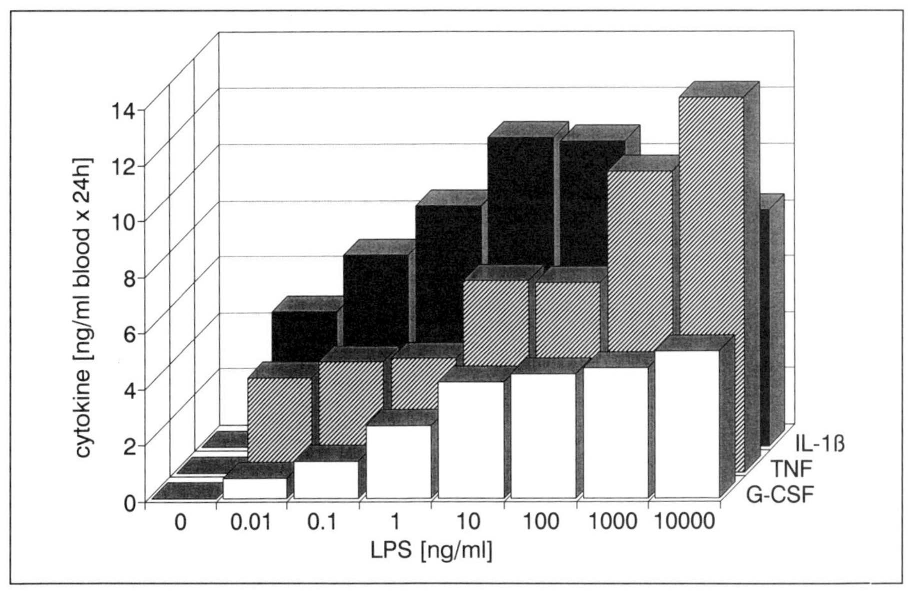

Fig. 2: **LPS concentration dependence of human whole blood cytokine release.** Whole blood (20% in RPMI 1640) from seven donors was incubated for 24 h in the presence of LPS at the concentration indicated. Data are mean cytokine release as measured by ELISA.

A variety of cytokines was released in response to the stimuli that we used. The concentration-dependence and the kinetics of release of the following cytokines were studied: interleukin-1ß (IL-1), IL-2, IL-6, IL-8, IL-10, IL-12, tumour necrosis factor-α (TNF-α), TNF-β, interferon-γ (IFN-γ), granulocyte colony-stimulating factor (G-CSF), granulocyte macrophage colony-stimulating factor (GM-CSF), macrophage inflammatory peptide 1α (MIP-1α) and leukemia inhibitory factor (LIF). Furthermore, the stimulated release of soluble TNF receptors (sTNF-R I and sTNF-R II) and the IL-1 receptor antagonist (IL-1ra) was assessed. Data in Figure 3 show the mean amounts (n= five blood donors) of some selected cytokines released within 24 h of incubation using the maximal concentration of the stimulus. There was no significant release of IL-3, IL-4 or IL-11 after any type of stimulation.

The overall result of these experiments was that each stimulus induced an individual pattern of cytokines following different dose-response curves and kinetics. The differences between individuals for every parameter were within one order of magnitude.

The macrophage in vitro model

LPS is not cytotoxic to hepatocyte cultures alone. In contrast, in the presence of at least 3% Kupffer cells, LPS induces a concentration- and time-dependent hepatocytotoxicity [1]. A variety of cytokines is formed by these cells as measured

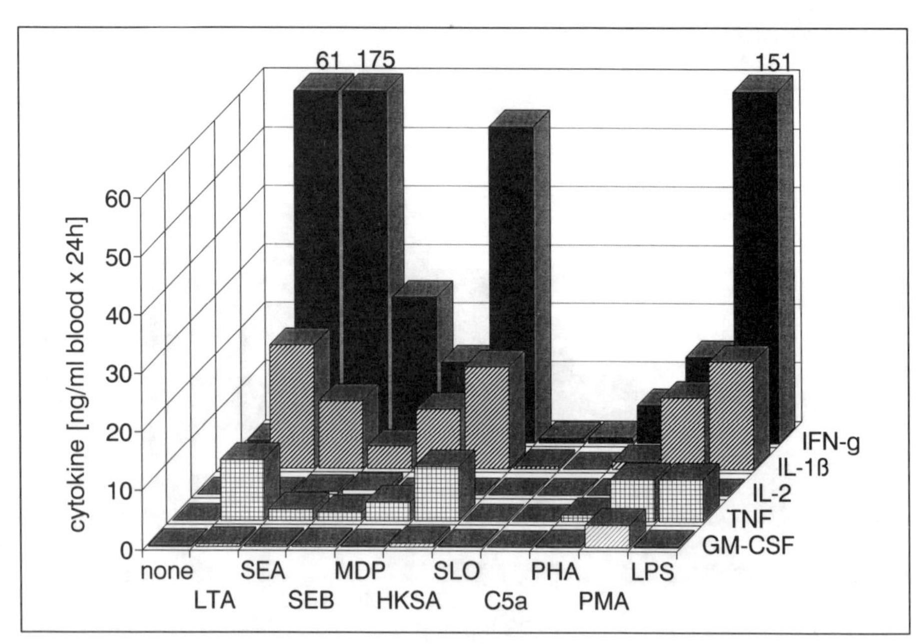

Fig. 3: **Cytokine pattern induced in human whole blood by various stimuli.** Whole blood (20% in RPMI 1640) from five donors was incubated in the presence of either stimulus for 24 h. Data are mean cytokine amount released in response to the maximal concentration used (see Methods).

by bioassay (Fig. 4). Since only very few specific neutralizing antibodies against rat cytokines are available, the relative contribution of either factor cannot be judged at present. In the case of TNF-α, co-incubation in the presence of a crossreacting antiserum against murine recombinant TNF-α resulted in partial protection [1]. Thus, TNF is one of the likely toxic mediators of macrophage-mediated hepatocytotoxicity.

Pharmacological inhibitors were used to study the contribution of further mediators. Pertinent examples are given in Table 1. Since a variety of substances known to interfere with thromboxane formation (e.g. aspirin and ibuprofen) and leukotriene D_4 synthesis or action (e.g. BW755C, FPL55712, I-660-117) were protective in this model, we conclude that these factors contribute to LPS-inducible cytotoxicity. In addition, a number of anti-oxidants attenuated LPS-induced cytotoxicity in this macrophage in vitro model, suggesting a contribution of reactive oxygen species.

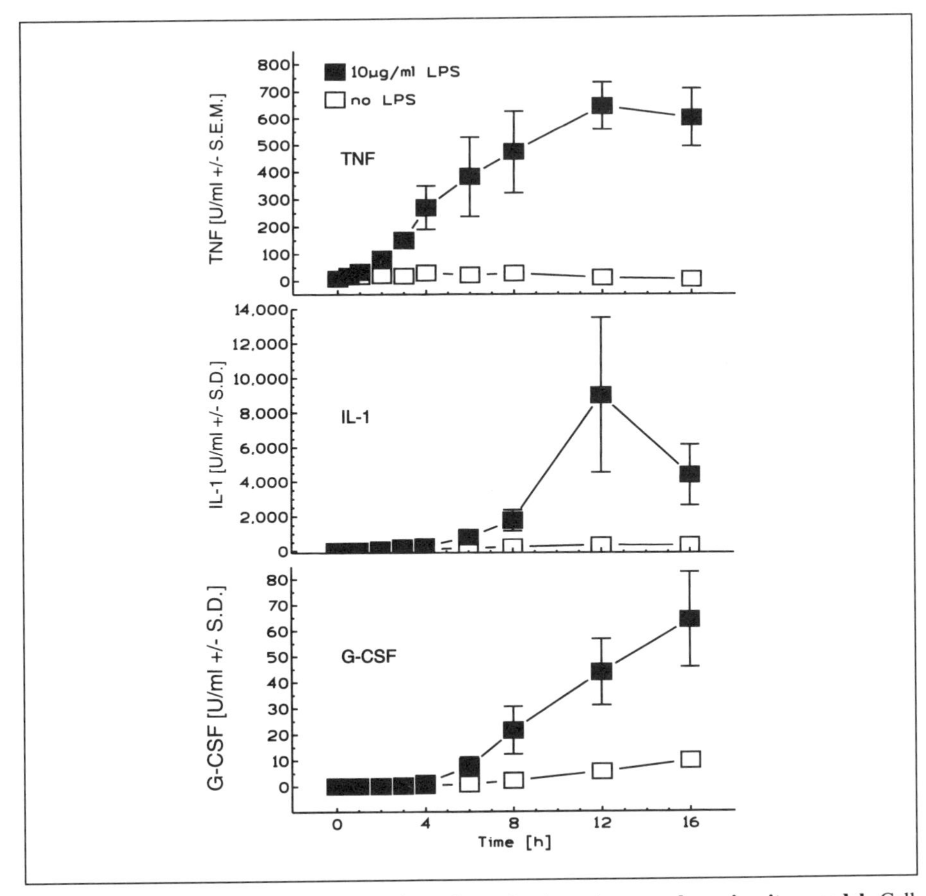

Fig. 4: **Kinetics of LPS-inducible cytokine release in the rat macrophage in vitro model.** Cell supernatants were taken at the time points indicated from rat liver cell co-cultures incubated in the presence or absence of 10 µg/ml LPS. Cytokines were measured by bioassays in cells from three independent preparations.

Table 1: Effect of various pharmacological inhibitors on LPS-inducible macrophage-mediated hepatocytotoxicity.

Pharmacological effector	**Inhibition [% ± S.D.]**
BW 755C [50 µM]	76 ± 5%
FPL 557 12 [20 µM]	66 ± 4%
I-660-177 [50 µM]	37 ± 2%
aspirine [30 µM]	58 ± 5%
ibuprofen [50 µM]	43 ± 4%
ascorbyl palmitate [20 µM]	82 ± 4%
dimethyl sulfoxide [5% v/v]	56 ± 2%

Co-cultures of rat Kupffer cells (3%) and hepatocytes were pre-incubated for 30 min. before incubation for a further 16 h in the presence of 10 µg/ml LPS. Cytotoxicity was assessed as lactate dehydrogenase release. Data are mean inhibition of LPS-inducible hepatocytotoxicity ± S.D. of three incubations.

To test whether this model translates to humans, we prepared an analogous liver cell co-culture from human surgical liver dissections. In five out of seven preparations obtained so far we have observed a LPS-inducible hepatocytotoxicity (Fig. 5). The small number of donors does not allow us at present to interpret the fact that liver cells from some patients were insensitive to LPS toxicity. We cur-

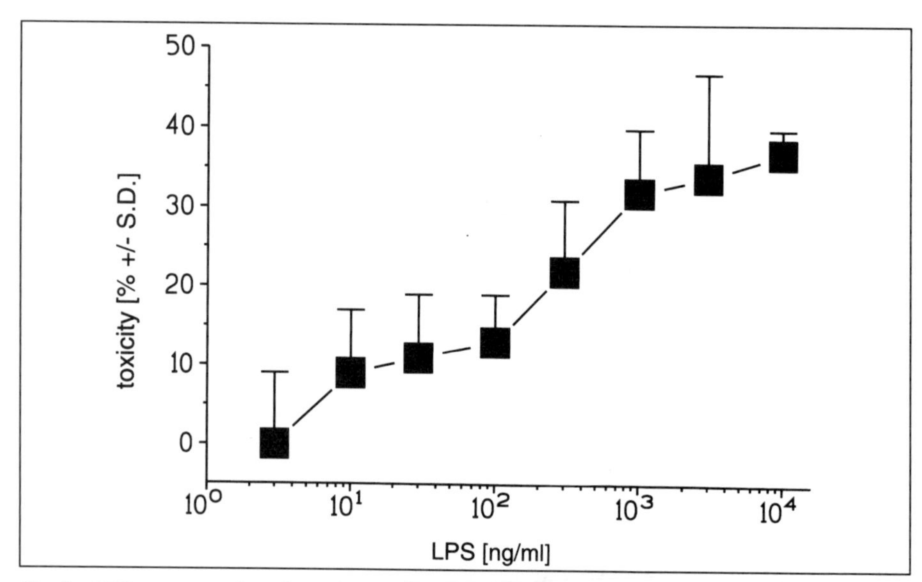

Fig. 5: **LPS concentration dependence of hepatocytotoxicity in human liver cell co-cultures.** Human liver cell co-cultures were incubated for 16 h in the presence of LPS at the concentrations indicated. Triple incubations were performed from a single donor.

rently characterize the mediation of LPS-inducible cytotoxicity in humans using a number of anti-cytokine-antibodies.

The macrophage/PMN in vitro model

To characterize the contribution of PMN to LPS-inducible liver injury further, we added PMN to the liver cell co-cultures. Addition of 7.5 PMN per hepatocyte increased the sensitivity of the cell system towards LPS by three orders of magnitude [3]. A selection of experiments using pharmacological and immunological inhibitors is shown in Table 2. Anti-murine-TNF-α-antiserum abrogated LPS-inducible cytotoxicity. Thus, PMN-dependent damage appears to be due to activation of PMN by TNF. In fact, recombinant human or murine TNF could replace LPS in inducing liver cell death in this model, whereas it had no effect in hepatocytes alone or in the macrophage in vitro model without addition of PMN. It is concluded that PMN were activated by Kupffer cell-derived TNF. We also observed a protective effect of serine protease inhibitors in the macrophage/PMN in vitro model while anti-oxidants failed to attenuate LPS-inducible PMN-dependent hepatocytotoxicity in this system. Thus, elastase primarily released by PMN degranulation is involved in an as yet not fully characterized step of cell death upon LPS exposure.

Discussion

To collect information on the human cytokine response to microbial stimulation, whole blood incubations represent a technically convenient and simple model. Compared to isolated leukocyte populations, this approach minimizes preparation

Table 2: Effect of various pharmacological inhibitors on LPS-inducible PMN-amplified hepatocytotoxicity.

Pharmacological effector	Inhibition [% ± S.D.]
anti-TNFα antiserum [1% v/v]	84 ± 6%
Eglin C [100 μg/ml]	88 ± 5%
α_1-antitrypsin [2 mM]	92 ± 8%
ascorbyl palmitate [100 μM]	18 ± 5%
ascorbate [1 mM]	0 ± 5%
catalase + SOD [1 mg/ml/1 mM]	1 ± 7%

Human PMN were added (7.5 per hepatocyte) to co-cultures of rat Kupffer cells (3%) and hepatocytes. Putative inhibitors were added and pre-incubated for 30 min. before incubation for a further 16 h in the presence of 10 μg/ml LPS. Cytotoxicity was assessed by the MTT reduction assay. Data are mean inhibition of LPS-inducible hepatocytotoxicity ± S.D. of three incubations.

artefacts and permits cellular interactions between different leukocyte populations involved in inflammatory reactions. Since in blood from healthy donors only minimal amounts of cytokines are detected, ex vivo stimulation of blood is an alternative to assess the cytokine production capacity in measurable quantities. In addition to the most commonly used initiator endotoxin (LPS), a number of further immunostimulators are capable of initiating an in vitro cytokine release.

We therefore extended these stimulatory agents to include the following: lipoteichoic acid (LTA), muramyl dipeptide (MDP), heat killed Staphylococcus aureus (S.aur.), streptolysin O (SLO), Staphylococcus enterotoxin A (SEA) and B (SEB). In addition, the general immunostimulators phytohaemagglutinin (PHA) and phorbol myristate acetate (PMA), as well as the endogenous stimulus complement factor 5a (C5a) were used. Most stimuli were able to induce pyrogenic factors such as IL-1β, TNF, IL-6 and PGE_2 (data on PGE_2 not shown). Thus, the spectrum of putative pyrogenic stimuli greatly exceeds those detectable by the limulus assay. The system was extremely sensitive e.g. as little as 10 pg/ml LPS induced significant amounts of these factors. We suggest that BCR might be an alternative to pyrogenicity testing in rabbits due to the advantage of measuring a human response. Its use appears to be advantageous compared to the limulus assay as the specificity is not restricted to LPS.

Our data show that a variety of cytokines and soluble receptors is released from human whole blood in response to different stimuli representing Gram-negative, Gram-positive as well as general immunostimulators. Some stimuli were extremely powerful inducers of cytokine release, in particular LPS, LTA, SEA, SEB and heat killed S. aureus. Among the compounds tested, PMA was the only inducer of TNFβ and LIF within 24 h of incubation; it was also a more potent inducer of GM-CSF than other stimuli. Lipoteichoic acid is increasingly discussed as a Gram-positive equivalent of LPS. Nevertheless, LTA appears to be more than an LPS mimetic: In some cases (G-CSF, IL-10), LTA was an even stronger inducer of cytokine release than LPS. As expected, the lymphokine IL-2 was found especially after stimulation with superantigens (SEA, SEB, PHA). The example indicates that by appropriate selection of stimulus and endpoint, the response of an individual cell population may also be studied in whole blood.

BCR also appears to be suitable for assessing immunomodulatory treatment ex vivo. We have recently employed BCR to monitor the effect of granulocyte colony-stimulating factor (G-CSF) showing marked changes in cytokine release capacity of different leukocyte populations after G-CSF injection [8].

To obtain in vitro data on the contribution of individual cytokines in pathological processes, more complex models are required. Here, two cell systems of different complexity were presented which share several features with the in vivo situation. Using human cells (PMN and in some experiments Kupffer cells as well as hepatocytes) the variety of immunological tools available for humans can be used to characterize these inflammatory processes. This information might be helpful in defining critical mediators concerning inflammation, thereby adding to the information gained from cytokine release models such as BCR.

Conclusions

The overall result of the study on blood cytokine responses was that each stimulus resulted in a unique cytokine pattern following individual concentration-dependence and kinetics. This set of data might allow suitable stimuli and endpoints to be defined to assess the cytokine release capacity of different leukocyte populations as well as their cooperative potential. The system appears to be useful for in vitro immuno-pharmacology and -toxicology, drug screening, monitoring of immunomodulating treatment as well as assessment of individual immunocompetence in man. The reservation has to be made that this system does not allow simple conclusions on the cytokine network in human blood because cellular interactions such as elimination and inactivation of cytokines are not taken into account. The relatively small individual variations and the possibility of standardizing the protocol of whole blood cytokine response allows a basic immune response of healthy volunteers to be assessed. If this pattern is known, the altered immunocompetence can be characterized in different states of disease and the efficacy of immunomodulatory treatment can be monitored ex vivo. The use of BCR in immunopharmacology and -toxicology might help to minimize animal experiments. To introduce this model for industrial use, we have established this model under GLP guidelines in cooperation with ANAWA Inc. (Munich, Germany). This project aims to test the suitability of BCR for drug screening and risk assessment of biomedical products and medical devices.

The liver cell models appear to be useful for studying basic mechanisms of inflammatory liver failure. Thus central mediators could be identified. In addition, these models might reduce animal experiments in drug screening. Their use in the pharmaceutical industry is under development in cooperation with MADAUS (Cologne, Germany) supported by the Ministry of Research and Technology.

In conclusion, three in vitro models of basic immune functions are presented which appear to be suitable for the testing of immunomodulatory properties of compounds. While the most simple model (BCR) is also useful for studying ex vivo effects of drug treatment, the more complex liver cell models could elucidate the biochemical pharmacology of some inflammatory processes. The use of the latter models is encouraged by the striking similarity of animal experiments to the in vivo situation.

Acknowledgements

The technical assistance by S. Otte and A. Gerth is thankfully acknowledged. The human liver cell project was performed in cooperation with Prof. Dr. D. Rühland (Hospital of Singen, Dept. of Surgery). The liver cell projects were supported by Deutsche Forschungsgemeinschaft (SFB 156), BGA / ZEBET and BMFT.

References

1 Hartung T, Wendel A: Endotoxin-inducible cytotoxicity in liver cell cultures. Biochem Pharmacol 1991;42:1129-1135.

2 Hartung T, Tiegs G, Wendel A: The role of leukotriene D4 in septic shock models. Eicosanoids 1992;5:S42-S44.

3 Sauer A, Hartung T, Wendel A: Cooperation of granulocytes and Kupffer cells in endotoxin-induced liver cell killing. J Leukoc Biol, submitted

4 Espevik T, Nissen-Meyer J: A highly sensitive cell line, WEHI 164 clone 13, for measuring cytotoxic factor/tumor necrosis factor from human monocytes. J Immunol Methods 1986;95:99-105.

5 Beuscher HU, Günther C, Röllinghoff M: IL-1β is secreted by activated murine macrophages as biologically inactive precursor. J Immunol 1990; 144: 2179-2183.

6 Shirafuji N, Asano S, Matsuda S, Watari K, Takaku F, Nagata S: A new bioassay for human granulocyte colony-stimulating factor (hG-CSF) using murine myeloblastic NFS-60 cells as targets and estimation of its levels in sera from normal healthy persons and patients with infectious and hematological disorders. Exp Hematol 1989;17:116-119.

7 Wilson BMG, Severn A, Rapson NT, Chana J, Hopkins P: A convenient human blood culture system for studying the regulation of tumor necrosis factor release by lipopolysaccharide. J Immunol Meth 1991;139:233-240.

8 Hartung T, Döcke W-D, Gantner F, Krieger G, Sauer A, Stevens P, Volk HD, Wendel A: Effect of G-CSF treatment on ex vivo blood cytokine response in human volunteers. Blood 1995;85:2482-2489.

Prof. Dr. A. Wendel, University of Konstanz, Biochemical Pharmacology,
D-78434 Konstanz, German

Brown F, Cussler K, Hendriksen C (eds): Replacement, Reduction and Refinement of Animal Experiments in the Development and Control of Biological Products.
Dev Biol Stand. Basel, Karger, 1996, vol 86, pp 97-102

Reducing the Use of the Target Animal Batch Safety Test for Veterinary Vaccines

B. Roberts[1], *R.N. Lucken*[2]

[1] National Office of Animal Health Ltd., Enfield, England
[2] Veterinary Medicines Directorate, Addlestone, England

Abstract: The need to submit each batch of every veterinary vaccine to a target animal safety test is questioned. It is proposed that a risk/benefit analysis should be conducted, on a product by product basis, to determine whether the continued application of this test to each batch of a product is beneficial and justified.
For an established product, the analysis should consider:
the number of batches manufactured;
the length of time for which the product has been manufactured;
the testing experience and the incidence of reported adverse reactions;
the level of GMP compliance and the standard of QA practised by the manufacturer;
the method and conditions of manufacture;
the use of animals for other batch tests;
the recommendations for the use of the product.
For a newly developed product the analysis should take into account:
the safety data generated during development;
the inherent risks in the product and its manufacture;
the use of other animal tests.
It is suggested that the continued use of the test could be reduced on a phased basis e.g. after 10 consecutive satisfactory tests the frequency could be reduced by limiting it to the first batch of each production campaign or to every n^{th} batch. Furthermore, the possibility of discontinuing the test altogether, for the product in question, could be considered. A return to full testing frequency would be required in the event of:
a test failure;
a batch related adverse reaction;
introduction of a new seed;
a change of manufacturer;
a process modification;
a specification amendment;
or an interruption of batch continuity.
When a manufacturer has established consistency of production and testing, an alternative approach to the need for batch testing, with a view to avoiding or minimising the use of animals, is both desirable and possible.

Introduction

A discussion paper was issued by the National Office of Animal Health (NOAH) in the UK in 1992. This paper proposed ways of reducing the use of animals, particularly for the routine control of established veterinary vaccines. It received both supportive and constructive responses from a wide field. These responses have been reviewed in consultation with the Veterinary Medicines Directorate (VMD), UK to try to find ways whereby some practical progress might be made towards the objective of reducing the use of animals in the testing of immunological products for veterinary use.

It is apparent that expressions of support for the international concern to minimise the use of laboratory animals are abundant [1-9]. Although this concern extends over the whole field of laboratory animal usage, particular attempts to address the issue have been directed specifically at the production and control of veterinary biological products. There are numerous reports of alternative in vitro tests, developed for controlling and measuring particular characteristics of individual types of products [10-13], as well as declarations from some regulatory agencies of intentions to minimise animal usage for this purpose [14]. However there is little evidence of any successful attempt to focus these various approaches and to derive any practical benefit in terms of reduced animal use.

It is also apparent that difficulties arise in trying to generalise. Any attempt to eliminate a particular usage of animals from the testing of a group of biological products or for measuring a type of parameter that is common to a range of products, will almost certainly encounter products or situations which raise specific difficulties or where such a move is not practicable. If progress is to be made, the problem must, therefore, be approached on a product-by-product basis.

A receptive attitude on the part of the regulatory authorities is also an essential part of the process and these authorities may themselves contribute directly either to work designed to provide alternative test procedures or to change established attitudes. However, the main responsibility for proposing and validating alternative in vitro procedures rests with the manufacturers, with the regulatory authorities co-ordinating the efforts of the industry as a whole. There is much benefit to be derived from the authorities encouraging manufacturers to assess the need for, and the usefulness of, any animal tests used or proposed for the control of their products.

There will always be competition for resources with which to develop alternative tests. Shortage of suitable resources is obviously a potential impediment to progress which will be made most rapidly if ways and means can be found which make a minimum demand on resources without compromising the value of the achievement. NOAH, with the support of the VMD, believes that there is scope for encouraging some amendments to certain established testing requirements and would challenge the need for certain mandatory animal tests and their value in individual cases. This approach relies essentially on making better use of existing information rather than generating anything new, placing the onus on manufacturers to propose alternative ways of verifying the quality of product batches.

Proposal

Particular criticism is made of the target animal batch safety test. This routine batch test involves administering either a double dose (killed vaccine) or a 10-fold dose (live vaccine) to susceptible animals of the species for which the product is intended and looking for adverse local or systemic reactions. Although the need for this test has been defended [15], NOAH and the VMD regard its use as particularly suitable for rationalisation. The number of animals routinely used, two per batch for large animals or 10 per batch for chickens, may not be considered large but any use of animals which can be reduced or eliminated is to be welcomed and may in turn point the way to further reductions with other tests.

It is proposed that the continued need for this test for each batch of each product should be challenged on the basis of a risk/benefit analysis. The analysis should be conducted in order to assess the value of the target animal safety test for the purposes of batch release. It should be conducted by the manufacturer and the results used to justify to the competent authority either a continued requirement for the test or that use be its discontinued for each of the manufacturer's products. It is stressed that this approach can only be followed on a product-by-product basis and each product must be assessed individually. For each established product such an analysis should take into account, for example but not exclusively, the factors indicated in Table 1.

In the case of new products where the manufacturing experience is short, such an analysis will not be possible but the requirement for target animal safety testing should still be reviewed by taking into account:

1. the safety data generated during development; this information will have demonstrated safety following a variety of dosing regimes and may indicate the margin available between the recommended dose and the toxic dose;

Table 1.

Factor	Risks
1. The number of batches manufactured	More batches means greater confidence in the quality of subsequent batches
2. The frequency with which batches are manufactured	Infrequent manufacture may increase the risk of variation between batches
3. The length of time over which the product has been made	Long successful manufacture indicates robust production and control
4. The occurrence of test failures or adverse reactions	Continuity of satisfactory testing provides confidence in process reliability
5. The level of GMP compliance practised by the manufacturer	High level compliance ensures low risk and batch-to-batch variation
6. The effectiveness of the manufacturer's QA system	Well documented, well validated and competently executed QA procedures increases batch security
7. The method and conditions of manufacture	Type and complexity of the process, nature of the starting materials, standard of process containment, design and age of the equipment
8. The recommendations for use	The higher the risk with product use, the more secure the controls need to be: parenteral administration of a live vaccine to a fully susceptible young animal is inherently more risky than a booster inoculation with a killed vaccine to an already primed individual

2. the inherent risks in the product and the process of manufacture; the degree of attenuation of a live vaccine seed, the security of the inactivation process for an inactivated vaccine, the risk attached to the use of materials of biological origin, and the degree of validation of the testing methods should all be critically assessed;
3. the use of animals for other batch tests; where animals are used to conduct other batch tests, such as for potency, the design of these tests should be considered to determine whether they might also fulfil the function of a safety test without using more animals.

Where the outcome of such an analysis shows that the risk of a safety test failure is significantly less than the benefit of discontinuing the animal test, a course of action either to reduce and eventually eliminate or to immediately discontinue the batch safety test should be initiated. A possible scheme, illustrating how the test might be phased out is given in Table 2.

If, after proceeding with reduced testing under the provisions of Stage 2, the results of subsequent tests fall out of compliance with regulatory requirements, or if significant changes to the manufacture of the product are introduced, then testing should revert to Stage 1 until consistency has again been demonstrated. Situations which would necessitate consideration of a return to Stage 1 would include but would not necessarily be limited to:

- a safety test failure;
- a batch related adverse reaction in the field;
- the introduction of a new seed;
- a change of manufacturer or manufacturing location;

Table 2.

Testing level	**Test frequency**	**Criteria**
Stage 1	Test every batch for the first (10?) consecutive batches	If all tests satisfactory and no adverse reaction in the field:– proceed to Stage 2
Stage 2	For products produced on a continuous basis test every (10th?) batch For products produced on a campaign basis test the first and last batch of the campaign	If all tests satisfactory and no adverse reaction in the field:– proceed to Stage 3
Stage 3	Consider combining the safety test with other animal tests (possibly not the target species) Consider whether the continued use of the test is justified	If the test can be replaced or if it cannot be justified:– discontinue the test

- a process modification;
- an amendment to the specification;
- an interruption in production continuity.

Once continuity of production has been re-established, or an investigation of the adverse incident has been carried out, or the experience gained with the amended procedure has been reviewed, consideration could be given to restoring limited testing, or discontinuing it altogether, by progressing through Stages 2 and 3. In many cases it may be possible to justify by–passing the intermediate stage of reduced testing and move directly to discontinuing testing on the basis of the risk assessment.

Discussion

It is emphasised again that application of this approach to the target animal safety test should only be contemplated on a product-by-product basis. However, where there is no accumulated evidence that a significant risk exists of an unsafe batch of the product being released for use, the need for an animal test as a pre-requisite to batch release cannot be justified. The opportunity exists for each manufacturer to review the testing of each product made and to justify the continued need for the target animal safety test for batch release purposes. Such a review would be conducted by a critical examination of documentation to determine the production and testing record of the product concerned, together with an assessment of the incidence of reported field reactions. There would be no demand for experimental work or for laboratory resources and the cost of such a review would be low. By arguing the case for reducing and ultimately eliminating the need for this test, supported by appropriate documentation and batch-testing records, this use of animals could be reduced at very little expense.

Conclusions

The adoption of the approach to target animal batch safety testing outlined in this paper offers an opportunity to reduce the routine use of animals for batch release of veterinary vaccines. The proposed action is inexpensive and relies on the initiative of the vaccine manufacturer to argue on a product-by-product basis for the reduction and eventual elimination of the need for this test.

This approach to the target animal safety test could be extended to other tests using animals. Abnormal toxicity tests and in vivo inactivation tests are obvious examples but consideration should also be given to applying a similar rationale to in vivo potency tests. The opinion expressed by Stainer et al [6] that when a manufacturer has established consistency of production and testing, alternative methods for ensuring the potency of those vaccines can be entertained, is relevant here. The issues may be more complex but nevertheless the need to test every batch in vivo could be reduced (given that an alternative in vitro test is available). As such tests typically use more animals than safety tests, the potential for reducing animal usage is proportionately greater.

References

1 Bleby J: Reduction of animal usage: some considerations. Dev Biol Stand 1986;64:17-18.

2 Carlsen B: A survey of the reduction of animal use in projects funded by the National Board for Laboratory Animals (CFN) 1979-1987. Acta physiol Scand suppl 1990;592:141-166.

3 Hyde RLW: Reduction in in vivo testing at the National Veterinary Services laboratories of the United States Department of Agriculture. Dev Biol Stand 1986;64:149-151.

4 Kamp M van der: Ways of replacing, reducing or refining the use of animals in the quality control of veterinary vaccines. Lelystad, Institute of Animal Science and Health (ID–DLO), 1994, 1-107.

5 Rabouhans M–L: Reduction of animal usage: British Pharmacopoeia Commission policy. Dev Biol Stand 1986;64:11-16.

6 Stainer DW: Reduction in animal usage for potency testing of diphtheria and tetanus toxoids. Dev Biol Stand 1985;65:241-244.

7 White VJ, Sojka MG: Immunoelectrophoresis in quality control of veterinary clostridial products. Dev Biol Stand 1986; 64:119-127.

8 Wood GW, Thornton DH: Application of polyacrylamide gel electrophoresis of genome fragments to control of reovirus products. Dev Biol Stand 1986; 64:213-218.

9 Wood JM, Mumford J, Schild GC, Webster RG, Nicholson KG: Single-radial-immunodiffusion potency tests of inactivated influenza vaccines for use in man and animals. Dev Biol Stand 1986;64:169-177.

10 Fitzgerald EA: Use of the single radial immunodiffusion test as a replacement for the NIH mouse potency test for rabies vaccine. Dev Biol Stand 1986;64:73-79.

11 Knight PA: Are potency tests for tetanus vaccines really necessary? Dev Biol Stand 1986;64:39-45.

12 McKenzie NM, Pinder AC: The application of flow microfluorimetry to biomedical research and diagnosis: a review. Dev Biol Stand 1986;64:181-193.

13 Thornton DH, Nicholas RAJ, Wood GW: Development of in vitro tests for detection of extraneous agents. Dev Biol Stand 1986;64:195-198.

14 Magrath DI: Summaries and conclusions (rapporteur). Dev Biol Stand 1986;64:303-309.

15 Knight PA: Summaries and conclusions (rapporteur). Dev Biol Stand 1986;64:309.

Dr. B. Roberts, Mallinckrodt Veterinary Limited, Breakspear Road South, Harefield, Uxbridge, Middx, UB9 6LS, England

Brown F, Cussler K, Hendriksen C (eds): Replacement, Reduction and Refinement of Animal Experiments in the Development and Control of Biological Products.
Dev Biol Stand. Basel, Karger, 1996, vol 86, pp 103-109

The Reduction of the Use of Challenge Testing to Provide Evidence of Efficacy in Tests of Immunological Veterinary Medicinal Products (IVMPs)

C.J. Webster

SmithKline Beecham Animal Health, Wyevale, Ont., Canada

INTRODUCTION

The demonstration of the efficacy of IVMPs is a prerequisite of their licensing in EU Member States (MS), as each MS has reflected, or will reflect, in national legislation the requirements of Directive 92/18/EEC. At present, the great majority of the data that are submitted to licencing authorities in support of applicants' efficacy claims is generated in experiments in which the immunity of vaccinated animals is challenged by their exposure to a dose of the virulent pathogen sufficient to cause clinical signs, and sometimes death, in unvaccinated control animals. This is because such challenge testing is explicitly preferred by the Directive. On the other hand, Directive 86/609/EEC, which lays the requirements for the humane use of experimental animals in the EU, requires that investigators, before conducting an experiment in animals, shall be able to demonstrate not only that no acceptable alternative to the use of animals exists, but also that no more humane method of conducting the experiment exists, even though it may still involve the use of animals. Thus, with regard to the testing of IVMPs by virulent challenge, these two Directives often appear to be in conflict. FEDESA's view is that, while it is necessary to the scientific rigour of some efficacy tests that they should involve a virulent challenge, it is increasingly possible to demonstrate efficacy by serological or immunological methods that do not require a challenge. In this paper, we will:

- discuss the reasons why opportunities to avoid the use of virulent challenges should be actively sought and taken;
- propose, in general terms, the circumstances in which a requirement to test efficacy by challenge might be abandoned;
- propose a provisional list of veterinary pathogens for which our existing knowledge suggests that the in vivo protective effects of vaccination can be correlated

directly with changes in immune effector activities measurable in vitro; and a scenario for the development of the critical data, derived from in vitro end-points, which could be accepted as providing satisfactory evidence of efficacy of IVMPs.

Reasons to Attempt to Replace Challenge Testing

In the course of generating the data necessary to support claims of efficacy for a marketing authorisation, efficacy end-points must be determined for each vaccinal valency on several occasions, as follows:

1. determine end-point in immunogenicity/minimum immunising dose study;
2. determine end-point after administration of primary vaccination regime;
3. determine end-point at end of claimed period of protection (duration of immunity);
4. for multivalent products, or different IVMPs administered simultaneously, determine end-point of each valency, and relate to end-point of the same valency given alone, to demonstrate lack of immunological interference;
5. in addition, it may be necessary to determine efficacy end-points under other conditions according to the claims made for the particular product, e.g. to support a claim of immunisation in the face of residual maternally-derived antibody.

Thus, FEDESA's view is that there could often be several opportunities, within particular registration dossiers, for the replacement of challenge testing. At present, there are not infrequent differences between national licencing authorities in their acceptance of efficacy end-points supported by in vitro data. Some authorities seem to insist that challenge testing is the only reliable method while others, considering the same product, are prepared to accept in vitro data for at least some of the end-points listed above. The fact that these differences occur seems to indicate that there already exists some scope, among the members of the Committee for Veterinary Medicinal Products (CVMP), for discussion of these issues with FEDESA in order to reach a more explicitly harmonised position.

The reasons for interest, on the part of IVMP manufacturers, regulatory authorities and society at large, in the replacement of challenge testing by in vitro methods are as follows:

a) Challenge testing is relatively expensive and time consuming.

Depending on the pathogen and the challenge method used, testing requires some or all of the following:

- killing experimental animals;
- purpose-built containment buildings and experimental handling facilities; special protective clothing for investigators and technicians;
- special administrative procedures to support the legal conduct of work involving live pathogens;
- preliminary experimentation, often quite prolonged and requiring resources because of its inherent complexity, to define and calibrate an authentic challenge model;

- the process of challenging treated and control animals, then waiting for the appearance of clinical signs in controls, and perhaps other analyses to confirm initial conclusions, is certainly more laborious and may take a longer time than in vitro testing. In the scheme of product development, elapsed time has its own, high cost.

In practice, the synergy between these factors makes the Directive's preference for challenge testing a very expensive burden on the industry.

b) Challenge testing represents a hazard to animals which are not part of the experiment and, possibly, to people.

- Cases of accidental infection resulting from the use of live pathogens in challenge experiments are not rare.

c) When a satisfactory alternative method exists, challenge testing is inhumane and unnecessary.

- By definition, the end-point of challenge testing involves the induction of clinical signs, often associated with profound distress, in the control animals and sometimes in treated animals as well (if, for example the challenge infection is more virulent than expected, or the product's efficacy is not such as to prevent totally the appearance of clinical signs).
- The requirements of Directive 86/609/EEC have been mentioned above. The point here is that (rightly, in FEDESA's view) the degree of distress caused to experimental animals is itself to be considered a material factor in judging the acceptability of experimental methods. In this regard, challenge testing must usually be judged less acceptable than other methods of end-point determination, provided that conclusions about the efficacy of the product are sound.

Circumstances under which a Requirement to Test Efficacy of an IVMP by Virulent Challenge Might be Replaced

Directive 92/18/EEC, Title II, Part 8, Section C, paragraph 1 states : «In principle, demonstration of efficacy shall be undertaken under well controlled laboratory conditions by challenge ...». The immunological basis of the protection conferred by vaccination is always multifactorial; even in those infections in which protection is known to be correlated closely with antibody titres, actual elimination of the pathogen also requires the effective functioning of other immune mechanisms. Thus we understand that the reason for the requirement for challenge testing is that this is the most expeditious and direct way for the various effects of vaccination, integrated into a single biological end-point, to be assessed. However, it is FEDESA's view that it is unnecessary to demonstrate the total effect of all the mechanisms involved in immune protection on every occasion upon which an efficacy end-point must be determined to support the claims for the product. On the contrary, FEDESA believes that, in infections in which it is clear that a direct causal relationship has been established between the presence of antibody (or any other effector activity measurable in vitro) and the immune effect in the animal that will be the basis of the product claim, it should not be acceptable to determine efficacy end-points by other than in vitro testing, because challenge testing does not increase significantly the certainty that animals are, or are not, protected.

If this approach is to be accepted, it is necessary to demonstrate unequivocally a direct relationship between the chosen in vitro end point and the protective effect of the vaccine in vivo. Certainly, some challenge testing will be necessary. It will also be necessary to define precisely the conditions under which the in vitro testing is performed. For infections in which it is possible to establish this relationship, FEDESA's view is that the necessary data can normally be generated most conveniently, for each vaccinal valency, in the course of the study, to establish the minimum immunising dose (MID) of the vaccine. The protocol for this study must allow the correlation of immune effector activities determined in vitro with protective effects assessed by virulent challenge. It must also accommodate the characterisation of responses to heterologous challenge strains, if these are relevant. FEDESA foresees two possible types of correlation between challenge protection and in vitro data, as follows:

1. Internal correlation

Minimum Immunising Dose (MID) is found by challenge for a particular vaccine, and in vitro threshold value is established above (greater than) the mean value of the in vitro variable that was observed in the protected vaccinates that had received the MID. This threshold value is used as an end-point in other efficacy studies with the same vaccine, but not for other vaccines containing the same antigen.

2. External correlation

As for 1. above, but the threshold value established by correlation with challenge protection with one vaccine (the standard) is used also as an efficacy end-point for other vaccines of the same type (i.e. antigen, strain, titre/adjuvant). No challenge testing is carried out to demonstrate the efficacy of vaccines other than the standard.

In some studies, the immune effector that would be measured in vitro is not physically present indefinitely after vaccination (e.g. duration of immunity studies). In these cases, FEDESA proposes that the demonstration of an anamnestic response to revaccination could be accepted as evidence of the continuation of primary vaccinal effect. The study of anamnestic responses, as markers of immune priming, could also be used in other situations, e.g. the ability of a vaccine to protect a young animal in the face of maternally-derived immunity.

Proposal of a Provisional List of Veterinary Pathogens for which Correlation of in Vivo Protective Effects of Vaccination with Immune Effector Activities Measured in Vitro may be Possible

Pathogen	Vaccinal protection observed in vivo	Immune effector measured in vitro	Assay method
In bovines: Bovine Viral Diarrhoea	According to applicant's claims	Serum antibody	SN test
Bovine Herpesvirus 1	Reduction of clinical signs	Serum antibody	SN test
Bovine Respiratory Syncytial virus	Reduction of viral shedding	Serum antibody	SN test
Parainfluenza 3	Reduction of viral shedding	Serum antibody	SN test
Rabies	Absence of clinical signs	Serum antibody	SN test
Leptospira spp	Reduction of clinical signs	Serum antibody	Microagglutination
Clostridial myositides and enterotoxaemias	Reduction of clinical signs	Serum antibody	Immune precipitation of purified toxins
In ovines/caprines: Clostridial myositides and enterotoxaemias	Reduction of clinical signs	Serum antibody	Immune precipitation of purified toxines
Rabies	Absence of clinical signs	Serum antibody	SN test
In porcines: Porcine parvovirus	Reduction of clinical signs	Serum antibody	SN, HI test
Swine influenza	Reduction of clinical signs	Serum antibody	HI test
Rabies	Absence of clinical signs	Serum antibody	SN test
Leptospira spp.	Reduction of clinical signs	Serum antibody	Microagglutination
In canines: Canine Distemper	Reduction of clinical signs	Serum antibody	SN test
Canine Adenovirus 1 & 2	Reduction of clinical signs	Serum antibody	SN test
Canine Parvovirus	Reduction of clinical signs	Serum antibody	SN, HI tests
Canine Parainfluenza	Reduction of clinical signs	Serum antibody	SN test

Pathogen	Vaccinal protection observed in vivo	Immune effector measured in vitro	Assay method
Rabies	Absence of clinical signs	Serum antibody	SN test
Leptospira spp.	Reduction of clinical signs	Serum antibody	Microagglutination
In felines: Feline Parvovirus	Reduction of clinical signs	Serum antibody	SN, HI tests
Feline Herpesvirus	Reduction of clinical signs	Serum antibody	SN test
Feline Calicivirus	Reduction of clinical signs	Serum antibody	SN test
Rabies	Absence of clinical signs	Serum antibody	SN test
In equines: Equine influenza	Reduction of clinical signs	Serum antibody	HI test
Rabies	Absence of clinical signs	Serum antibody	SN test
In avians: Avian encephalomyelitis virus	Reduction of clinical signs	Serum or egg antibody	SN test, ELISA
Infectious bronchitis virus	Reduction of virus isolation	Serum antibody	SN test
Infectious bursal disease virus	Reduction of clinical signs	Serum antibody	SN test, ELISA
Infectious laryngotracheitis virus	Reduction of clinical signs	Serum antibody	SN test, ELISA
Newcastle disease virus	Reduction of clinical signs	Serum antibody	HI test, ELISA
Mycoplasma gallisepticum	Reduction of clinical signs	Serum antibody	Microagglutination, ELISA

Key: SN test = Serum neutralisation test; HI test = Haemagglutination inhibition test.

Conclusion

The table is not intended to be definitive nor comprehensive, but it shows that in vivo responses to many of the antigens included in modern, commercial vaccines can be quantified in vitro; in almost every case, the tests mentioned have been used in this way in published work. Further, it is known that serum antibody is an important immune effector in each of these infections, to the extent that it is possible for a quantitative correlation between in vivo protection and serum antibody concentration to be found, at least for each individual product (internal correlation). For some pathogens, the use of standard methods and reagents would allow these correlations to be extended generically (external correlation), and FEDESA supports, and would like to see accelerated, the work of the European Pharmacopoeia Commission in biological standardisation of IVMPs.

As progressively more becomes known about the immune mechanisms and correlates of protection in each infection, by the application of more specific and sensitive immunological methods, the possibilities for replacement of challenge testing by in vitro assays will increase. The benefits of the establishment of standard methods and accepted end-point values could also be significant in the context of international harmonisation of regulatory standards. In the meantime, it is FEDESA's view that it would be reasonable and extremely helpful if agreement could be reached quickly with CVMP on a list of pathogens, perhaps shorter than that proposed above, for which European licencing authorities accept that the internal correlation described above is normally possible, so that applicants for marketing authorisations do not encounter such differences in the regulatory treatment of the same data between MS as often occurs at present. FEDESA commends this approach to CVMP and remains ready to enter into technical discussions that would lead to agreement.

Dr. C.J. Webster, C.J. Webster & Co., Ltd., Wymboldwood Beach, RR1, Wyevale, Ontario LOL, 2T0, Canada

Brown F, Cussler K, Hendriksen C (eds): Replacement, Reduction and Refinement of Animal Experiments in the Development and Control of Biological Products.
Dev Biol Stand. Basel, Karger, 1996, vol 86, p 111

SESSION V

Viral Vaccines

Chairmen: A. Osterhaus (Rotterdam, The Netherlands)
L. Bruckner (Mittelhäusern, Switzerland)

Brown F, Cussler K, Hendriksen C (eds): Replacement, Reduction and Refinement of Animal Experiments in the Development and Control of Biological Products.
Dev Biol Stand. Basel, Karger, 1996, vol 86, pp 113-120

Use of Animals in the Development and Control of Viral Vaccines

P.D. Minor

National Institute of Biological Standards and Control, Potters Bar, Hertfordshire, U.K.

INTRODUCTION

Animals have been used at many stages in the development and routine control of vaccines against viral disease. This includes their use as models of pathogenicity and protective efficacy in the development of vaccines. In addition they have been used in routine assessment of live vaccine batches for virulence, in immunogenicity studies for potency, and in examining vaccines for untoward effects associated with contaminants such as extraneous agents. The use of animals has been essential at some stage in all viral vaccine development and use. The purpose for which they have been used varies with the vaccine however, and will be illustrated by considering the development of live and killed poliovaccines, live mumps vaccines and possible future measles vaccines.

POLIOVACCINES

Studies of the pathogenesis of poliomyelitis illustrate both the necessity and drawbacks of animal models. Poliomyelitis began to be a significant health problem at the turn of the century, when the previous pattern of a few endemic cases was replaced by the occurrence of major epidemics, chiefly because improvements in hygiene resulted in the exposure of children to the virus when maternal antibody had declined to non-protective levels. The virus was first transmitted to monkeys in 1909 by Landsteiner and Popper [1], by intracerebral inoculation of material derived from the faeces of a fatal case. This finding was taken up by Flexner and his co-workers [2], who showed that monkeys developed clinical signs closely resembling human poliomyelitis after instillation of virus suspensions into the nose. Despite epidemiological and virological studies by Swedish workers [3], which indicated that most infections were not paralytic and that virus could be demonstrated in the faeces and intestinal tract, the view developed that the virus was neurotropic in nature, entering the CNS via the olfactory lobes after entering the nose, and moving to the spine where it caused extensive damage. This view was consistent with the progression of the disease in the monkey model. It was so strongly held that progress towards the current view that the infection is largely enteric in nature, with a systemic phase before the invasion of the central nervous system, was effectively prevented for thirty years. The currently accepted model gives greater hope for intervention in the form of immunisation, while the neural model suggests that an immune response would probably be of limited value in preventing disease.

The neural model was ultimately discarded when it was shown that the histopathology of human poliomyelitis differed from that in animals infected by the nasal route, particularly in that lesions were absent from the olfactory lobes in the human disease but well represented in the animal model [4]. At the same time studies of cases of poliomyelitis showed that the virus was found in intestinal contents, the oropharynx and to some degree in lymphatic tissues in amounts comparable to or greater than those found in the CNS, suggesting other replication sites [5]. The models of pathogenesis were thus gradually changed to encompass the idea of an orally transmitted virus which occasionally found its way to the spinal cord where it produced the damage typical of poliomyelitis. Both monkeys and chimpanzees were used extensively in these studies, and were essential to the eventual development of the vaccines which have controlled poliomyelitis in many parts of the world. The history of poliomyelitis therefore provides excellent illustrations of both the value of animal models and the extent to which they can be misleading.

The intellectual basis for the development of vaccines against poliomyelitis was provided by a study of the prophylactic effect of gammaglobulin on the development of poliomyelitis in a massive trial in approximately 55,000 children conducted in the USA in 1952 [6]. Half the children received gammaglobulin, and half placebo, and it was demonstrated that there was a detectable level of protection for eight weeks following the administration of passive antibody, although previous studies had shown that antibody had no effect on severity once the disease was diagnosed. Two types of vaccines capable of inducing neutralising antibody were developed, and their relative merits are still debated today.

The consequences of infection with a virulent strain of poliovirus are potentially so serious that vaccine consisting of killed but immunogenic virus is the most obvious choice. After a false start in 1935, when a clinical trial of an allegedly killed vaccine resulted in cases of poliomyelitis, a vaccine was developed by Salk in the early 1950s, based on virus preparations carefully inactivated by treatment with formalin to destroy infectivity while retaining antigenicity. The Francis trial of 1954 concluded that such a preparation was both safe and effective in human use, but its protective efficacy had already been repeatedly demonstrated in chimpanzees and monkeys. The vaccine was approved for use on 12th April 1955 when the results of the Francis trial were finally released and the first case of vaccine-associated poliomyelitis in what has become known as the Cutter Incident was admitted to hospital on 25th April 1955, having received the vaccine on 16th [7]. A total of 60 cases in recipients and 89 in contacts were directly related to the vaccine, involving two bulk preparations which had been incompletely inactivated. Production procedures were immediately reviewed, and no case of poliomyelitis attributable to killed poliovaccine has been reported since. The cause of the incident was the existence of virus clumps in which virions at the centre of the aggregate were protected from the formalin treatment. As a result the production process now involves filtration both before and after the inactivation step to remove aggregates, which may harbour live virus.

Live virus could be isolated from the implicated batches of Cutter vaccine, but the infectious process was more prolonged, either because of the low numbers of infectious particles or because individual particles were less infectious due to the surrounding debris. However it was found that infectivity was readily detected by intraspinal inoculation of monkeys. While most countries now accept that tissue culture methods and process monitoring provide total assurance of the inactivation of the virus, the US code of Federal Regulations still specifies this test for inactivated poliovaccine to demonstrate lack of infectivity.

A consequence of the introduction of filtration steps was that some of the viral antigen was removed, so that the potency of the vaccine could no longer be expressed in terms of the amount of infectious virus before formalin treatment. Alternative tests were therefore required and currently involve immunogenicity studies in either guinea pigs or chicks, as specified by the European Pharmacopoeia, or the immunisation of monkeys, which must be shown to have acquired a certain titre of neutralising antibodies, as specified in the US code of Federal Regulations. Neither test is totally satisfactory; alternative tests involving immunisation of rats have been developed but are not fully accepted at present. An in vitro possibility would involve ELISA or similar methods for the quantification of antigen content [8]. Possible drawbacks to this are illustrated in Table 1 where the potency of a vaccine was assessed in a capture ELISA using specific monoclonal antibodies. The inactivated vaccine had different potencies with respect to the live virus from which it had been derived, depending on the antibody used, almost certainly because the formalin treatment modified the site recognised by the monoclonal antibody. It is likely that this also affects the protective potency of the virus in that antibodies induced against modified sites will not react with and neutralise live virus. It is thus not clear how to translate such potency estimates into clinically relevant potency specifications. The standardisation of inactivated poliovaccine potency is an area requiring further study and currently requires immunogenicity studies in animals if only to validate in vitro methods.

The live attenuated vaccine strains of Sabin are the most widely used poliovaccine, infecting the recipient and inducing immunity without causing disease. They are the end result of extensive studies involving large numbers of primates and other animals to isolate and identify strains which would grow well in the gut but poorly, if at all, in nervous tissue. Some indication of the care required is given by the caution in interpreting the results of neurovirulence assays of viral populations inoculated intrathalamically into primates [9]. The population may consist of spinal variants unable to grow in the brain but which may diffuse to the more sensitive spinal areas where replication can occur freely. The virus may be unable to grow in any neural site, resulting in damage solely due to an abortive infection at the site of inoculation. Strictly neurotropic virus may develop which is unable to grow in vitro, but can be passaged from animal to animal. A mixture of neurotropic and spinal variants can develop in a population, and finally viruses with truly differing degrees of virulence may be selected. Examples of all five phenomena were quoted by Sabin [9], and careful studies are required to distinguish them, including passage at different doses

Table 1: Potency of Type 3 component of inactivated poliovaccine compared to live infectious control virus of the same strain.

Antibody	Potency of live virus	Potency of Vaccine
214	1.0	0
440	1.0	0
471	1.0	0
113	1.0	0
194	1.0	0
439	1.0	0.02
441	1.0	0.7
442	1.0	0.57
134	1.0	0.64
132	1.0	0.36
495	1.0	1.09
875	1.0	0
877	1.0	0.63
1281	1.0	1.11

Potencies were measured by capture ELISA using a parallel line assay.

in animals and re-isolations. The process is necessary to ensure the safety of the vaccine strains obtained which should be of uniformly low neurovirulence rather than a mixed population containing viruses of low and high virulence which is attenuated in test systems because of the relative amounts present. Sabin used large numbers of animals in isolating and characterising the strains now used in production of the live vaccines, which have controlled poliomyelitis in many parts of the world.

However in a limited number of cases, estimated at one per 530,000 primary vaccinees or 1 in 2×10^6 vaccinees overall, the live attenuated vaccine strains have been shown to cause poliomyelitis [10]. While the risk is extremely small, to the extent that it is difficult to quantify, there is no doubt that it is real. The consequence however is that the components of every batch of vaccine must be tested for neurovirulence, and while there are certain encouraging developments relating in vitro molecular biological markers to virulence in animals [11,12], none has been satisfactorily validated.

The only currently acceptable test involves primates and has been rationalised somewhat to improve the quality of the data and to use animals more economically [13]. Alternative in vivo tests under investigation include the use of transgenic mice which carry the gene encoding the human receptor for poliovirus and are therefore susceptible to infection [14]. Again the system has not yet been validated.

Animal use, especially of primates, has therefore been central to the development of vaccines against poliomyelitis and continues to be essential for certain aspects of their routine control.

Mumps Vaccines

In contrast to poliomyelitis, mumps is generally regarded as a trivial if unpleasant disease of childhood. It results in few deaths, but can be responsible for just under 40% of hospitalisations for meningitis or encephalitis in unimmunised populations. It may also cause orchitis in adult males which may result in sterility and mastitis in adult females. The usual consequences of infection with a wild type mumps virus are undesirable but not so potentially disastrous as an infection with a wild-type poliovirus and this has been reflected in the development and control of mumps vaccines.

In 1946 Enders and co-workers [15] showed that mumps virus could be attenuated for monkeys by serial passage in hens' eggs, leading to the suggestion that the resulting virus might also be attenuated for human recipients. Three vaccine strains of mumps virus have been of particular interest within Europe, namely Jeryl Lynn, Urabe and Rubini.

The Jeryl Lynn strain was isolated from a female patient with unilateral parotitis in 1963, by amniotic inoculation of embryonated hens' eggs [16]. It was passaged in eggs and in cell cultures from chick embryos. The virus strain was assayed at two passage levels in a variety of cell culture systems and compared with a wild type strain. While most cells failed to distinguish between the Jeryl Lynn preparations and a virulent strain of virus, primary human amnion cell cultures were able to grow the virulent control virus but neither of the Jeryl Lynn preparations. Monkeys were inoculated with the two Jeryl Lynn batches intracranially, intraspinally and intramuscularly; 17 of 20 animals developed neutralising antibodies which were of higher titre for the lower passage level preparation. No significant histopathological abnormalities were observed. Routine tests for extraneous agents were made in culture and in animals, and the preparations tested in children. The lower passage preparation gave a slighly higher seroconversion rate than the other, but also produced parotitis and virus shedding in a few children.

The Urabe Am9 strain was obtained by serial passage in the amniotic cavity of chick embryos, followed by cloning in quail embryo fibroblasts [17]. The uncloned material gave a good antibody response and no clinical signs in monkeys. The biological characteristics of the virus were studied in human embryo kidney and chick embryo fibroblasts. Differences in titre on the two cell systems ranged from 10-fold to no difference although there were also differences in plaque size. Six cloned preparations were tested in children, where it was concluded that those preparations giving small plaques on chick embryo fibroblasts gave higher seroconversion rates. The Urabe Am9 strain was derived from one such preparation.

The Rubini vaccine strain was isolated from the urine of a child by isolation in the human diploid cell WI38 and then passaged into the amnionic or allantoic cavity of embryonated hens' eggs before adaptation to the human diploid cell MRC-5, which is the substrate for production [18]. It has a distinctive small plaque morphology on Vero cells compared to Jeryl Lynn and Urabe, and inoculation of baby and adult cynomolgous monkeys produced seroconversion without side effects. After testing for extraneous agents in animals and cell culture the virus was then given a clinical trial in children.

Each of these three vaccine strains has been used extensively in children and it is clear that, compared to either live or killed poliovaccines, animal models played only a small role in their development or routine control as safe and effective vaccines other than in extraneous agent testing. However the widespread use of the Urabe strain in Canada, Japan and the UK led to the observation that mumps virus derived from the Urabe strain could be isolated from the cerebral spinal fluid of children admitted to hospital with suspected meningitis. Assuming that the virus caused the symptoms, the estimates of the incidence ranged from about 1 in

100,000 to 1 in 3,000 vaccinees [19,20]. The main reason for the wide range appears to be the frequency with which csf samples taken clinically were given a detailed virological analysis after it was established that they did not contain bacteria. As a result of these observations, vaccines containing the Urabe strain are no longer used in the United Kingdom. It is possible however that the virus is not the cause of the clinical symptoms, but that by coincidence children who have been vaccinated experience febrile convulsions or other symptoms caused by other circumstances which prompt neurological investigation; by this view the virus would frequently be found in csf even in healthy vaccinees. Mumps virus is reported in csf in cases of clinical mumps at a high frequency even where there are no neurological symptoms. There is no suitable animal model for mumps to investigate this possibility and it is clearly ethically unacceptable to do so in children.

The Jeryl Lynn strain of mumps vaccine is the only vaccine currently used in the UK and has never been implicated in possible neurological consequences of vaccination or isolated from csf. However it has been shown that the vaccine contains a mixture of two strains, probably representing two distinct isolates [21]. The significance of this observation is difficult to evaluate in the absence of an animal model but it is possible that the ratio of the two strains is a significant factor in the properties of the vaccine, for example, if a slow-growing highly attenuated strain competes with a faster growing more immunogenic strain to reduce its virulence. The fact that increasing passage appeared to reduce clinical reactogenicity and immunogenicity [16] could mean that changes in the ratio of the strains were affecting the properties of the vaccine.

Finally the Rubini strain, originating from a Swiss child of the same name is identical to the laboratory strain Enders, originating from the USA decades previously as far as the sequence of the parts of its genome currently known are concerned (M Afzal, personal communication [22]). It is therefore possible that it arose by laboratory contamination at some stage. There are in vitro biological differences between Rubini and Enders, (M Afzal, personal communication) which may relate to their qualities as a vaccine, although this cannot be tested easily.

The lack of a satisfactory animal model therefore makes it difficult to assess the significance of clinical observations such as the occurrence of Urabe-derived virus in csf of some vaccinees. The uncertainty surrounding the existing vaccines with respect to the mechanism by which they induce immunity with minimal possible side effects makes further scientifically based development of mumps vaccines difficult. While the safety and quality of the vaccines in use are excellent, they are largely based on previous clinical experience and tight control of the production process to ensure that the vaccines are the same as those made previously, not on objective tests of the end product.

Measles

Measles, like mumps, is regarded as a trivial childhood disease in developed countries, although the death rate in the UK used to approach 1 in 3,000 cases in an epidemic. In developing countries it is the major preventable cause of childhood death, killing an estimated 1.4 million children annually, chiefly from diarrhoea, pneumonia or encephalitis. This figure takes into account the extensive global vaccination strategies of WHO which are believed to protect approximately

80% of the children born. Modifications in the measles vaccine would be desirable. Currently the live vaccine strain is given by intramuscular injection so that the presence of maternal antibody prevents virus replication and therefore immunisation in a significant number of children. Thus vaccination is delayed until nine months in developing countries when the take rate is acceptable because the recipients are then susceptible. Unfortunately they are also susceptible to wild type measles which has a particularly high mortality in the very young. For these and other reasons a vaccine which could be given earlier would be an advantage. Development of novel measles vaccines or even vaccination strategies are handicapped by past experience in this field. A killed measles virus could be effective in the presence of maternal antibody, but in the late sixties such a preparation was associated with severe adverse consequences when recipients with waning immunity were exposed to infection with a wild type strain [23]. Atypical measles developed, consisting of a sudden onset of very high fever, a high incidence of pneumonitis and an atypical rash. While there were no deaths in the developed countries in which it was used, the frequency of hospitalisation was very high and it is likely that deaths would occur in less developed areas.

A second more recent phenomenon was observed when vaccinees were given high amounts of the measles vaccine at six months to overcome effects of the maternal antibody [24]. While measles was prevented there was an excess mortality rate in the recipients of the high titre vaccine compared to low titre recipients. Measles has long been known to have a disruptive effect on the immune system, and other undesirable consequences of vaccination are conceivable.

Measles vaccine development is thus inhibited by the ethical difficulties of testing novel vaccines or novel strategies in people, although such developments hold out the prospect of reducing death rates. It is difficult to see how improvements in vaccination can be made in the absence of good animal models and several groups are working to develop them to examine a number of different aspects of the problems raised.

Summary and Conclusions

Animal models were central to the development of poliovaccines and remain essential in some form in the routine quality control of both live and killed vaccines. The necessity of an animal model is illustrated by the examples of mumps and measles vaccines where the existing materials, while satisfactory, have a number of drawbacks and where changes in current practice raise concerns for safety and efficacy.

References

1 Landsteiner K, Popper E: Übertragung der Poliomyelitis acuta auf Affen. Ztschr f Immunitätsforsch u exper Therap Orig 1909;2:377-390.

2 Flexner S, Clark PF: A note on the mode of infection in epidemic poliomyelitis. Proc Soc Exp Biol Med 1912;10:1-2.

3 Kling C, Wernstedt W, Pettersen A: Recherches sur le mode de propagation de la paralysie infantile épidémique. z Immunitätsforsch 1912; 12:316-323; 657-670.

4 Sabin AB: The olfactory bulbs in human poliomyelitis. Am J Dis Child 1940;60:1313-1318.

5 Sabin AB, Ward R: The natural history of human poliomyelitis I. Distribution of virus in nervous and non-nervous tissues. J Exp Med 1941;73:771-793.

6 Hammon WD, Coriell LL, Wehrle PF, Stokes J: Evaluation of Red Cross gammaglobulin as a prophylactic agent for poliomyelitis. 4. Final report of results based on clinical diagnosis. JAMA 1953;151:1272-1285.

7 Nathanson N, Langmuir AD: The Cutter Incident. Poliomyelitis following formaldehyde inactivated poliovirus vaccination in the United States during the spring of 1955. Am J Hyg 1963;78:16-81.

8 Wood DJ, Heath AB, Sawyer LA: A WHO Collaborative study on assays of the antigen content of inactivated poliovaccines. Biologicals 1994, 23:83-94.

9 Sabin AB: Characteristics and genetic potentialities of experimentally produced and naturally occuring variations of poliomyelitis. Ann NY Acad Sci 1955;61:924-938.

10 Nkowane BU, Wassilak SG, Oversteen WA, Bart KJ, Schonberger LB, Hinman AR, Kew OM: Vaccine associated paralytic poliomyelitis in the United States: 1973 through 1984. JAMA 1987;257:1335-1340.

11 Minor PD: The molecular biology of poliovaccines. J Gen Virol 1992;73:3065-3077.

12 Chumakov KN, Powers LB, Newman KE, Robinson IB, Levenbook IS: Correlation between amount of virus which altered nucleotide sequence and the monkey test for acceptability of oral poliovirus vaccine. Proc Nat Acad Sci USA 1991;88:199-203.

13 WHO Requirements for poliomyelitis vaccine (oral) (Revised 1982) Technical Report Series. Geneva: World Health Organization. 1983;687:107-174.

14 Ren R, Moss EG, Racaniello VR: Identification of two determinants that attenuate vaccine-related type 2 poliovirus. J Virol 1991;65:1377-1382.

15 Enders JF, Levens JH, Stokes Jr. J, Maris EP, Berenberg W: Attenuation of virulence with retentions of antigenicity of mumps virus after passage in the embryonated egg. J Immunol 1946;54:283-296.

16 Buynak EB, Hilleman MR: Live attenuated mumps virus vaccine. 1. Vaccine development. Proc Soc Exp Biol Med 1966;123:768-775.

17 Yamanishi K, Hosai H, Ueda S, Takahashi M, Okuno Y: Studies on live attenuated mumps virus vaccine. II. Biological characteristics of the strains adapted to the amniotic and chorio-allantoic cavity of developing chick embryos. Biken J 1970;13:157.

18 Glück R, Hoskins JM, Wegman A, Just M, Germanier R: Rubinia new live attenuated virus strain for human diploid cells. Dev Biol Stand 1986;65:29-35.

19 Hockin JC, Furesz J: Mumps meningitis possibly vaccine related. Ontario Canadian Diseases Weekly Report 1988;14:210.

20 Colville A, Pugh M: Mumps meningitis and measles, mumps and rubella vaccine. Lancet 1990;340:786.

21 Afzal MA, Pickford AR, Forsey T, Heath AB, Minor PD: The Jeryl Lynn vaccine strain of mumps virus is a mixture of two distinct isolates. J Gen Virol 1993;74:917-920.

22 Kunkel U, Schreien E, Siegl G, Schulze D: Molecular characterization of mumps virus strains cirulating during an epidemic in Eastern Switzerland 1993/93. Arch Virol 1994;136:433-438.

23 Fulginiti VA, Eller JJ, Downie AW, Kempe CH: Altered reactivity to measles virus. JAMA 1967;202:1075-1080.

24 Aaby P, Knucten K, Whittle H, Lisse IM, Thaarup J, Poulsen A, Sodermann M, et al.: Long term survival after Edmonston-Zagreb measles vaccination in Guinee Bissau: Increased female mortality rate. J Pediatr 1993;122:904-908.

Dr. P.D. Minor, National Institute of Biological Standards and Control, Blanche Lane, South Mimms, Potters Bar, Hertfordshire EN6 3QG, UK

Brown F, Cussler K, Hendriksen C (eds): Replacement, Reduction and Refinement of Animal Experiments in the Development and Control of Biological Products.
Dev Biol Stand. Basel, Karger, 1996, vol 86, pp 121-127

Rat Immunogenicity Assay of Inactivated Poliovirus

J.M. Bevilacqua, L. Young, S.W. Chiu, J.D. Sparkes, J.G. Kreeftenberg

Connaught Laboratories Limited, Willowdale, Ontario, Canada

Key words: Assay, immunogenicity, IPV, polio, potency, rat.

Abstract: A rat immunogenicity assay for IPV potency was validated and applied to routine vaccine testing as a potential alternative to the CFR Monkey Potency Assay. Potencies of pure trivalent polio, various combinations and experimental vaccines were tested with a view to producing a single dilution assay.

Introduction

The North American «gold» standard for the determination of Inactivated Polio Vaccine (IPV) potency and lot release is the 21 CFR 630.3, the N.I.H. Monkey Antigenicity assay (U.S. Health Regulations 1964). Vaccines which pass the minimum standard for this assay are considered to be efficacious and appropriate for human use. The difficulty with the assay as a release test is that it requires large numbers of expensive primates (Cynomolgous monkeys), employs a reference serum (rather than a reference vaccine) and is therefore at best, qualitative. In searching for a replacement, our rationale was to obtain an assay that reduced the use of primates and was sensitive, reproducible, quantitative and cost effective. An in vitro potency assay would be preferable; however, our experience with current methods and published reports [1] do not show good agreement between antigenicity and immunogenicity for vaccine preparations stored at 4°C for long periods. This determination is critical for vaccine stability evaluation. In addition, other components such as Pertussis and adjuvants may have an impact on the immunogenicity but not on antigenicity. We therefore proceeded with a rat model for immunogenicity similar to that previously described by van Steenis et al [1].

This paper will describe the IPV Rat Immunogenicity in terms of its validation, which includes the selection and standardization of the assay reference (Connaught's CLL Reference). In addition, the applicability of the rat potency test to pure poliovirus and its combinations, IPV from different methods of manufacture (Vero or MRC-5 cells) and the adaptability of the test to a single dilution assay will be examined.

Materials and Methods

Vaccines tested

The Reference vaccine is a commercially prepared trivalent blend of enhanced potency IPV formulated to contain polio types 1, 2, and 3 in the proportions of 40-8-32 DU/0.5 ml produced in MRC-5 cells (CLL Ref). The assigned potency of this reference vaccine is one unit for all three types. Test vaccines were also commercially prepared by the same manufacturer and included trivalent vaccines with pure polio blends, combinations (DPT-IPV, DT-IPV, Td-IPV) and experimental combinations with component pertussis and HIB vaccine. In addition, a Vero cell vaccine was evaluated (PMsv, Lot A).

Rat immunogenicity assay

Five serial two-fold dilutions of reference and test vaccine were made and the following volumes were injected intramuscularly into groups of 10 rats (Charles River Laboratories): 1 ml undiluted (2 × 0.5 ml), or 0.5 ml undiluted, 1/2, 1/4, and 1/16.

Blood was collected by jugular slit at 21 days post inoculation processed and heat inactivated at 56°C for 30 minutes. The serum was tested in duplicate in a standard virus neutralization assay against Mahoney (Type 1), MEF (Type 2) and Saukett (Type 3) strains of poliovirus (approx. 100 $TCID_{50}$, neutralization period three hours at 37°C, followed by 18 hours at 4°C).

Statistical methods

Data were analysed either as log2 titres, using a parallel line method or, by Probit Analysis [2] where log2 titres of ≥2, ≥8, or ≥4.5 for poliovirus types 1, 2, 3 respectively were determined to be cut-off levels.

Results

Reference vaccine

During development of the rat immunogenicity assay no North American or International reference vaccine for IPV was available. It was therefore necessary to select and validate an appropriate CLL reference material. The optimal material would be a vaccine preparation which had been tested and proved efficacious in a human clinical trial. Unfortunately, a current lot meeting such criteria was not available. Therefore, a vaccine lot produced by our standard manufacturing method, shown to result in efficacious vaccination in humans, was used. The selected lot had been proved to be immunogenic in the CFR Monkey Potency assay and the Chick Potency assay [3]. The rationale for this approach is that such a vaccine would be considered to be efficacious in humans according to the current regulations, thereby validating its use as an internal reference preparation and providing a link to the monkey assay.

The reference lot, CLL Ref., exhibits linear dose response profiles for all three poliovirus types as indicated by mean titre responses in Figure 1. To limit the number of animals used in the assay, only one set of serial dilutions was used for all

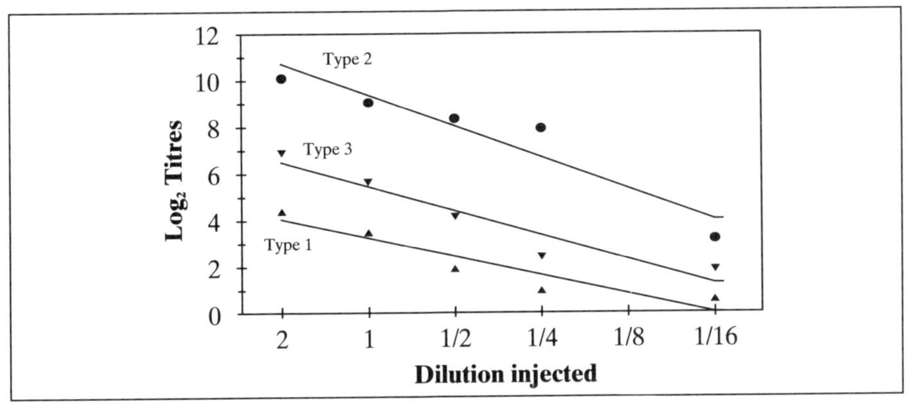

Fig. 1: Type 1, 2 and 3 mean titre (log_2) response for CLL Ref. Titres are mean (n=15) responses of sera collected 21 days post intramuscular inoculation.

three polio types. The 1/8 dilution was omitted from the series, in favour of a 1/16 dilution, so as to ensure an endpoint for type 2 virus. From the accumulated data (approximately 15 assays) it was possible to generate test validity criteria for the Probit method:

1. No significant non-linearity;
2. No significant non-parallelism (Ref. vs vaccine);
3. Confidence limits of Reference ED_{50} 50% to 200% of estimated value.

Significance for criteria 1 and 2 are tested at the 5% level. Observed confidence limits for CLL Ref. ED_{50} were within the range 65% to 153% for all three polio types.

Vaccine Potency

With validity criteria in place, it was possible to do actual potency testing on pure trivalent blends, DPT IPV, DT-IPV, Td-IPV and experimental combinations (CLL manufactured vaccines). Representative dose response profiles for a trivalent blend and DPT-IPV combinations are shown in Figure 2. The profiles for all combinations are not shown; however, in all cases (pure, combination and experimental blends), the data exhibit good linear dose-response curves as well as parallelism to the reference. These findings support the use of a single trivalent reference for relative potency determination of all IPV combinations examined.

Table 1 shows the relative potencies of the test vaccines calculated by parallel line and Probit analysis. Vaccines stored at elevated temperatures are designated S (Stability). There appears to be reasonable agreement between the two methods of calculation for all three polio types. In addition, the results are relatively consistent for each vaccine combination. The Probit gives consistently more valid analysis and is therefore the preferred method at this time. It is further noted that the combination vaccines tend to achieve higher polio potencies than are observed in the pure polios, for all three polio types.

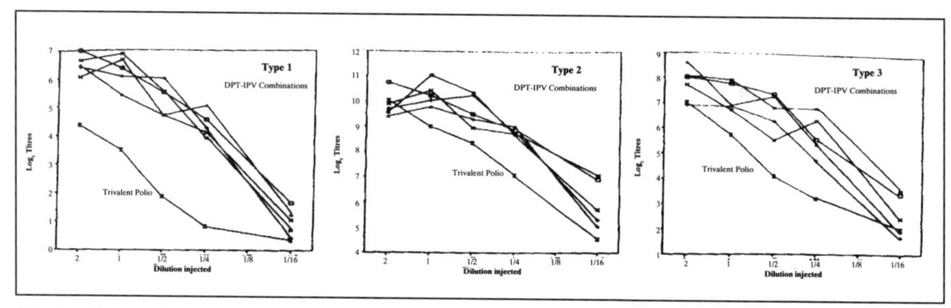

Fig. 2: Comparison of DPT-IPV combination vaccine $\log_2$ titre responses compared to representative pure trivalent polio. Individual DPT-IPV combination lots are represented by a single line.

Examination of the feasibility of a single reference material was extended to include the potency testing of IPV vaccine prepared by an alternative method of manufacture (PMsv, Vero Lot A). CLL Reference Vaccine (MRC-5) and Vero Lot A were shown to have dose response curves that are approximately linear and parallel [Fig. 3]. It seems that there is similarity in dose response to support the use of a single reference to analyse MRC-5 as well as Vero IPV containing vaccines. However, further investigations will be necessary.

Pass/fail criteria

If the IPV Rat potency assay is to be used routinely for lot release, criteria to determine acceptable vaccine potency limits are necessary. In theory, the relative

Table 1: Relative potency of representative vaccines as calculated by parallel line or probit analysis.

	Relative potency*					
	Parallel line			Probit		
Vaccine lot #	**Type 1**	**Type 2**	**Type 3**	**Type 1**	**Type 2**	**Type 3**
IPV						
1	0.9	1.0	1.2	0.8	1.1	0.8
2S	0.4	0.7	1.1	0.5	3.9	2.0
DPT-IPV						
1	5.7	3.3	4.9	5.3	5.4	5.5
2	3.6	3.2	1.4	4.3	1.9	1.6
3	4.4	2.2	2.9	6.2	3.0	1.4
4	1.5	3.5	1.3	1.0	3.7	2.2
DT-IPV						
1S	2.2	1.1	1.7	1.8	1.1	2.2
2S	0.8	1.9	4.0	0.9	1.2	0.8
Experimental combinations						
1S	2.1	1.0	0.5	1.7	1.6	0.7
2S	0.8	0.7	2.2	1.2	1.2	2.1

* Expressed in units against the CLL Ref.

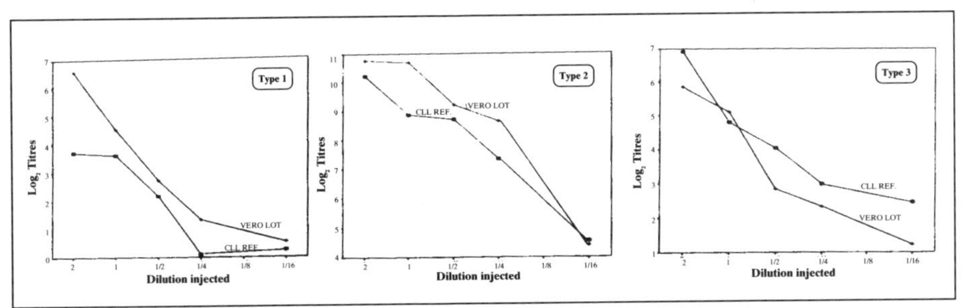

Fig. 3: Comparison of mean titre response of polio types 1, 2 and 3 for CLL Ref. and Vero Lot A. Each dose response curve is a mean of two assays.

potency of a test vaccine should not be significantly less than the potency of the reference. We reviewed the 95% limits of our potency data and found that the lower limits relative to the CLL Ref. generally exceeded 0.50. We decided to retest the sample if the estimated potency is less than 0.5.

Single dilution assay

The possibility of reducing our rat IPV immunogenicity assay to a single dilution assay was also examined. The test results of 35 vaccine samples, tested during the past year, were analysed for a statistically relevant vaccine dilution which could estimate vaccine potency. The relative potency of each test vaccine was estimated using the full dose response curve of the CLL Ref. and a single vaccine dilution (1/4). These potencies were compared to the potencies calculated using full dose response for both the test and reference vaccines (Table 2). There is good general agreement between the potencies calculated by both methods; however, the absolute potency values differ. The single dilution assay as a screening test could be used to identify vaccine lots of low potency. Lots of borderline potency could be retested using the single dose assay or even with a full, five-dose assay. Accumulation of potency results for our refined method will increase confidence in this approach.

Discussion

The rat potency assay for IPV appears to be sensitive, reproducible and cost effective (compared to the monkey assay). Part of the difficulty with the interpretation of any animal assay is the inherent variability in the test system i.e. animals, route of inoculation and serological testing. It is preferable to use a quantitative measure (antibody titres) to evaluate the relative potency and other test parameters; however, our experience with parallel line analysis of the data is that this approach still gives problems with linearity, parallelism and homogeneity of the test and reference vaccine, and consequently needs some fine tuning. Probit analysis was employed to overcome these difficulties and relatively equivalent potencies have been demonstrated. The development of a standardized computer program method is necessary.

Table 2: Relative potency of representative vaccines as calculated by full dose or single dilution probit analysis.

Vaccine lot #	Relative potency*					
	From dilution method			Single dilution method		
	Type 1	Type 2	Type 3	Type 1	Type 2	Type 3
IPV						
1	0.8	1.1	0.8	2.4	0.5	≤0.3
2S	0.5	3.9	2.0	0.4	1.1	1.0
DPT-IPV						
1	5.3	5.4	5.5	≥8.0	4.7	7.1
2	4.3	1.9	1.6	≥8.0	2.5	3.6
3	6.2	3.0	1.4	6.0	8.7	51.0
4	1.0	3.7	2.2	1.4	3.6	1.2
DT-IPV						
1S	1.8	1.1	2.2	1.1	0.9	0.9
2S	0.9	1.2	0.8	0.9	0.8	0.3
Experimental combinations						
1S	1.7	1.6	0.7	1.2	0.8	≤0.3
2S	1.2	1.2	2.1	1.3	1.2	1.1

* Expressed in units against the CLL Ref.

The immunogenic activity of the DPT-IPV combination tends to be higher than the reference or DT-IPV combinations. This suggests that pertussis has an adjuvant effect. Similar trends have also been observed in monkey potency data.

Conclusions

Our data indicate that the Rat Immunogenicity Assay for IPV is a good model and potential alternative method to either/both of N.I.H. Monkey and Eu. Pharm. Chick Potency Assays. This conclusion is based on the performance of the assay with vaccines from different manufacturing methods, various combinations and vaccine kept for long periods at 4°C which is antigenically but not immunogenically stable. The anticipated release of an International Reference IPV vaccine will facilitate the standardization of rat potency assays and calculation methods from different laboratories. However, it will be important to investigate whether relative potencies, measured against an international reference, reflect the potency in man for the various products. Until that time, we consider that the use of a homologous reference to demonstrate consistency in manufacturing in comparison with a reference vaccine of known quality is the best short-term approach for use of the rat test as a release test for IPV.

These concepts are under discussion with the Canadian authorities. Our goal is to have the rat potency assay accepted as a release assay and to pursue the application of the single dilution assay as a screening test. This should result in a significant reduction in animal use.

Acknowledgements

We wish to thank Sema Firestone for administrative assistance.

References

1 Van Steenis G, van Wezel AL, Sekhuis VM: Potency testing of killed polio vaccine in rats. Dev Biol Stand 1981;47:119-128.

2 Finney DJ: Probit Analysis 3rd Edition Cambridge University Press; 1971.

3 European Pharmacoepia Monograph 214 Vaccinum Poliomyelitidis Inactivum 1983;6:96.

J. Bevilacqua, Connaught Laboratories Limited, 1755 Steeles Avenue West, Willowdale, Ontario M2R 3T4, Canada

Brown F, Cussler K, Hendriksen C (eds): Replacement, Reduction and Refinement of Animal Experiments in the Development and Control of Biological Products.
Dev Biol Stand. Basel, Karger, 1996, vol 86, pp 129-135

Hepatitis A-Vaccines: a Comparison between Three Methods of Antigen Determination

F. Burkhardt, R. Glück, B. Finkel-Jimenez, S. Brantschen

Swiss Serum and Vaccine Institute, Berne, Switzerland

Keywords: Hepatitis A-vaccine, virosomal vaccine, hepatitis A-antigen, HAAg solid phase RIA, HAAg solid phase EIA, HAV mouse potency test.

INTRODUCTION

In 1978 it became possible to diagnose hepatitis A virus (HAV) infections serologically by determining IgM and IgG antibodies to HAV with solid-phase immunoassays. However, the antigen preparations used for these tests were purified faecal extracts from human and later marmoset monkey origin. It took a few more years until HAV could be cultivated in cell cultures of different species: For reasons yet poorly understood, primary isolation of HAV from faecal samples may take up to 210 days [1]. The propagation and later attenuation of HAV in human diploid cell cultures [2] gave way to the concept of manufacturing vaccines against Hepatitis A.

It is therefore not surprising that so far only two hepatitis A-vaccines are available. In 1992, Smith, Kline, Beecham brought Havrix to the market, an alum-absorbed inactivated vaccine, and this year, our company, the Swiss Serum and Vaccine Institute, launched Epaxal Berna, an inactivated vaccine with a new design: HAV was coupled to Influenza A Virosomes [3]. High immunogenicity has been shown in several clinical studies [3,4]. On the basis of that short historical review, it is easy to understand why several problems in potency testing have so far not been solved in a uniform way:

1. Only a few producers are testing the hepatitis A Antigen (HAAg) content with different methods, be it solid phase immunoassays or immunogenicity tests in mice.
2. No European Pharmacopoeia Monograph exists so far on inactivated hepatitis A vaccines, only a draft.

3. No international standard reference preparation is available at present.
4. Antigen units are still arbitrary: a preliminary collaborative study has brought WHO experts to the conclusion that for the moment, antigen content should be declared as in-house determined units and not as ng/dose until standardization would allow such a declaration in the form of a future WHO standard.
5. The big difference in the design of both existing vaccines does not allow a direct comparison in antigen content.

Before we present the methods and results of antigen determination, a production schedule for Epaxal Berna is summarized in Figure 1.

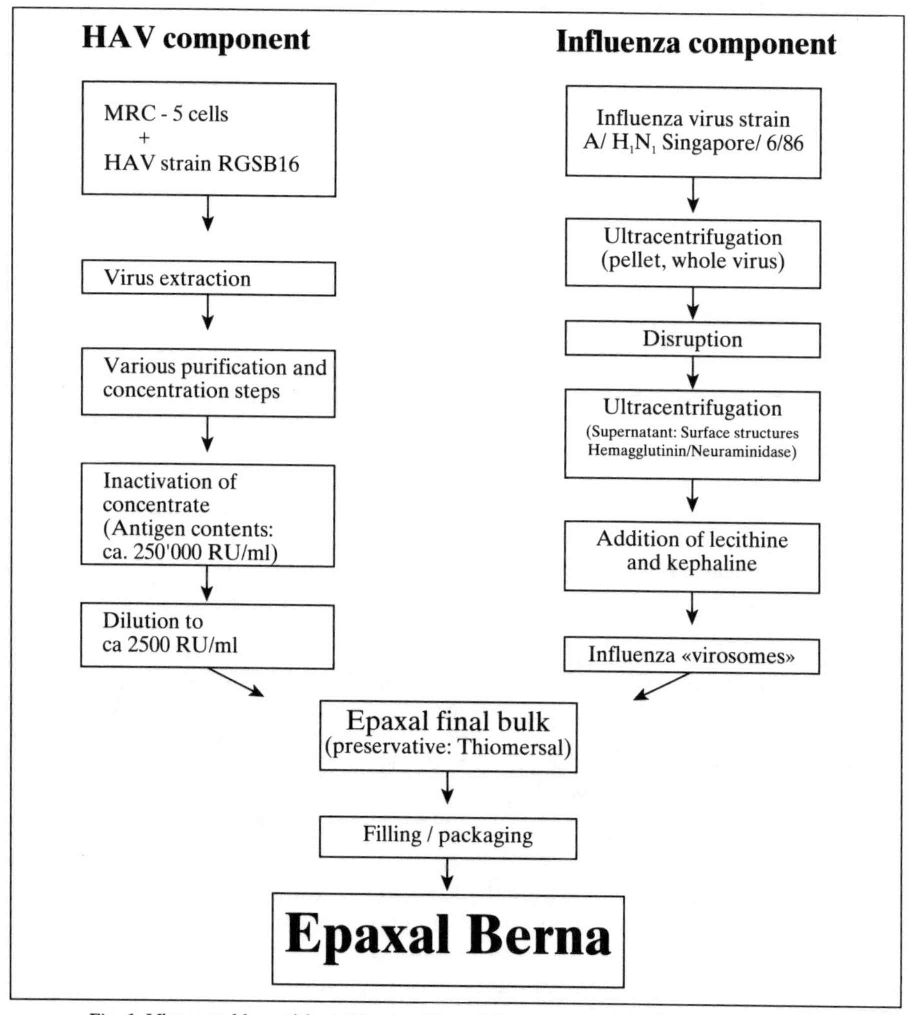

Fig. 1: Virosomal hepatitis A-Vaccine Epaxal Berna summarized production steps.

1. HAV component

HAV strain RGSB16 is propagated on MRC-5 cells. Viral extraction is followed by various purification and concentration steps, before formalin inactivation.

The concentrate usually contains around 250,000 RIA Units (RU)/per ml (see below) and is then diluted 100-fold down to roughly 2,500 RU/ml.

2. Influenza component

Influenza virus strain A/H1N1/Singapore/6/86 is pelleted by ultracentrifugation, and the resuspended pellet disrupted in several different steps. After a second ultracentrifugation, the haemagglutinin and neuraminidase are collected from the supernatant and re-aggregated to so-called virosomes after addition of lecithin and kephaline.

3. Epaxal Berna

Coupling the product of the last preparation step to the virosomes results in Epaxal final bulk. After preservation with Thiomersal, filling and packaging, the virosomal vaccine is presented as Epaxal Berna. The model of this virosomal vaccine is shown in Figure 2.

Materials and Methods

The first four consecutive production lots of Epaxal Berna (FB 07, 08, 09 and 10) as well as the Havrix 1440 production lot VHA 401B6 were submitted to three different tests for assessment of antigen content or immunogenicity in mice.

1. Solid phase Immunoassay for direct detection of HAAg

1a) Solid phase Sandwich RIA

As shown in Figure 3, the solid phase (polystyrene bead) is coated with a highly diluted post-infectious serum specimen with an antibody titre of >1:2500, as measured by Abbott Havab test. The sera were collected from a hepatitis A outbreak in a Swiss village in 1979 [5] and were then also used by one of the authors of the study to develop the antigen detection test.The solid phase is then reacted with the HAAg containing specimen and standard in appropriate dilutions, incubated and washed. In the third step, radiolabelled antibody of human origin from the Abbott Havab test kit is added. After incubation and washing, samples are measured in a Gamma counter and the antigen content is calculated by regression analysis. This method has been used and validated during the entire development phase of our vaccine and is part of our registration documentation.

1b) Solid phase Sandwich EIA

The assay shown in Figure 4 is of the same principle as the one described above, with the following technical differences:

1. The test is performed on microtitre plates, coated with a mouse monoclonal antibody to HAV.
2. HRPO-conjugated mouse monoclonal antibody to HAV is used as tracer.
3. H_2O_2/TMB are added as substrate/co-enzyme in a fourth step. The OD (450) is then measured.

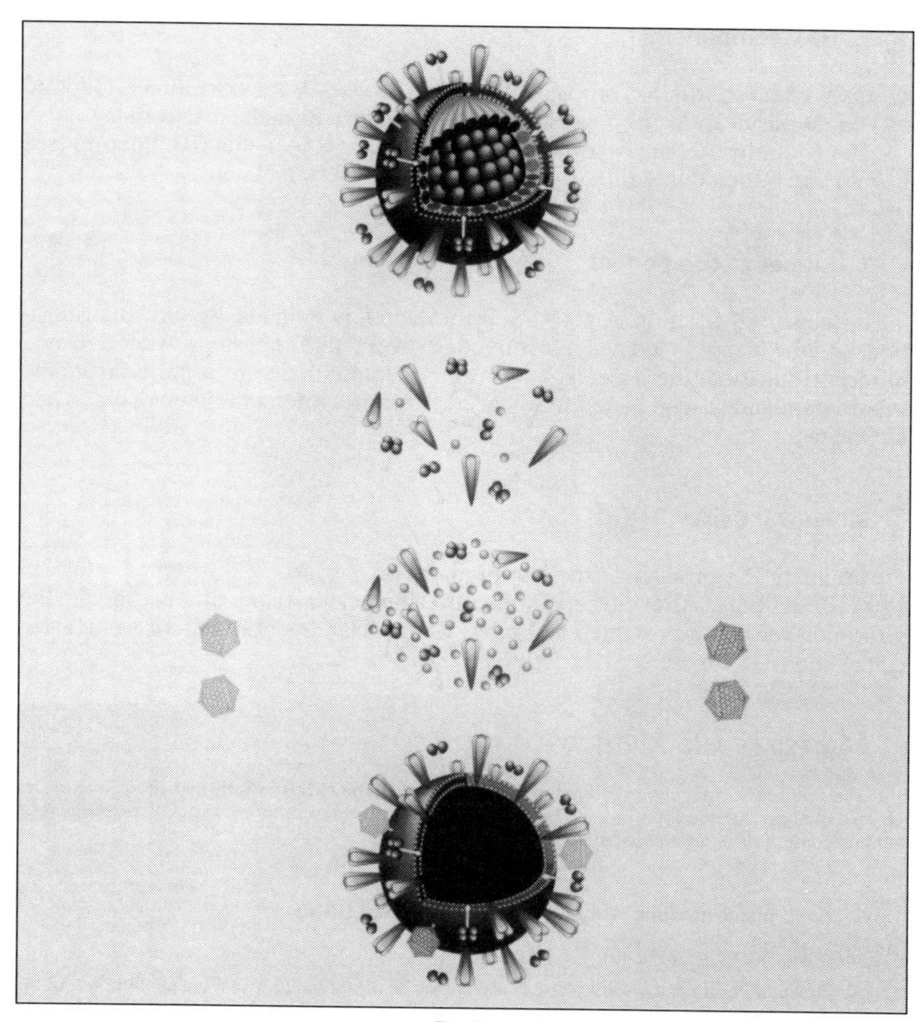

Fig. 2.

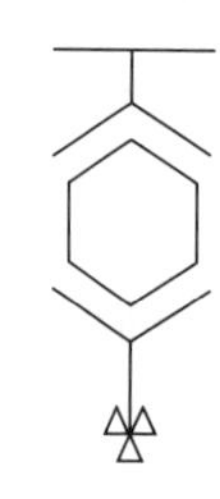

Solid phase coated with diluted post-infection serum with high antibody titre to HAV

HAAg containing sample in appropriate dilutions

radiolabelled antibody to HAV (human) from Havab (Abbott) test kit

Fig. 3: Solid phase radio-immunoassay (RIA) for the detection of hepatitis A-Antigen (HAAg).

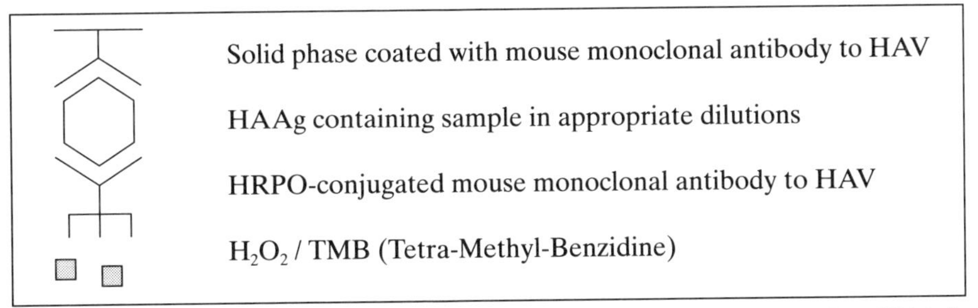

Fig. 4: Solid phase enzyme-immunoassay (EIA) for the detection of hepatitis A-Antigen (HAAg).

The calculation method is the same as for the RIA, and the RU values of the standard are taken as EU values in a 1:1 ratio. The assay has been recently developed to parallel the RIA.

2. Immunogenicity test in mice

Five-week-old mice, as proposed by the EP draft dated June 1993, were immunized intraperitoneally with 0,5 ml of the following dilutions of the vaccine lot to be tested: 1:10, 1:40, 1:160, 1:640, 1:2,560 and 1:10,240.

Dilutions were made in PBS. Ten mice were each immunized with 0.5 ml PBS and served as negative controls in the antibody assay. After 28 days, all test animals were bled and the sera were submitted to a solid phase antibody EIA as shown in Figure 5. The solid phase again consisting of a microtitre plate as in the previously described EIA, is coated with hepatitis A antigen and reacted with the individual sera of immunized mice at an appropriate dilution (same dilution for all mice). After incubation and washing of the plates, HRPO-conjugated antibody to mouse IgG of goat origin is added. After incubation and washing, H_2O_2/OPD are added as substrate/co-enzyme. The OD 490 is measured for all samples and reactive versus non-reactive specimens are recorded for each dilution.

Results are calculated in two ways:

1. ED_{50} according to Spearman-Kärber.
2. EP_{50} according to Reed and Muench.

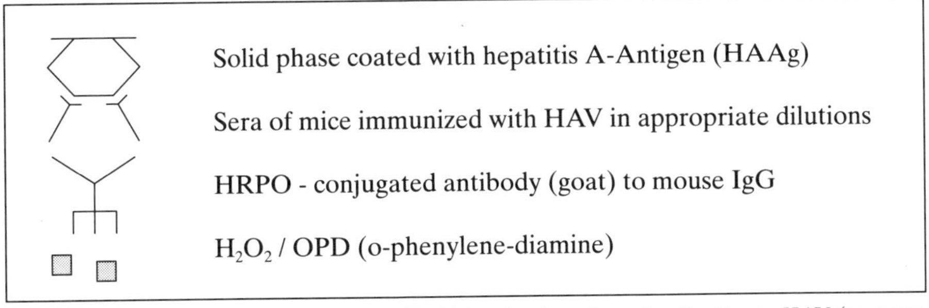

Fig. 5: Solid phase enzyme-immunoassay (EIA) for the detection of antibodies to HAV (to assess ED_{50} value after potency testing in mice).

RESULTS

The results of all tests for the specimens described are compiled in Table 1.

DISCUSSION

Apart from the fact that our solid phase RIA was validated by experts of the Institute for Mathematical Statistics of the University of Berne, the results presented in this paper are not abundant. This is due to the fact that our product has only recently been introduced and only four lots have been produced on an industrial basis. However, we conclude from the current data:

1. There is a consistency of 95% to 98% for Lots FB07, 08 and 09 between the solid phase assays for Epaxal Berna and 87% for lot FB10.
2. The variations in ED_{50} as well as EP_{50} are acceptable from test to test for the same lot (Epaxal Berna FB 07 or Havrix 1440). There was an **almost six-fold difference** in the mouse potency test between Lot FB08 and FB09 as compared to virtually identical results in RIA for the same lots. This is beyond acceptable limits, particularly so because FB08 and FB09 are two filling lots of an identical final bulk. A similar comment can be made for FB 10 in comparison with the other lots submitted to both kinds of assays. Some more parallel testing will be needed to confirm this statement.

Table 1: Determination of hepatitis A-Antigen (HAAg) in Epaxal Berna.

	Direct antigen detection by solid phase immunoassay		**Immunogenicity tests in mice**		
Vaccine lot	RIA (Results in RU/ml) 1)	EIA (Results in EU/ml) 2)	ED50 (Spear-mann-Kärber)	Test Nr.	EP50 (Reed and Münch)
FB 07	**1955** 3) SD abs.: 189 95% C.I. abs.: ±370 rel.: ±18.9% (1585-2325)	**1892** 3) SD abs.: 233 95% C.I. abs.: ±457 rel.: ±24% (1435-2349)	**5.38 RU** **2.70 RU** **7.09 RU** **6.18 RU**	1/94 2/94 3/94 4/94	**1:363** **1:724** **1:268** **1:389**
FB 08	**2940** 4)	**2788**	**3.08 RU**	3/94	**1:1259**
FB 09	**2939**	**2884**	**0.53 RU**	3/94	**1:5754**
FB 10	**2320**	**2668**	**9.67 RU**	2/94	**1:187**
Havrix 1440 VHA 401 B6	**xxx** 5)	**xxx** 5)	**xxx** **xxx** **xxx**	1/94 3/94 4/94	**1:69** **1:209** **1:105**

Legend:
1) Values compared to an in-house standard (Std F) containing 2185 RU/ml (SD abs.: 187 RU, SD rel.: 8,6% (validated)).
2) EIA units (EU) were adapted 1:1 to RIA units (RU) see 1).
3) Mean of 20 (RIA) resp. 10 (EIA) independent tests.
4) FB 08-10: Results represent the mean of four independent tests for RIA and EIA.
5) RU resp. EU/ml could not be determined for Havrix due to alum component; therefore ED_{50} values also cannot be given.

3. At present, it is not possible to compare in vitro antigenicity between our product and Havrix, since we have no technical information about the in vitro assays that SKB perform on their product. A closer cooperation in this field would be of great help to both producers.

Conclusion

Looking at the results of our solid phase antigen assays on the one hand and at the mouse potency tests on the other, we would suggest that at least as far as our hepatitis A vaccine Epaxal Berna is concerned, there are reasons for continuing mouse potency tests to prove immunogenicity of the next five to ten lots, but then abandon them and stay with the in vitro tests for lot release, because the mouse potency assays have no value in adjusting the final bulk from the concentrate.

References

1 Flehmig B: Hepatitis A, in Baillière's Clinical Gastroenterology, 1990, Vol. 4, Nr. 3, pp 707-720.

2 Glück R, Althaus B, Berger R, Just M, Cryz SJ Jr: Development, safety and immunogenicity of new inactivated hepatitis A vaccines: effects of adjuvants, in Lobel HO, Steffen R, Kozazarsky PE (eds): Travel Medicine 2. Atlanta, Proc 2nd Conf Ind Travel Med,1992, pp 135-136.

3 Glück R, Mischler R, Brantschen S, Just M, Althaus B, Cryz SJ Jr: Immunopotentiating reconstituted influenza virus virosome vaccine delivery system for immunization against hepatitis A. J Clin Invest 1992;90:2491-2495.

4 Loutan L, Bovier P, Althaus B, Glück R: Inactivated virosome hepatitis A vaccine. The Lancet 1994;343:322-324.

5 Schilt U, Burkhardt F, Siegl G: Virological and serological data of a hepatitis A outbreak in Wasen i.E., Switzerland. Experientia 1982;38:1366.

Dr. F. Burkhardt, Med. Chem. Labor Thun, Postfach 126, CH-3608 Thun, Switzerland

Brown F, Cussler K, Hendriksen C (eds): Replacement, Reduction and Refinement of Animal Experiments in the Development and Control of Biological Products.
Dev Biol Stand. Basel, Karger, 1996, vol 86, pp 137-145

Development and Evaluation of Alternative Testing Methods for the in vivo NIH Potency Test used for the Quality Control of Inactivated Rabies Vaccines

E. Rooijakkers[1], J. Groen[2], J. Uittenbogaard[3], J. van Herwijnen[4], A. Osterhaus[1,2]

1 Institute of Virology, Erasmus University, Rotterdam, The Netherlands
2 Department of Clinical Virology, University Hospital Rotterdam, Rotterdam, The Netherlands
3 Laboratory of Vaccine Development and Immunemechanisms, National Institute of Public Health and Environmental Protection (RIVM), Bilthoven, The Netherlands
4 European Veterinary Laboratory b.v., Woerden, The Netherlands

Abstract: The potency control of rabies vaccines is routinely performed in a vaccination-challenge test (NIH test), which induces substantial distress and suffering in laboratory mice. Although, according to the recommendations by the WHO, partial replacement of this in vivo potency test by in vitro antigenicity testing is permitted, no internationally accepted alternative testing method is yet available. In the present study, we focussed on the use of monoclonal antibody-based ELISA systems for the quantitative detection of rabies glycoprotein in vaccines for human and veterinary use. Results in the newly developed competitive binding ELISA for glycoprotein quantification showed good correlation with NIH potencies. Also, the applicability for vaccines based on different virus strains was demonstrated. Quantification methods were remodelled to convenient ELISA kits for interlaboratory evaluation of the sensitivity and reproducibility in an international collaborative study. The relevance of the data generated with these assays is currently assessed by parallel studies evaluating vaccine-induced B and T cell-mediated immunity. We speculate that the assays developed will provide, at least in part, a suitable replacement for the NIH potency test.

INTRODUCTION

In Western Europe, regulations for assessment and assurance of rabies vaccine potency are dictated by control authorities, according to the recommendations by the World Health Organization (WHO) [1] and European Pharmacopoeia (Eur. Ph.)

[2]. In general, vaccine manufacturers have adopted the in vivo vaccination-challenge procedure in mice (NIH test, [3]) for testing the potency of rabies vaccines. However, for both practical and ethical reasons, the application of in vivo tests for potency control has been widely critisized [4]. Alternative methods for potency estimation, e.g. enzyme-linked immunosorbent assays (ELISAs) [5-8], the immunodiffusion assay (SRD) [9] and the antibody binding test (ABT) [10], have been proposed. The alternatives presented so far showed practical and technical drawbacks and lacked scientific justification, which hampered their general acceptance.

In this report we describe enzyme-linked immunosorbent assays for quantitative determination of the rabies glycoprotein (GP) in rabies vaccines. We demonstrate their applicability to in-process control and final-lot control procedures of adjuvanted and non-adjuvanted rabies vaccines. The antigenicity data generated with these assays are compared with potency values in the NIH test. We also present preliminary data of a collaborative study and the additional evaluation that has been carried out to validate the ELISAs.

Considering the «Three R concept» for replacement, reduction and refinement of animal testings, the impact of introducing antigenicity testing in potency control of rabies vaccines is discussed.

Materials and Methods

Rabies vaccines

Freeze-dried inactivated rabies vaccines for human use, produced with the Pitman-Moore virus strain on primary dog kidney cell cultures (PM-DKC vaccines) and liquid inactivated vaccines for veterinary use, obtained from different manufacturers, were stored refrigerated until use in these studies. The in vivo potency tests and ELISAs were calibrated against an in-house reference rabies vaccine (R0-41A, containing 4.2 IU of rabies vaccine per ampoule).

Monoclonal antibodies

Two glycoprotein specific murine monoclonal antibodies, MAb 6-15 and 2-22 [11], were used in these studies. Mab 2-22 was labelled with biotin (NHS-d-Biotin, Sigma Chemicals) to increase the sensitivity of the ELISA.

Potency test

The in vivo potency test for rabies vaccines was either carried out following a standardized operating procedure, according to WHO recommendations for human vaccines [3] or to Eur. Ph. regulations for veterinary vaccines [2].

Enzyme-linked immunosorbent assays

ELISAs for the detection of rabies glycoprotein (GP) were based on the sandwich principle (two-antibody ELISA) or competition principle (antigen-competition ELISA). The two-antibody ELISA was carried out as follows: MAb 6-15 was coated in excess on ELISA plates. After washing and blocking the plates, serial twofold dilutions of vaccine samples were prepared and incubated on the coated plates. Again after washing, biotin-conjugated MAb 2-22, in excess, was incubated on the plate. Subsequently, horse radish peroxidase (HRPO)-bound streptavidin (SBC-HRPO, Amersham International, Amersham, UK) was allowed to bind to biotin, followed by development with a tetramethylbenzidine (TMB) substrate solution.

The antigen-competition ELISA was carried out as follows: serial twofold dilutions of vaccine samples were prepared on low-binding microtitre plates, followed by the addition of equal volumes of a titrated dilution of biotin-conjugated MAb 2-22. After 1 hour, the antigen-antibody mixture was transferred to ELISA plates previously coated with a standard preparation of rabies viral antigen. The bound MAb was detected with SBC-HRPO and subsequently developed with a TMB substrate solution.

Lymphocyte stimulation assays

Antigen-induced proliferation of lymphocytes, isolated from blood of human donors previously immunized with rabies vaccine, was determined in lymphocyte stimulation assay (LST). Briefly, lymphocytes were separated by gradient centrifugation from whole blood. Subsequently, threefold dilutions of rabies vaccine samples were prepared in triplicate in microtitre culture plates, followed by adding the isolated lymphocytes (100k cells per well). After culturing for three to five days, [3H]-labelled thymidine ([3H]-Trd) was added to the culture medium and incubated for 16 hours. Subsequently, proliferation of the cells was measured, after harvesting the proteins of the cells on filter paper, by counting the radioactivity in a ß-plate reader.

Calculation procedure for ELISA antigenicity values

A reference curve of the positive standard was constructed by plotting optical density (OD) values, measured at serial dilutions of the standard, as a function of the corresponding antigenic values. The antigenicity of a sample was determined in the plot by reading the antigenic value corresponding to the OD value measured. The results were expressed in equivalent units (EU) of GP, related to the positive standard used.

Results

Antigenicity testing

The content of glycoprotein (GP) in culture supernatants of two vaccine production runs was determined in the two-antibody GP-ELISA. The concentration of GP in successive harvests showed a time and batch-dependent course (Fig. 1). The quantitative detection of the GP content in this ELISA offers the possibility of efficient monitoring of the production process within three hours.

For a panel of nine final-lot human rabies vaccines, the GP-specific antigenicity was measured in both the two-antibody and the antigen competition ELISA. Comparison of the ELISA data with potency values determined in the NIH test showed good correlation, with correlation coefficients of r=0.8860 (two-antibody ELISA vs NIH test) and r=0.9600 (competition ELISA vs NIH test) for the vaccine samples tested (Fig. 2A).

Furthermore, the GP content of a panel of six commercial veterinary rabies vaccines of different manufacturers was tested in the competition ELISA. Irrespective of the origin of the vaccine – with regard to cell substrate, vaccine strain and adjuvant – the antigenic values in GP-ELISA showed good correlation (correlation coefficient, r = 0.9659) with potency values determined in the NIH test carried out according to the Eur. Ph. protocol (Fig. 2B).

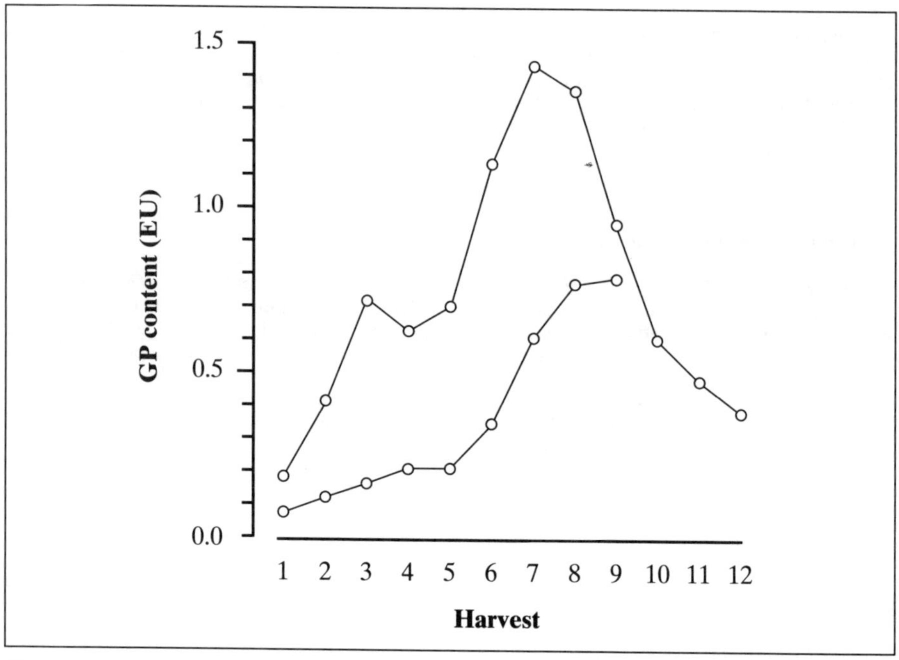

Fig. 1: Quantification of GP content in successive harvests of two PM rabies vaccine production runs in DK cells. Cell culture supernatants were tested in the two-antibody ELISA, using the GP-specific MAbs 6-15 (coated) and 2-22 (biotinylated). The results are expressed in ELISA units (EU).

Collaborative testing of ELISA

The antigen competition GP-ELISA was remodelled to a convenient kit for testing the robustness of the assay in a collaborative study. Nine participants, from production and control laboratories in Europe, carried out the GP-ELISA on a panel of selected rabies vaccines in a single trial. Valid ELISA data were generated by seven participants for the positive standard and by five participants for a panel of three vaccine samples. The results for the positive standard showed an inter-laboratory variation of 20% (CV) for the slope of the titration curve (data not shown). The antigenic values, generated by five participants with the GP-ELISA kit, showed a variation of 20, 14 and 26% (CV) for the three vaccine samples tested (Fig. 3).

Immunogenicity testing

The capacity of vaccine samples to induce a secondary immune response in lymphocyte cultures was investigated in lymphocyte stimulation assays, by measuring the radioactivity of incorporated [3H]-labelled thymidine. The lymphocytes showed an antigen-specific, dose-dependent proliferative response (data not

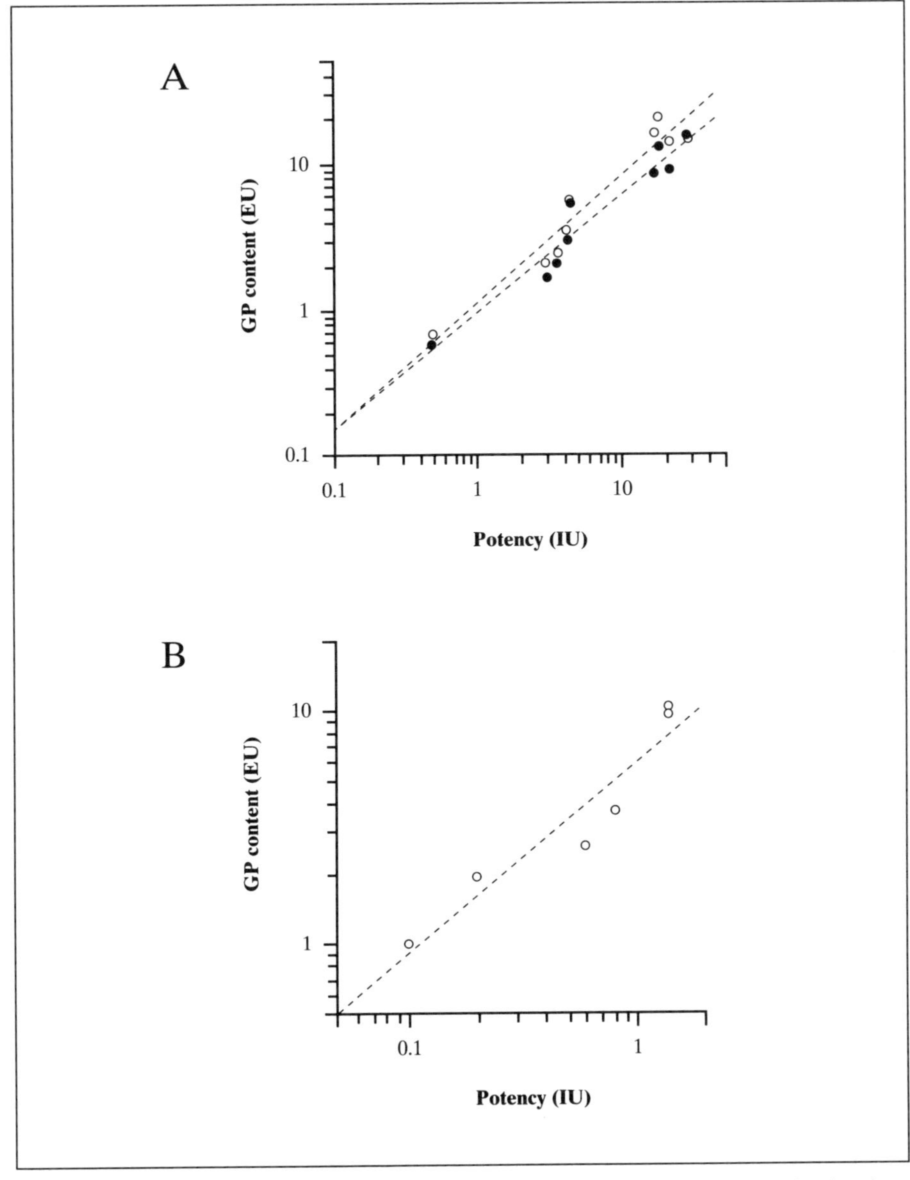

Fig. 2: The relationship between the potency determined in the NIH test and the antigenic values in ELISA.

A. For a panel of nine human PM-DKC vaccine lots the antigenic values, determined in the two-antibody ELISA (○) and the antigen competition ELISA (•), are compared with the potency values in the NIH test (according to the WHO recommendations).

B. For a panel of six commercial veterinary rabies vaccines the antigenic values, determined in the antigen competition ELISA, are compared with the potency values in the NIH test (according to the Eur. Ph. regulations).

Regressions are indicated by dotted lines (correlation coefficients: see text in results section).

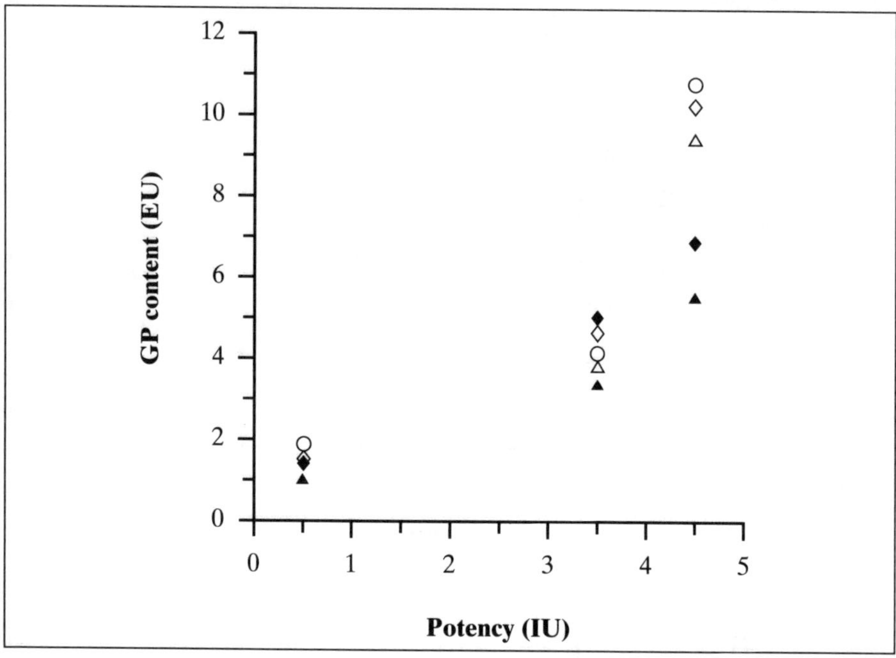

Fig. 3: Comparison of glycoprotein content (GP-ELISA kit) and potency (NIH test) for three rabies vaccines, tested in a collaborative trial. The antigenic values for three lots of PM-DKC vaccines were determined by five participants of the trial. Individual ELISA results are indicated by different symbols. The interlaboratory variation (%CV) in antigenic values for the three vaccines (from low to high potency) was respectively 20%, 14% and 26%.

shown). The proliferation of lymphocytes correlated well with vaccine potency, with a correlation coefficient of r=0.9709 (Fig. 4).

Discussion

In Europe, millions of rabies vaccine doses are produced annually for human and veterinary use. Manufacturers are obliged to comply with the quality recommendations or requirements outlined by the international control authorities, WHO and Eur. Ph. In agreement with the requirements for potency, each batch of vaccine has to pass the in vivo potency test before release. The reliability of this animal test has been discussed since its introduction in 1953; however it is still recommended by both the WHO and Eur. Ph. The test is not only laborious and poorly reproducible but also unethical in terms of animal welfare. Consequently, studies were initiated to develop alternative methods for potency testing. So far, the proposed methods for in vitro antigenicity testing, immunodiffusion, antibody binding and enzyme-linked immunosorbent assays, suffered from limited applicability due to strain specificities of the antibodies and interference by adjuvants [5-10, 13].

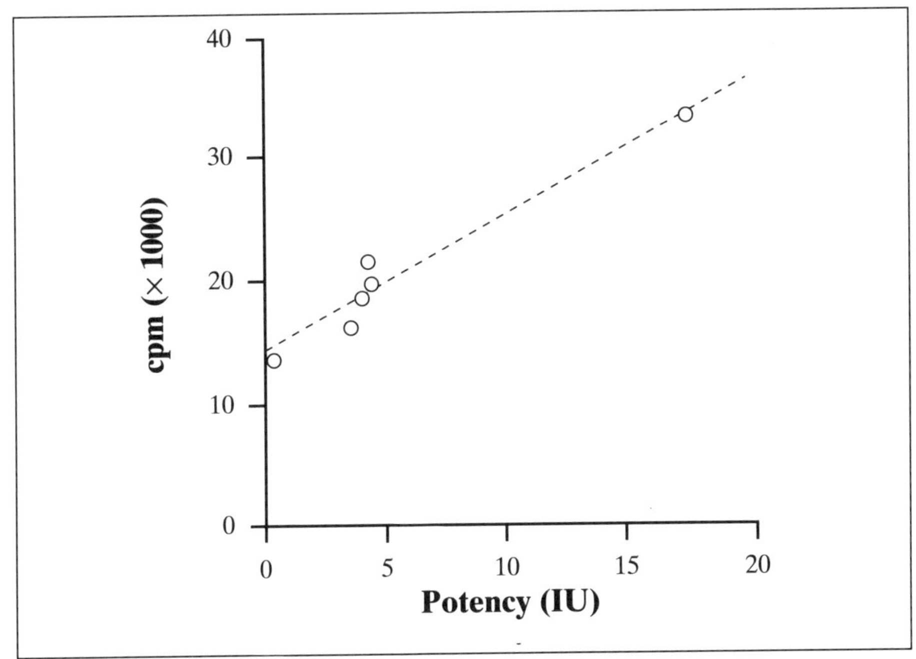

Fig. 4: Comparison of vaccine potencies (NIH test) and vaccine-induced proliferative responses in human lymphocyte cultures (LST) for a panel of six successive batches of PM-DKC vaccines. The regression is indicated by a dotted line (correlation coefficient r=0.9709).

In the current study we focussed on ELISAs for GP quantification, suitable to the various types of human and veterinary rabies vaccines, presently applied in Europe. The newly developed GP-ELISA proved to be applicable for testing vaccines based on PM, PV, Flury-LEP and SAD virus strains. Compared to assays based on polyclonal antibodies, the use of broadly reactive monoclonal antibodies has advantages in terms of specificity and unlimited supply. Another major improvement, compared to SRD, ABT and direct ELISAs is the possibility of using this ELISA in the presence of adjuvants (e.g. aluminium salts), that are widely applied to veterinary vaccines. The applicability of a similar ELISA system for NP quantification, recently developed by us, is still restricted to unadjuvanted vaccines and needs further evaluation.

Comparison of the potency values determined in the GP-ELISA and NIH test for human and veterinary rabies vaccines showed good correlation. However, additional information about the immunogenicity of the quantified GP not dependent on the NIH test would obviously support the acceptance of the ELISA for potency estimation. For this purpose, we demonstrated antigen-induced proliferative responses in cultures of lymphocytes from vaccinees. The data indicated that another relevant parameter of immunogenicity would be the in vitro induction of a secondary immune response. The quality of this response could be further investigated by analysing the epitope specificity of the induced antibodies in a recently described inhibition ELISA [14]. The advantage of this approach is that the use of lymphocytes from human rabies vaccinees yields information directly relevant for protection in the target species and probably offers the opportunity to abolish the in vivo antibody induction assay applied for immunogenicity testing.

Finally, the practical use of the GP-ELISA was studied in an international collaborative effort. The results of the trial showed an acceptable level of variation. Moreover, based on our experience and that of participants, we expect even higher reproducibility after familiarization with the system. The robustness of the ELISA kits used may be improved with minor modifications, which will be tested in a follow-up multi-participant trial.

Conclusion

The correlation between the results in GP-ELISA, LST and NIH test obtained with a panel of PM-DKC rabies vaccines clearly indicates the potential of the GP-ELISA for estimating the potency of rabies vaccines. The applicability of the ELISA for potency testing of vaccines based on other vaccine strains and production procedures should be further investigated in collaboration with vaccine manufacturers. These collaborative efforts should lead to standardized procedures, preferably performed with commercially available test kits to facilitate the acceptance of the proposed methods. Since the mechanisms which lead to vaccine-induced immunity to rabies are not fully understood, additional research of B and T cell-mediated immunity should provide more information about the immunological mechanisms for protection against rabies. Since virus-neutralizing antibodies induced by the GP of the virus offer direct in vivo protection against rabies, control authorities should be convinced of the reliability of the GP-ELISA method to replace at least in part the NIH test.

Acknowledgements

The authors wish to thank the participants in the collaborative trial for testing the ELISA kits: Dr. L. Brückner (Institute for Virology and Immunoprophylaxis, Mittelhäusern, CH), Dr. M. Ferguson (National Institute for Biological Standards and Control, Hertfordshire, UK), Dr. P. Flore (Solvay Duphar bv, Weesp, NL), Dr. R. Glück (Swiss Serum and Vaccine Institute, Berne, CH), Dr. M. Langlois (Agence du Médicament, Lyon, F), Dr. M. Mainka (Paul Ehrlich Institute, Langen, D), Dr. P. Marbehant (SmithKline Beecham Animal Health S.A., Louvain-la-Neuve, B), Dr. L. Peyron (Pasteur Mérieux Serums et Vaccins, Marcy l'Etoile, F) and Dr. N. Visser (Intervet International bv, Boxmeer, NL).

References

1 WHO Expert Committee on Rabies, 8th Report. WHO Technical Report Series, Geneva, 1992, vol 824.

2 Vaccinum rabiei inactivatum ad usum veterinarium. In: Monograph European Pharmacopoeia, 2nd edition, 1985, part II/9:451.

3 Seligmann EB Jr: The NIH test for potency, in Kaplan MM, Koprowski H (eds): Laboratory Techniques in Rabies, 3rd edition, World Health Organization, Geneva, 1973, pp 279-286.

4 Barth R, Diderrich G, Weinmann E: NIH test, a problematic method for testing potency of inactivated rabies vaccine. Vaccine 1988;6:369-377.

5 Van der Marel P, Van Wezel AL: Quantitative determination of rabies antigen by ELISA. Dev Biol Stand 1981;50:267-275.

6 Thraenhardt O, Ramakrishnan K: Standardization of an enzyme immunoassay for the in vitro potency assay of inactivated tissue culture rabies vaccines: Determination of the rabies virus glycoprotein with polyclonal antisera. J Biol Stand 1989;17:291-309.

7 Lafon M, Perrin P, Versmisse P, Sureau P: Use of a monoclonal antibody for quantitation of rabies vaccine glycoprotein by enzyme immunoassay. J Biol Stand 1985; 13:295-301.

8 Osterhaus A, Groen J, UytdeHaag F, Bunschoten E, De Groot I, Van der Meer R, Van Steenis G: Quantification of rabies virus glycoprotein by ELISA with monoclonal antibodies: correlation with single radial diffusion and the induction of anti-viral antibodies and protection in mice, in Thraenhart O, Koprowski H, Bögel K, Sureau P (eds): Progress in rabies control, IMVI Essen, Rochester, Staples Printers Rochester LTD., 1990, pp 324-341.

9 Ferguson M, Seagroatt V, Schild GC: A collaborative study on the use of single radial immunodiffusion for the assay of rabies virus glycoprotein. J Biol Stand 1984;12:283-294.

10 Barth R, Gruschkan H, Milcke L, Jaeger O, Weinmann E: Validation of an in vitro assay for the determination of rabies antigen. Dev Biol Stand 1986;64:87-92.

11 Bunschoten H, Gore M, Claassen I, UitdeHaag F, Dietzschold B, Wunner W, Osterhaus A: Characterization of a new virus-neutralizing epitope that denotes a sequential determinant on the rabies virus glycoprotein. J Gen Virol 1989;70:291-298.

12 Dietzschold B, Gore M, Casali P, Ueki Y, Rupprecht C, Notkins A, Koprowski H: Biological characterization of human monoclonal antibodies to rabies virus. J Virol 1990;64:3087-3090.

13 Ferguson M, Heath A: Report of a collaborative study to access the determination of glycoprotein antigen content of rabies vaccines for human use. Biologicals 1992;20:143-154.

14 Huynh A, Rooijakkers E, Nguyen T, Groen J, Osterhaus A: Methods for the purification of equine rabies immunoglobulin: effects on yield and biological activity. Biologicals 1994;22:1-6.

Prof. A.D.M.E. Osterhaus, Institute of Virology, Erasmus University, Dr. Molewaterplein 50, P.O. Box 1738, NL-3000 DR Rotterdam, The Netherlands

Brown F, Cussler K, Hendriksen C (eds): Replacement, Reduction and Refinement of Animal Experiments in the Develoepment and Control of Biological Products.
Dev Biol Stand. Basel, Karger, 1996, vol 86, pp 147-156

Serological Responses in Calves to Vaccines against Bovine Respiratory Syncytial, Infectious Bovine Rhinotracheitis, Bovine Viral Diarrhoea and Parainfluenza -3 Viruses

M. Tollis[1], L. Di Trani[1], P. Cordioli[2], E. Vignolo[1], I. Di Pasquale[1]

[1] Istituto Superiore di Sanità, Rome, Italy
[2] Istituto Zooprofilattico Sperimentale, Brescia, Italy

Key words: Efficacy, vaccine, bovine, viral respiratory disease.

Abstract: The Istituto Superiore di Sanità (ISS), the National Veterinary Services Laboratory in Italy, is in charge of assessing the quality, safety and efficacy of veterinary vaccines before and after licencing. To evaluate the relative potency of several vaccines against bovine respiratory syncytial virus (BRSV), infectious bovine rhinotracheitis virus (IBRV), bovine viral diarrhoea virus (BVDV) and parainfluenza-3 virus (PI3V), the serological responses in vaccinated calves were studied. Vaccination with any of the vaccines under study induced specific antibody titres against the different viral antigens. The differences of the mean antibody titres within and among the test group vaccines were statistically significant. The results confirm and support those obtained by other authors in similar studies, suggesting that serological responses in vaccinated calves can be used as a helpful means of assessing the relative potency of vaccines against viral respiratory diseases of cattle. The criteria allowing such an evaluation are discussed.

Introduction

Two types of vaccines against viral respiratory diseases of cattle, either attenuated or inactivated, available as single or combined products, are applied in the field. The variety and complexity of factors which affect such pathologies make it difficult to assess the efficacy of these vaccines and, as a consequence, to discriminate between the activiy of the different products. The lack of reference virus strains and experimental models capable of reproducing clinical signs of some diseases, thus allowing a clear differentiation between vaccinated and control animals (mainly for BRSV and PI3V), constitute the major obstacles to the definition of clinical efficacy [1]. Moreover, the widespread occurrence of these infections makes it very difficult to find seronegative calves for the evaluation of alternative parameters (e.g. demonstration of vaccine «take», reduction of the challenge virus excretion, ability to boost a serological response) for the activity of the vaccines [2, 3, 4]. Seroconversion in vaccinated calves has been shown to be not necessarily correlated with protection in the field [5, 6, 7, 8]; therefore the limits of a significant antibody response have not been clearly defined [9, 10]. Nevertheless the serological activity of the vaccine, in terms of the ability to induce specific antigenic seroconversion in vaccinated animals can be regarded as an objective criterion to discriminate the quality of different products [1, 11, 12]. This criterion has been adopted for years at the ISS; therefore seroconversion in vaccinated animals has been routinely verified during the pre-licencing control phase of vaccines against viral respiratory diseases of cattle. Moreover, specific and sensitive techniques capable of improving the detection of serum antibody responses have been developed and validated to allow a comparison of the relative potency of the different vaccines tested.

Materials and Methods

Vaccines

Fourteen vaccines containing modified live (MLV) of killed (K) viruses agains BRSV, BVDV, IBRV and PI3V were tested. They were coded from A to N. All the vaccines, with the exception of vaccine L, were presented as single or combined virus products (Table 1). All killed viruses had been inactivated with binary-ethylenimine (BEI); all the inactivated antigens, with the exception of vaccine I (adsorbed on alum), had been adsorbed on aluminium hydroxide. Vaccines G, H and I also contained saponin; a proprietary adjuvant had been added to vaccines D and E; aluminium hydroxide and paraffin were contained only in the MLV vaccine N. According to manufacturers' indications, only vaccine M was inoculated intranasally; all others were administrered intramuscularly. Manufacturers presented data on challenge experiments and/or field trials on target species for all the antigenic components of the vaccines.

Animals

All the trials were carried out at the Agriculture Research Station of ISS situated within the Azienda Tomassini (Palombara Sabina-Roma). The farm is a closed semi-intensive system dairy farm maintained under strict health control by ISS veterinarians where mandatory testing for brucellosis, tuberculosis and leucosis are carried out on a regular basis. No prophylactic vaccination is carried out and no outbreak of the major infectious diseases has been recorded in the last ten years. Ten Friesian calves, five to seven months old and weighing about 300 kg, were randomly selected and used in each vaccine trial. Among the ten animals, seven calves were vaccinated; the others were kept, in close contact, as unvaccinated controls.

Serology

The vaccinated and control calves were subjected to pre- and post-vaccination serology. Calves were vaccinated twice, three or four weeks apart, according to the manufacturers' directions. Blood samples for sero-analysis were taken from the jugular vein of each of the calves at the time of each vaccination (T-0, T-1) and two weeks after the second vaccination (T-2). The sera, separated from the clots and stored at –20°C, were inactivated for 30 minutes at 56°C just before testing. Except for the sera of animals vaccinated with L and N vaccines (which were only tested respectively against BVDV and PI3V), sera were tested for the presenc of antibodies against each of the virus antigens present in the specific vaccines.

The potency of the vaccines to induce specific seroconversion was investigated using the hemagglutination inhibition (HAI) test for PI3V, the serum neutralization (SN) test for BVDV, the enzyme-linked immunosorbent assay (ELISA) for IBRSV and BRSV.

Table 1: Vaccine components.

Vaccine	Components
A	MLV IBRV
	MLV BVDV
	MLV PI3V
	MLV BRSV
B	MLV PI3V
	MLV BRSV
C	MLV BRSV
D	K IBRV
	K BVDV
	MLV PI3V
	MLV BRSV
E	K IBRV
	K BVDV
F	MLV IBRV
	MLV PI3V
	MLV BRSV
	K BVDV-Cyt
	K BVDV-NCyt
G	K BVDV
	K IBRV
	K PI3V
	K BRSV
H	K BRSV
I	K BRSV
J	MLV BVDV
K	MLV BVDV
L	K IBRV
	K BVDV
	P. mult.A+B
M	MLV IBRV
	MLV PI3V
N	MLV IBRV
	MLV PI3V

HAI and SN tests were carried out according to well-established procedures. Briefly, for the HAI test, after kaolin treatment, the sera were challenged in microtitre plates agains 4 HAU/25 µl of the SF4 strain of PI3V adapted to grow on Aubek cell cultures. The HAI titres of the sera were expressed as the reciprocal of the highest final serum dilution inhibiting guinea pig red blood cell agglutination. The limit for negative serum was an HAI titre of <2. The quantitative evaluation of antibodies against BVDV was carried out on microtitre plates against 25 µl of a virus suspension containing 100 TCID50 of the cytopathic BVD-CNR virus strain (a field BVDV isolate kept in the virus collection at the Istituto Zooprofilattico Sperimentale of Brescia), adapted to grow on Aubek cell cultures. SN titres were expressed as the reciprocal of the highest final serum dilution inhibiting 100% CPE in all inoculated cultures. The limit for a negative serum was an SN titre of <2.

A «trapping ELISA» was developed for the detection of antibodies to IBRV [13] and BRSV [14]. A positive correlation had been previously established between ELISA and SN for IBRV and between ELISA and an indirect immunofluorescence test (IIF) for BRSV, by collaborative assays coordinated by ISS.

The specificity and sensitivity of the test was guaranteed by the selection of monoclonal antibodies for both the two virus antigens and bovine IgG [15, 16]. For BRSV, a pool of three monoclonal antibodies specific for the nucleoprotein (N) and fusion protein (F) were selected for the test. For IBRV, the 4E5 neutralizing monoclonal antibody, a well-conserved epitope common to more than forty reference and field strains of IBRV, was used.

The procedure, similar for both viruses, is based on the capability of the specific monoclonal antibodies adsorbed on the solid phase, to capture specific virus antigens present in the supernatant of infected cell cultures. For BRSV, the virus antigen was BEI-inactivated crude supernatant of Aubek cell cultures infected with the RB-94 strain of BRSV; for IBRV it was prepared from the crude supernatant of Aubek cell cultures infected with a BHV1.1 strain (a field virus strain kept in the collection at the Istituto Zooprofilattico Sperimentale of Brescia), by inactivation with BEI.

Starting from 1:50, two-fold dilutions of the sera to be tested were then added, followed, after the usual incubation and washing procedures, by the constant addition of a pool of four antibovine IgG (IgG1/IgG2) monoclonal antibodies-HRP conjugate. Ortho-phenylenediamine (OPD)/H2O2 was used as chromogen. Reading was performed at 492nm using a Titertek ELISA reader (Flow Laboratories, USA). In the trapping ELISA the results are expressed as the difference of the optical density (Δ O.D.) between the average value of O.D. obtained from two determinations of the serum to be tested in the presence of the virus antigen and the O.D. obtained in the control well without antigen. Quantitatively, the value of the O.D. is directly proportional to the specific antibody concentration present in the serum. In the case of a specific reaction the cut-off +/- of Δ O.D. is 0.150.

Statistical methods

The differences among the mean antibody titres in vaccinated animals were evaluated for statistical significance using the analysis of variance 14 days after the second administration (T-2) of each test group of vaccines. The antibody titres were transformed as $\log_2$; the F test was applied to verify the statistical significance of the mean titres among the various groups (the differences were tested for significance at = 0.5%.) The geometric mean titres (GMT $\log_{10}$) corresponding to the reciprocal dilution were used for the graphic representation of the results.

Results

At the time of the first vaccination (T-0), all calves were seronegative to BRSV, IBRV and BVDV (GMT = 0). The GMT of HAI antibodies to PI3V was 0.42. None of the unvaccinated control calves experienced a rise in antibody titre, thus excluding intercurrent natural field exposure or interference due to the shedding of virus in the case of animals vaccinated with MLV vaccines. All calves vaccinated with either live or inactivated vaccines against BRSV, IBRV and PI3V already showed a significant increase of antibody titres at the time of the second vaccination (T-1). At T-1, the GMT to BRSV (Fig. 1) and IBRV (Fig. 2) were higher in animals vaccinated with MLV vaccines compared with those vaccinated with inactivated vaccines. At T-1, antibody titres in calves vaccinated with the only inactivated PI3V vaccine (G) were similar to those induced by the most immunogenic MLV vaccines (Fig. 3). Two groups of animals vaccinated with BVDV inactivated vaccines (D and E), did not seroconvert at T-1, whereas the other inactivated and MLV vaccines induced an antibody response in all the animals. At T-1 antibody titres to BVDV induced by MLV vaccines were higher than those induced by inactivated vaccines (Fig. 4).

A clear anamnestic response to the second vaccination (T-2) was recorded in almost all vaccinated animals. The very few exceptions concerned the monovalent MLV vaccine C and the MLV vaccines A, M and N; in the first case a drop in the mean ELISA value to BRSV was noted; in the second case the means of HAI antibody titres to PI3V declined in calves vaccinated with vaccine A yet remained constant in those vaccinated with vaccines M and N.

At T-2, antibody titres induced by all the vaccines containing inactivated BRSV were higher than those induced by MLV vaccines (Fig. 5).

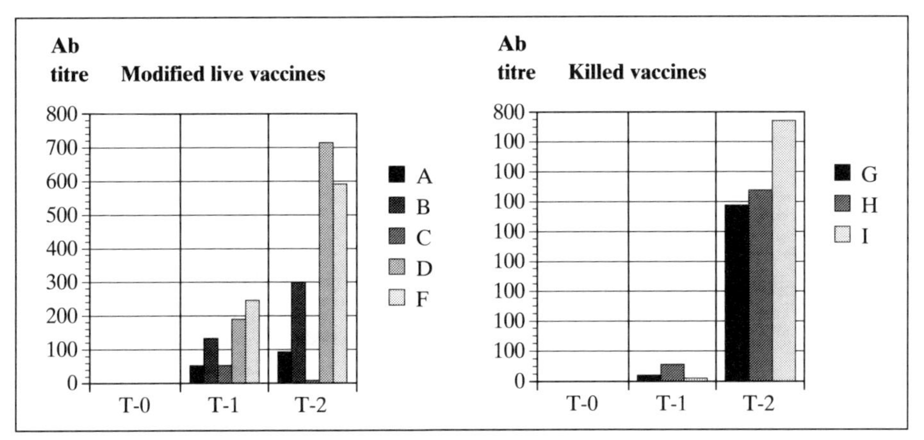

Fig. 1: Comparative ELISA antibody responses to BRSV in calves before and following administration of modified live and killed BRSV vaccines.

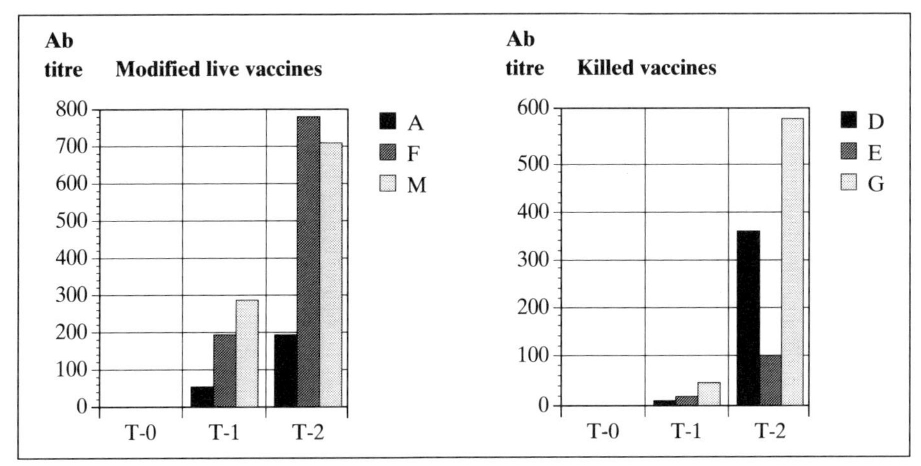

Fig. 2: Comparative ELISA antibody responses to IBRV in calves before and following administration of modified live and killed IBRV vaccines.

At T-2 the antibody titres to IBRV in calves vaccinated with MLV vaccines were still generally higher than those induced by inactivated vaccines (Fig. 6). At T-2 the inactivated vaccine G induced the highest antibody response to PI3V (Fig. 3) and BVDV. With the exception of vaccine G, MLV vaccines induced a better anamnestic response to BVDV compared with inactivated vaccines (Fig. 7).

Based on the data in Figure 8, it seems clear that antibody responses were more homogeneous first within the groups of calves vaccinated with BVDV vaccines (1.52), and then within the groups of calves vaccinated with the PI3V (1.87), BRSV (2.38) and IBRV (4.79) vaccines. The differences in antibody titres among the test group vaccines were statistically significant, decreasing from BVDV (52.8) to

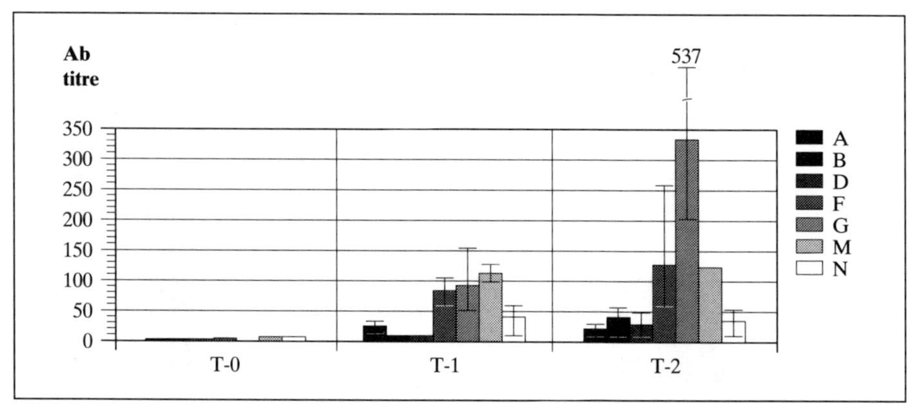

Fig. 3: HAI antibody response to PI3V in calves before and following administration of modified live and killed PI3V vaccines.

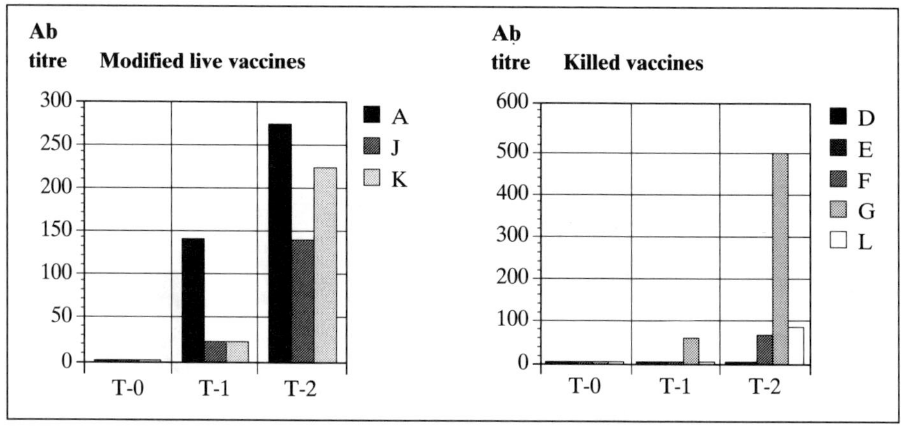

Fig. 4: Comparative serum neutralization antibody responses to BVDV in calves before and following administration of modified live and killed BVDV vaccines.

BRSV (51.3), IBRV (37.1) and PI3V (29.0) vaccines. The values of F test were 34.6, 21.5, 15.5, 7.7, respectively for BVDV, BRSV, PI3V and IBRV.

As an introductory approach to a more comprehensive study, a further statistical analysis was made taking into account only the groups of animals vaccinated with vaccines containing BRSV and, moreover, excluding those groups that had shown a high variability of mean antibody titres to obtain a more homogeneous set of results to test for statistical significance. Consequently, due to the high variability of mean antibody titres recorded in the homologous groups of vaccinated animals, the test group vaccines excluded from three further analysis of variance were first only vaccine C, then vaccines C and F and lastly vaccines C, F, G.

In all the three cases, the differences among the remaining test group vaccines were still significant. Furthermore, the analysis of variance was conducted on test group vaccines D, F, G, H, I, selected because of their similarity in the mean anti-

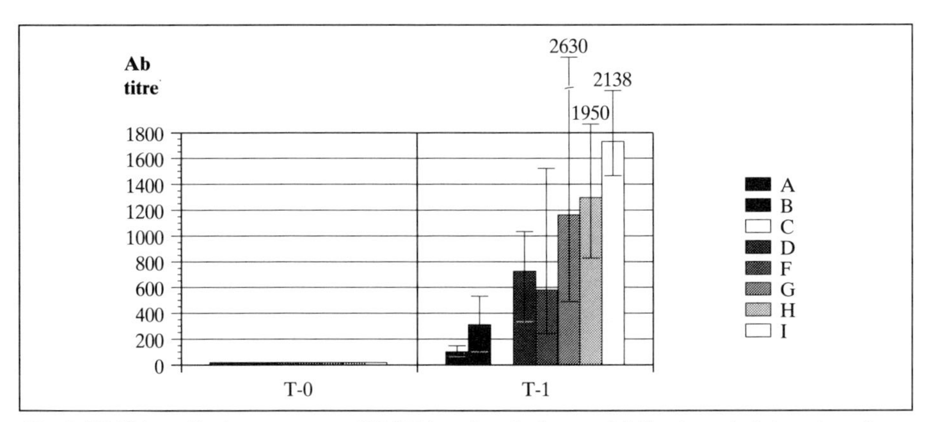

Fig. 5: ELISA antibody response to BRSV in calves before and following administration of test group vaccines.

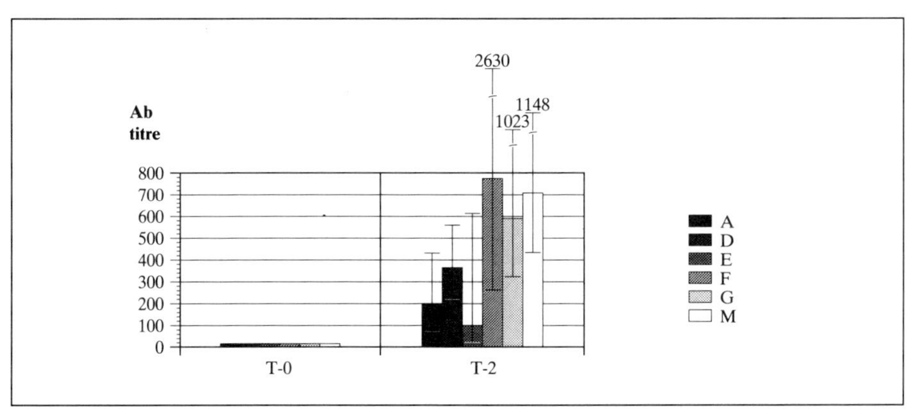

Fig. 6: ELISA antibody response to IBRV in calves before and following administration of test group vaccines.

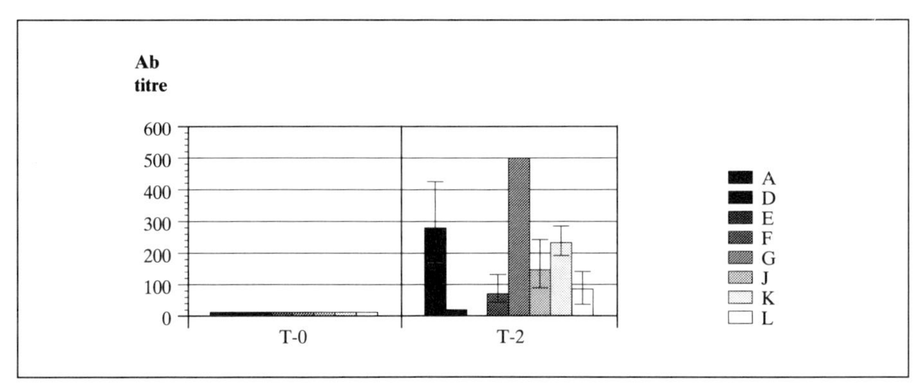

Fig. 7: Serum neutralization antibody response to BVDV in calves before and following administration of test group vaccines.

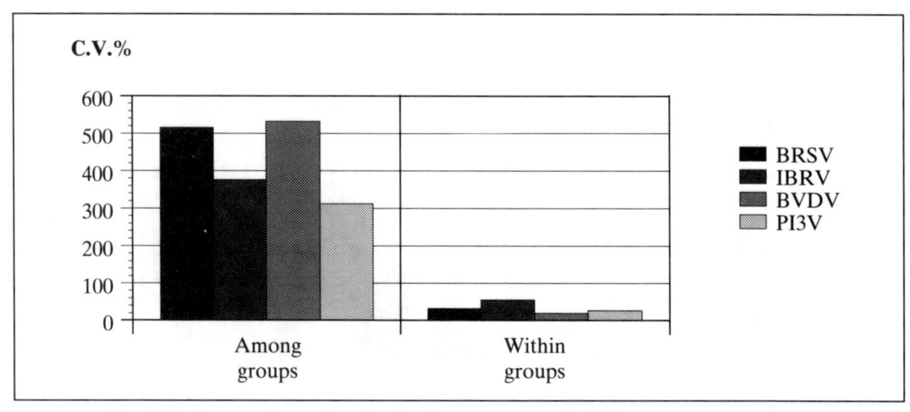

Fig. 8: Variance values of antibody titres among and within test group vaccines.

body titres and the highly significant differences with respect to those of test group vaccines A, B, C. Differences were still significant at =5% (F=2.5).

Discussion

The role of viruses in causing respiratory diseases of cattle and the use of specific vaccines in preventing them have been widely debated. The ultimate proof of a vaccine's efficacy is normally its ability to protect under significant challenge. Nevertheless, as the basis of protection against respiratory diseases of cattle is not yet fully understood and serious consequences of infections are not present, it might be questionable to demonstrate such an extreme effect of vaccination. It could be more useful to rely on the value of other immunomechanisms in the assessment of vaccine activity.

Immunity in cattle vaccinated against respiratory diseases is probably determined by a complex interaction of humoral and cellular parameters; the protective effect, depending on the immunogenic properties of each vaccine virus component, might generate single, additive or synergistic activities of antibody and effector T lymphocytes [17, 18]. In this respect the combination of different antigens in the formulation of vaccines must be regarded as a crucial point. Vaccination against respiratory diseases of cattle has been shown to be highly effective in: (a) reducing virus shedding and consequently livestock environment contamination; (b) preventing primary localization of viruses and, as a consequence, by inhibiting colonization by secondary bacteria and viruses, thus preventing more serious consequences not necessarily due to the primary viral infections. Even if the presence of serum antibodies is not able to protect calves from reinfection, nevertheless it can reduce severity or protect the animals from developing clinical signs of the diseases. Depending on the level of the health and economic impact that these limits may impose on livestock management, the value of performing challenge experiments in all circumstances appears to be questionable.

Besides the difficulties encountered in experimentally reproducing respiratory diseases of cattle or assessing clinical efficacy of vaccines from the results of field trials, the relationship between, on the one hand, the number of experimental animals and the associated costs of testing and, on the other, the limited information obtained (poor demonstration either of total protection or poor differentiation between vaccinated and control animals) has raised significant criticisms. Detection and titration of antibody in the sera of vaccinated cattle are useful means of determining the activation of the immune system even if often they do not clearly depict the extent and efficacy of the immune response. Nevertheless they remain objective, reliable and suitable procedures for assesssing the immunogenic properties of vaccines. In this respect, the development of diagnostic methods capable of detecting specific immune responses affecting the protective activity of vaccines and the adoption of reference sera and vaccines – for which serological response in animals has been evaluated and compared to the level of efficacy achieved in challeng experiments – are likely to be of fundamental importance.

This study presents the preliminary results in terms of statistically significant differences in immunogenic activity which seem to suggest the relative potency of vaccines against viral respiratory diseases of cattle. They can be regarded as a starting point for discussion among manufacturers, users, competent authorities and control laboratories to establish common criteria and to recognize the significacance, means and limits of producing, testing and using vaccines against respiratory diseases of cattle.

References

1 Ribble CS: Assessing vaccine efficacy. Canadian Veterinary Journal 1990;31:679-681.

2 Baker JC: Bovine respiratory syncytial virus; pathogenesis: clinical signs, diagnosis, treatment and prevention. Compendium Food Animal 1986;8:F31.

3 Verhoeff J, van Nieuwstadt APKMI: Prevention of bovine respiratory syncytial virus infection and clinical disease by vaccination. The Veterinary Record 1984;10:488-492.

4 Kahrs RF: Viral Diseases of cattle. Parainfluenza-3. Ames, Iowa, The Iowa State University Press, 1986, vol 13, pp 171-181.

5 Grotelueschen DM, Mortimer RG: Bovine viral diarrhea virus infections in beef cattle: clinical aspects and control. Agri-Practice-Virology 1988;9:25-27.

6 Frerichs GN, Woods SB, Lucas MH, Sands JJ: Safety and efficacy of live and inactivated infectious bovine rhinotracheitis vaccines. The Veterinary Record 1982;7:116-122.

7 Stott EJ, Thomas LH, Taylor G, Collins AP, Jebbett J, Crouch S: A comparison of three vaccines against respiratory syncytial virus in calves. Journal of Hygiene 1984;93:251-261.

8 Westenbrink F, Kimman TG, Brinkhof MA: Analysis of the antibody response to bovine respiratory syncytial virus proteins in calves. Journal of General Virology 1989;70:591-601.

9 Martin SW, Bohac JG: The association between serological titres in infectious bovine rhinotracheitis virus, bovine virus diarrhea virus, parainfluenza-3 virus, respiratory syncytial virus and treatment for respiratory in Ontario feedlot calves. Canadian Journal Veterinary Research 1986;50:351-358.

10 Martin SW, Bateman KG, Shewen PE, Rosendal S, Bohac JG: The frequency, distribution and effects of antibodies to seven putative respiratory pathogens, on respiratory disease and weight gain in feedlot calves in Ontario. Canadian Journal Veterinary Research 1989;53:355-362.

11 Van Donkersgoed J, van den Hurk Jan V., Mc Cartney D, Harland RJ: Comparative serological responses in calves to eight commercial vaccines against infectious bovine rhinotracheitis, parainfluenza-3, bovine respiratory syncytial, and bovine viral diarrhea viruses. Canadian Veterinary Journal 1991;32:727-733.

12 Ellis JA, Russell H, Cavender J, Haven TR: Bovine respiratory syncytial virus-specific immune responses in cattle following immunization with modified-live and inactivated vaccines. Analysis of the specificity and activity of serum antibodies. Veterinary Immunology and Immunopathology 1992;34:35-45.

13 Perrin B, Bitsch V, Cordioli P, Edwars S, Eloit M, Guerin B, Lenihan P, Perrin M, Ronsholt L, Van Oirschot JT, Vanopdenbosch E, Wellemans G, Wizigman G, Thibier M: An European comparative study of serological methods for the diagnosis of infectious bovine rhinotracheitis. Revue Scientifique et Technique, Office International des Epizooties 1993;12 (3):969-984.

14 De Simone F, Brocchi E, Archetti YL, Gamba D, Foni E: Use of monoclonal antibodies for the serological and virologic diagnosis of the respiratory syncytial virus in bovine. Atti Società Italiana Scienze Veterinarie 1986;40:903-906.

15 Capucci L, Gamba D, Archetti YL, Civardi A: Monoclonal antibodies to bovine IgG1, IgG2 and IgM: characterization and possible uses in the serological diagnosis of viral diseases. Atti Società Italiana Scienze Veterinarie 1986;40:898-901.

16 Archetti YL, Gamba D: Anticorpi monoclonali anti-isotopo: produzione e caratterizzazione di anti-IgG1 e anti-IgA bovine, anti-IgG e anti-IgM suine. Atti Società Italiana Scienze Veterinarie 1988;42:637-639.

17 Howard CJ: Immonological response to bovine virale diarrhea virus infections. Revue Scientifique et Technique, Office International des Epizooties 1990;1:95-103.

18 Ellis JA, Belden EL, Haven TK, Cavender J: Bovine respiratory syncytial virus-specific immune responses in cattle following immunization with modified live and inactivated vaccines. Analysis of proliferation and secretion of lymphokines by leuckocytes in vitro. Veterinary Immunology and Immunopathology 1992;34:21-34.

Dr. M. Tollis, Istituto Superiore di Sanita, V. le Regina-Elena 299, I-00161 Roma, Italy

Brown F, Cussler K, Hendriksen C (eds): Replacement, Reduction and Refinement of Animal Experiments in the Development and Control of Biological Products.
Dev Biol Stand. Basel, Karger, 1996, vol 86, pp 157-163

Replacement of Challenge Procedures in the Evaluation of Poultry Vaccines

C. Jungbäck, H. Finkler

Paul-Ehrlich-Institute, Federal Agency for Sera and Vaccines, Langen, Germany

Abstract: Vaccination of poultry flocks, especially parent flocks, is often performed with the intention of protecting the progeny via maternal antibodies during the first weeks of life. The efficacy of this vaccination schedule, more precisely described as induction of indirect protection, is normally proven by challenging the chickens. To avoid the challenge, some trials were performed, with the intention of establishing a correlation between antibody titres of vaccinated hens, embryonated eggs, and hatched chickens. The parent flocks were divided into several groups which were vaccinated either with Infectious Bursitis (IBD) vaccine or with Newcastle Disease (ND) vaccine. An acceptable correlation could be established. A well described standard or reference serum and a standardized and reproducible test procedure could be used in different laboratories. This approach allows serological test results to be compared. The possible parameters for the reference sera, including antibody titre, protein content and additional parameters, are discussed.

Introduction

Efficacy testing is an important part of the evaluation of vaccine quality. These trials require a large number of animals, which have to undergo stringent test conditions, especially in the challenge trials. These kinds of trials will always be necessary but should be reduced whenever possible. Our study presents the possibility of replacing some of the challenges. Two series of experiments on Newcastle and infectious bursitis vaccines act as examples.

Legal Prescriptions

The relevant legal prescriptions concerning efficacy testing are the following (Table 1).

European Union

- Directive 92/18, Annex, Title II, Part 8 and
- Guidelines for the production and control of avaian live and inactivated viral and bacterial vaccines.

Table 1: Legal prescriptions.

EU:	Directive 92/18, Annex, Title II, Part 8
	Guideline for the production and control of avian live and inactivated viral and bacterial vaccines
Eur. Phar.:	Vaccinum pseudopestis aviariae vivum cryodesiccatum
	Vaccinum pseudopestis aviariae inactivatum
	Vaccinum bursitidis infectivae aviariae vivum cryodesiccatum
	Vaccinum bursitidis infectivae aviariae inactivatum
USA 9 CFR:	113. 329 Newcastle disease vaccine, live virus
	113. 205 Newcastle disease vaccine, killed virus
	113. 331 Bursal disease vaccines, live virus
	113. 212 Bursal disease vaccines, killed virus

These provisions require testing of each category of target species recommended for vaccination, by each recommended route of administration and using the proposed schedule of administration. Any claims regarding onset and duration of protection shall be supported by data from trials. The animals must be vaccinated with the minimum titre. Challenges must be undertaken under well controlled conditions [1]. For vaccines intended for the passive protection of progeny of vaccinated dams via maternal immunity, data shall be presented showing the duration of protection of progeny hatched from eggs taken at the end of the period over which vaccine efficacy is claimed (usually the end of lay) [2].

European Pharmacopoeia Monographs

- Vaccinum pseudopestis aviariae vivum cryodesiccatum
- Vaccinum pseudopestis aviariae inactivatum
- Vaccinum bursitidis infectivae aviariae vivum cryodesiccatum.

  These monographs require chickens, vaccinated at the minimum age of life once with the minimum titre or a much lower titre. The challenge takes place 14 to 21 days after vaccination.
- Vaccinum bursitidis infectivae aviariae inactivatum.

  This monograph requires chickens, vaccinated at four weeks of age with one dose. The animals are serologically controlled four to six weeks after vaccination. The serology must be performed against a standard, defined by the monograph [3].

USA: 9 CFR

- 113. 329 Newcastle disease vaccine, live virus

  This monograph requires chickens, vaccinated at an undefined age, with a vaccination schedule recommended on the label. The challenge takes place 20 to 28 days after vaccination.

- 113. 205 Newcastle disease vaccine, killed virus

  This monograph requires chickens, vaccinated at two to six weeks of age as recommended on the label. The challenge takes place 14 to 21 days after vaccination.

- 113. 331 Bursal disease vaccine, live virus

  This monograph requires chickens, vaccinated at one day of age. The challenge takes place 17 days after vaccination.

- 113. 212 Bursal disease vaccine, killed virus

  This monograph requires chickens, vaccinated at 14 to 28 days of age. The challenge takes place 21 to 28 days after vaccination [4].

The Challenge Model

This efficacy testing requires different challenges: first the challenges after a single vaccination as required in the European Pharmacopoeia and in the 9CFR. The second series of challenges for the evaluation of the duration of protection requires the following procedure: chickens are vaccinated according to the vaccination schedule, given by the manufacturer. Challenges are performed on laying birds and on their progeny at several times during the laying period.

The Way to the Replaced-Challenge Model

Our procedure attempts to avoid some challenges for the evaluation of protection, to replace challenges by serological methods and to reduce the number of animals for the efficacy evaluation. Chickens are vaccinated according to the vaccination schedule. Thereafter the vaccinated animals, their progeny and eggs are controlled by serological investigations.

The ND-antibody titres were quantified by haemagglutination inhibition as described in the ND-directive [5], which differs only slightly from the HI-test, described in the EP-monograph for inactivated ND-vaccines. The IBD-antibody titres were tested by a defined SN-test.

Results

Our experiments with ND vaccination showed the following results (Fig. 1). The mean titres of all groups of hens, chickens and eggs appear between two $\log_2$ dilutions. In general the HI-titres of the yolk antibodies are one $\log_2$ degree higher than the serum titres from hens and chickens. The individual titres show a slightly wider range from $\log_2$ 6 to $\log_2$ 10, whereby the outer limits are only reached by single samples.

For the Gumboro vaccination the results are as follows (Fig. 2): the mean titres of all groupes of hens, chickens, and eggs appear within one log 10 dilution. Here between $\log_{10}$ 3,47 and $\log_{10}$ 3,2. The individual titres show a slightly wider range from $\log_{10}$ 2,6 to $\log_{10}$ 3,6, whereby the outer limits are only reached by single samples.

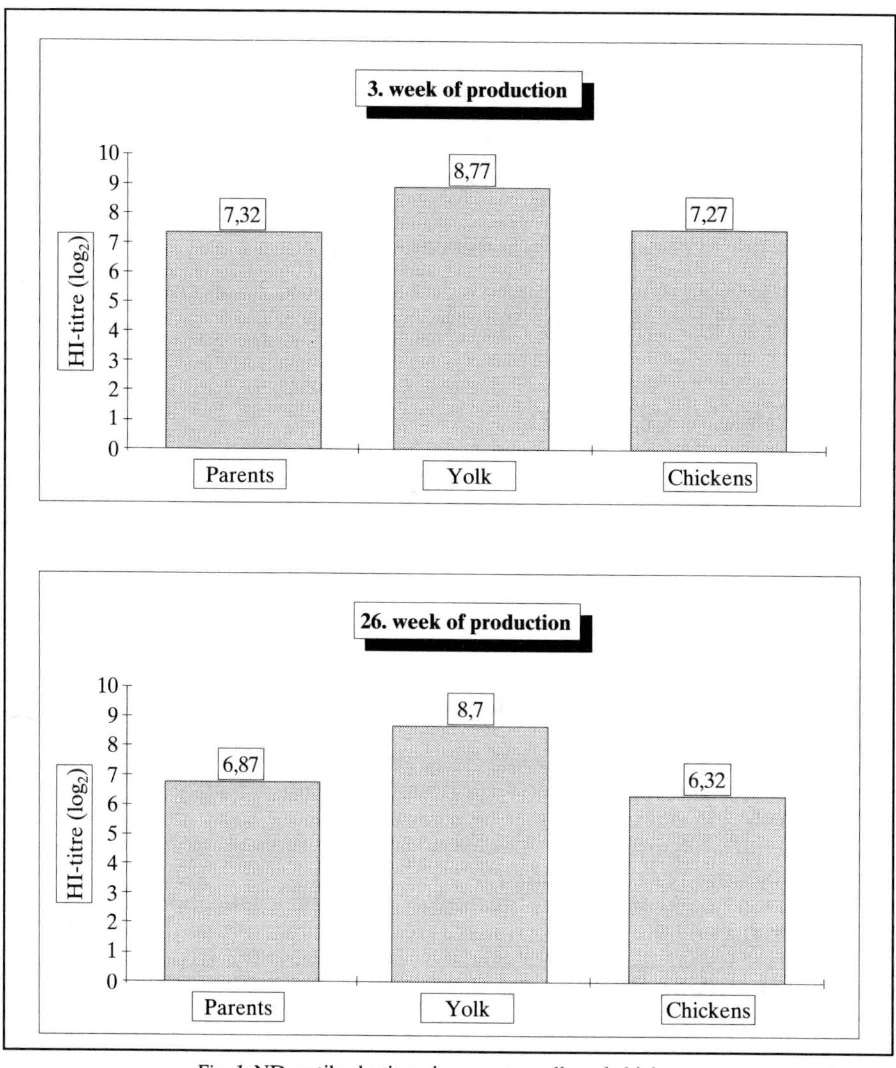

Fig. 1: ND-antibody-titres in parents, yolk and chickens.

Conclusions and Recommendations

Challenges of progeny from vaccinated hens to demonstrate the protection through maternal antibodies can be avoided if egg antibodies are quantified. In addition, the amount of blood samples taken from the hen can be reduced. The antibody titres in hens, eggs and chickens correlate if they are taken at corresponding times; presumably the serological tests are comparable. Moreover, all challenges in the trials concerning the duration of protection can be avoided if all serological tests are performed in an identical manner.

Therefore, a validated test procedure, as well as defined reference/standard substances, are necessary. A comparison of different SN-test procedures for the evaluation of TBD antibodies illustrates the need for explicit test descriptions. As demonstrated in Figure 3, changes in concentration of cells and/or virus lead to significant differences in SN-titres.

The recommendations for the test procedures and for the standard/reference substances should be as follows: the test method must be described in detail, including the description of the cells used, their source, passage, medium etc., the virus strain, passage number, host system, titre, the erythrocytes with all information on

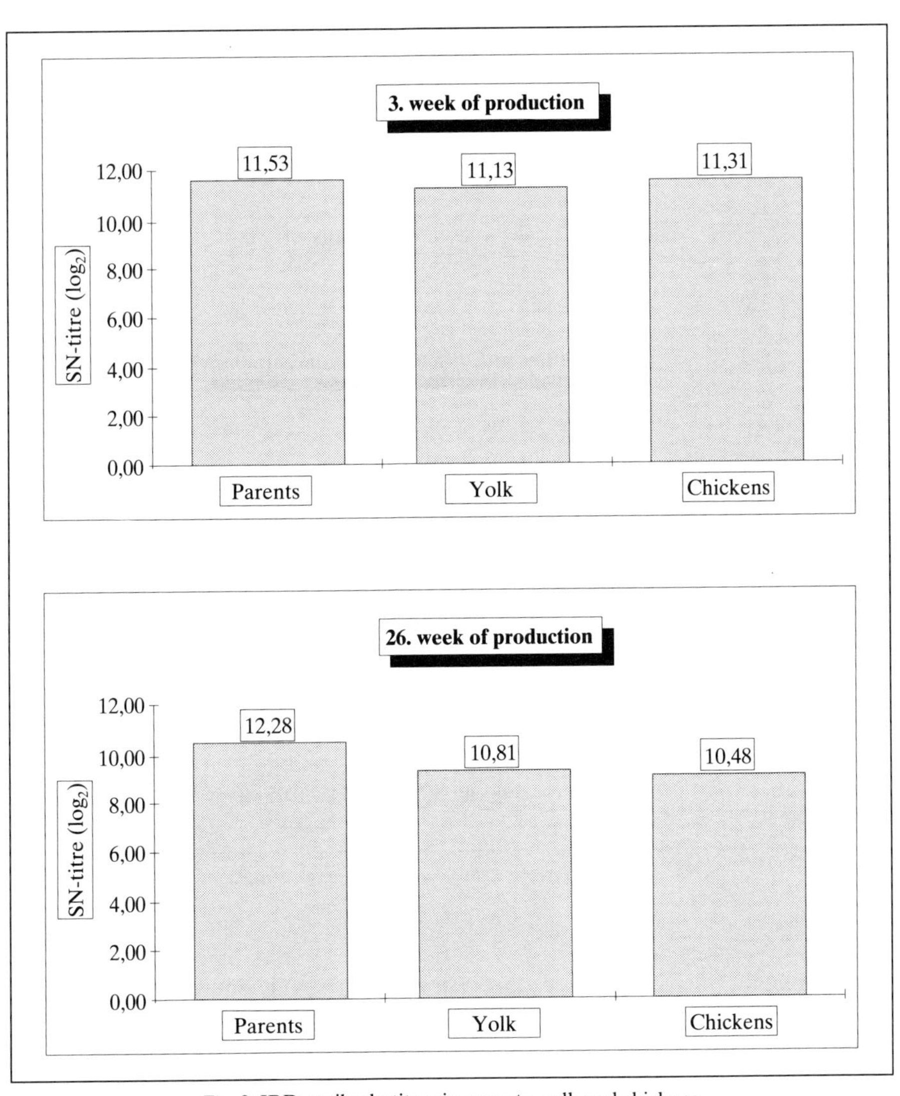

Fig. 2: IBD-antibody-titres in parents, yolk and chickens.

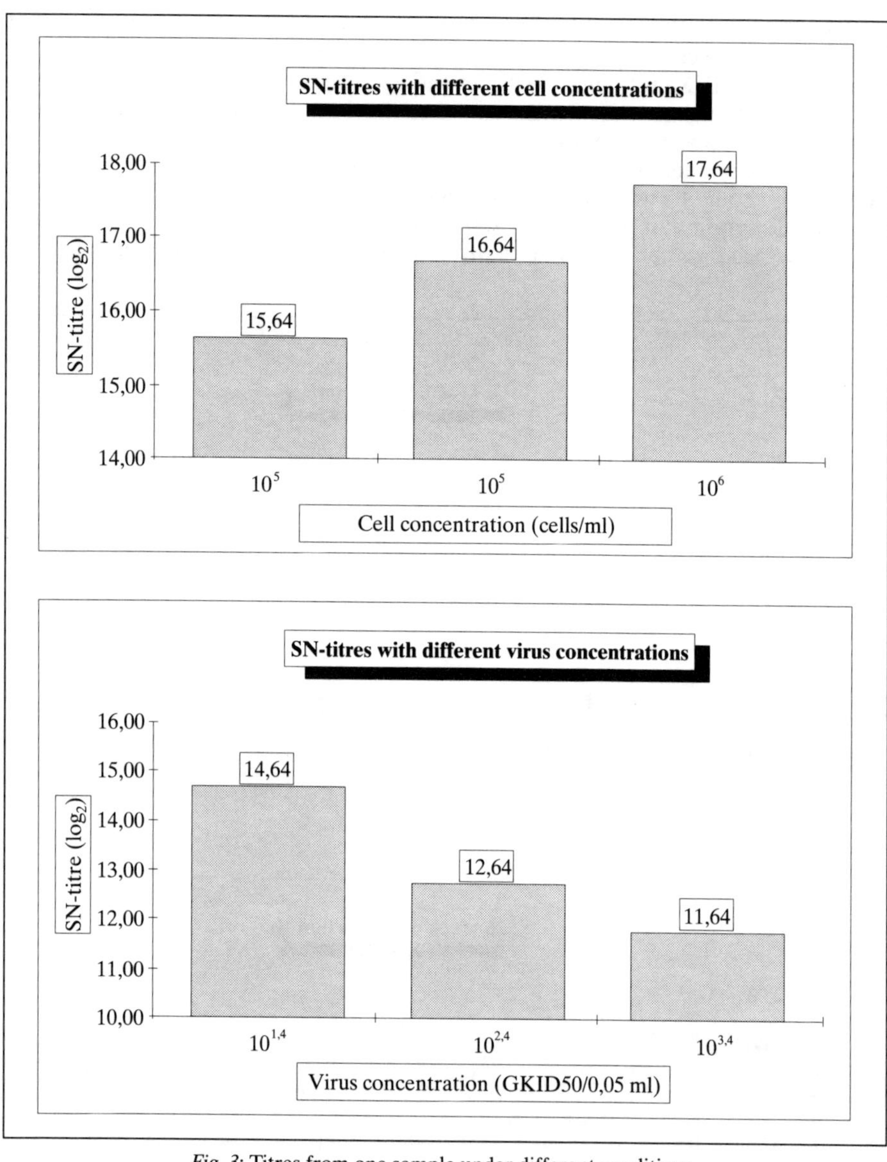

Fig. 3: Titres from one sample under different conditions.

donor animals, purification procedures, storage times and conditions, and concentrations. The amount of test substances and times of incubation must be fixed carefully, and criteria for the interpretation of the results, including a statistically evaluated correlation between antibody titres of hens, chickens and eggs, must complete the test description.

Table 2: Recommendations on standards.

Validated test method:	detailed test description cells virus erythrocytes reference sera SPF-quality of donor animals interpretation of results
Reference serum:	donor animals vaccination schedule sampling serum titre serum parameters

The reference materials, especially the sera must be characterized. The definition of the titre does not appear to be sufficient. It seems to be necessary to give information of the donor animals, the vaccination schedule, the time schedule for the serum sampling, the antibody titre, and some clinical serum parameters, to ensure that no pathological changes of the serum have occurred (Table 2).

Acknowledgement

The authors thank Eva Schirk and Ute Stückrath for their excellent assistance.

References

1 Commission Directive 92/18/EEC, 20.3.1990 modifying the Annex to Council Directive 81/852/EEC.

2 Guidelines for the Testing of Veterinary Medicinal Products, Vol. VII of the rules governing veterinary medicinal products, July 1994.

3 European Pharmacopoeia, 2nd Edition, 1993.

4 9 Code of Federal Regulations, 1994.

5 Council Directive 92/40/EEC, 22.6.1992.

Dr. C. Jungbäck, Paul-Ehrlich-Institute, Federal Agency for Sera and Vaccines,
Paul-Ehrlich-Str. 51-59, D-63225 Langen, Germany

Brown F, Cussler K, Hendriksen C (eds): Replacement, Reduction and Refinement of Animal Experiments in the Development and Control of Biological Products.
Dev Biol Stand. Basel, Karger, 1996, vol 86, p 165

SESSION VI

New Developments in Vaccine Production and Quality Control

Chairmen: F. Brown (New York, USA)
C. Hendriksen (Bilthoven, The Netherlands)

Brown F, Cussler K, Hendriksen C (eds): Replacement, Reduction and Refinement of Animal Experiments in the Development and Control of Biological Products.
Dev Biol Stand. Basel, Karger, 1996, vol 86, pp 167-173

An in Vitro Assay for Acute Pathogenicity of Immunodeficiency Viruses

M.T. Dittmar[1], *S. Wagener*[1], *P. Fultz*[2], *R. Kurth*[1], *K. Cichutek*[1]

[1] Dept. of Medical Biotechnology, Paul-Ehrlich-Institute, Langen, Germany
[2] University of Birmingham, Alabama, USA

Key words: Acute disease, apathogenicity, replication competence in non-stimulated peripheral blood mononuclear cells.

Abstract: As a model for AIDS, experimental infections of old-world monkeys with various simian immunodeficiency viruses (SIV) are frequently carried out to study mechanisms of pathogenicity. For example, $SIV_{smmPBj14}$ was isolated from a pig-tailed macaque (*Macaca nemestrina*) suffering from acute viral disease. The molecular virus clone $SIV_{smmPBj1.9}$, which displays close genetic homology to other related SIVs, was shown to induce an acute viral disease in vivo after infection of pig-tailed and rhesus macaques. The acute pathogenicity of $SIV_{smmPBj1.9}$ was correlated with its unique ability to replicate in non-stimulated peripheral blood mononuclear cells from pig-tailed macaques. We have exploited this in vitro assay to resolve putative pathogenic genetic determinants of another SIV, namely SIV_{agm3}, isolated from African green monkeys (*Cercopithecus aethiops*). Hybrid viruses encompassing subgenomic regions of $SIV_{smmPBj1.9}$ in place of comparable regions of molecular virus clone SIV_{agm3mc} were constructed and tested for their ability to replicate in non-stimulated PBMC from pig-tailed macaques and African green monkeys. Only those hybrid viruses comprising the U3 region of the viral LTR of $SIV_{smmPBj1.9}$ replicated in non-stimulated peripheral blood mononuclear cells. This in vitro assay will be used to determine the potential of SIV and of hybrid viruses between different SIVs to induce acute viral disease in vivo. It will help to avoid excessive experimental infections of monkeys with respective hybrid viruses for determining genetic determinants of acute pathogenicity of immunodeficiency viruses.

Introduction

As a model for AIDS, experimental infections of non-human primates with human (HIV) and simian immunodeficiency viruses (SIV) are frequently carried out. The animal experiments are aimed at understanding mechanisms of virus pathogenicity or immune response of the host. Pathogenic lentiviruses induce chronic infections of experimentally inoculated animals accompanied by a continuous loss of CD4-positive T-cells detectable in the peripheral blood and by other chronic impairments of the immune system. For several days or weeks animals may suffer from weight loss, fever, lymphadenopathy and subsequent infections with other bacterial or viral pathogens present in the animal population. We have used an in vitro system developed by Fultz et al [1-3] which allowed us to assess in vitro a correlate of the acute pathogenicity of a specific isolate of SIV_{smm} from sooty mangabey monkeys (*Cercocebus atys*) termed $SIV_{smmPBj14}$, and of hybrid viruses between $SIV_{smmPBj14}$ and SIV_{agm3} from African green monkeys (AGM; *Cercopithecus aethiops*). Consequent use of this test system will allow us to reduce the number of experimental infections necessary to study genetic determinants of specific biological properties of simian immunodeficiency viruses.

$SIV_{smmPBj14}$, which was isolated from an experimentally infected, moribund pig-tailed macaque suffering from severe immunodeficiency [4, 5], induces an acute disease in macaques. Full-length provirus 1.9 ($pSIV_{smmPBj1.9}$) [6] was subsequently cloned and also shown to induce an acute disease in pig-tailed macaques [6, 7]. The acute disease is characterized by rapid loss of the CD4- and CD8-positive T-cells in the peripheral blood of the infected animals, weight loss, diarrhoea, fever and lymphadenopathy. Most infected monkeys die within days to weeks and show visible lesions of the intestines, possibly due to extensive viral replication in intestinal epithelial cells. In contrast, SIV_{agm} is apathogenic in infected AGM. SIV_{agm} belongs to an evolutionary old and distinct group of lentiviruses [8-11]. Provirus $pSIV_{agm3}$ was molecularly cloned from isolate SIV_{agm3} obtained from a long-term infected healthy AGM [12]. SIV_{agm3mc} used for infections of AGM and pig-tailed macaques (*Macaca nemestrina*) did not cause acquired immunodeficiency during an observation period of at least four years [12, 13].

To define the genetic determinants underlying the acute pathogenicity of $SIV_{smm-PBj14}$ the biological properties of hybrid viruses encompassing complementary genomic fragments of both parental viruses were analysed. By use of the in vitro test developed by Fultz et al [1-3] we were able to show that some but not all of the hybrid viruses generated displayed the hallmark of $SIV_{smmPBj14}$ acute pathogenicity, namely its capacity to replicate in non-stimulated monkey peripheral blood mononuclear cells (PBMC).

Methods

Construction and generation of recombinant viral genomes

The construction and complete sequences of proviral plasmid clones $pSIV_{agm3}$ (SIV_{agm3mc}; Genbank M30931) and clones 1 and 9 of $SIV_{smmPBj14}$ which were combined for the construction of full-length provirus $pSIV_{smmPBj1.9}$ ($SIV_{smmPBj1.9mc}$; Genbank M31325) were described previously [6, 11]. Nucleotide numbering of $pSIV_{agm3}$ starts with nucleotide 1 of the R region of the LTR, designated here to show its derivation from SIV_{agm} as A1, whereas nucleotide 1 of $pSIV_{smmPBj1.9}$ is as described in ref. 6, designated here as P1. Briefly, $pSIV_{agm3}$ was obtained by insertion of the Eco R I fragment of pSIVMB1 [13] into plasmid pEX2.4 which contains the Eco R I (A6990)-Xho I (A334) 3′ half fragment of pSIVMB1 inserted into Eco R I and Sal I of pUC8 (Pharmacia, Freiburg, Germany). The construction of the hybrid proviruses will be published elsewhere [14].

Transfections and reverse transcriptase assays

For transfection, 10 µg of each of two plasmids containing the respective 5′ or 3′ genomic half of the full-length hybrid provirus to be formed were digested by treatment with the appropriate restriction enzymes (see above) and ligated in a total volume of 100 μl using T4 DNA-ligase (NEB, Schwalbach, Germany). Each ligation mixture or 20 μg of plasmids containing full-length proviruses with 2 LTRs were then transfected into $1x10^6$ C8166 cells using the DOTAP transfection reagent according to the manufacturer's procedure (Boehringer, Mannheim, Germany). Cell supernatants were harvested from virus-producing cultures, viruses were pelleted and virion-associated reverse transcriptase (RT) activity was assayed as described previously [11, 14].

Infections of PBMC

Virus stocks were diluted in three-fold steps using RPMI 1640 medium and eight 50 μl aliquots of each sample were used for the titration of viruses infectious for C8166 cells. Immunoperoxidase assay (IPA) was used for the detection of SIV antigen expressing cells using sera from an SIV_{agm3} infected pig-tailed macaque and an HIV-2 infected human, respectively [15]. PBMC were isolated from SIV-seronegative AGM and pig-tailed macaques by Ficoll-Hypaque density gradient centrifugation. These were either infected immediately, i.e. non-stimulated, or incubated for two days in the presence of 5μg/ml PHA (Wellcome, Dartford, UK) and 100U/ml recombinant IL-2 (Eurocetus, Frankfurt/Main, Germany) before infection (m.o.i. = 0.01). After two days the cells were washed and cell supernatants were removed for titration after an incubation period of 15 days (see above).

Results and Discussion

Acutely pathogenic $SIV_{smmPBj1.9}$ but not apathogenic SIV_{agm3mc} can replicate in non-stimulated PBMC

As an in vitro correlate of its capacity to induce acute disease in infected pig-tailed macaques, $SIV_{smmPBj14}$ was shown to replicate in non-stimulated PBMC from pig-tailed macaques [1-3]. To analyse in vitro the correlate of acute pathogenicity of $SIV_{smm-PBj1.9}$ versus that of SIV_{agm3mc} (for proviral structures see Fig. 1), infections of stimulated and non-stimulated PBMC isolated from pig-tailed macaques and African green monkeys were carried out. First, proviruses were transfected into C8166 T-cells permissive for the replication of both molecular virus clones and virus-containing supernatants produced within the first five days after transfection were used for a second round of infection of C8166 T lymphocytes. RT activity released into the supernatants was monitored for 25 days (data not shown). Virus containing supernatants harvested from the infected cells were pooled and subsequently used for infection of PBMC from pig-tailed macaques and AGM stimulated for two days in the presence of interleukin-2 (IL-2). As depicted in Figures 2A and B, both parental molecular virus clones were able to replicate in these PBMC. To test the capacity of both parental viruses to replicate in non-stimulated PBMC, infections were carried out using freshly isolated PBMC which were not stimulated in the presence of IL-2 before infection. Figure 2C shows that $SIV_{smmPBj1.9}$ replicated in PBMC isolated from pig-tailed macaques whereas SIV_{agm3mc} was unable to grow in these cells. Surprisingly, non-stimulated PBMC from AGM also supported replication of $SIV_{smm-PBj1.9}$ (Fig. 2D), but not of SIV_{agm3mc}. Thus, a clear distinction could be made between the acutely virulent and the apathogenic parental virus by assessing their capacity to replicate in non-stimulated PBMC from pig-tailed macaques and AGM.

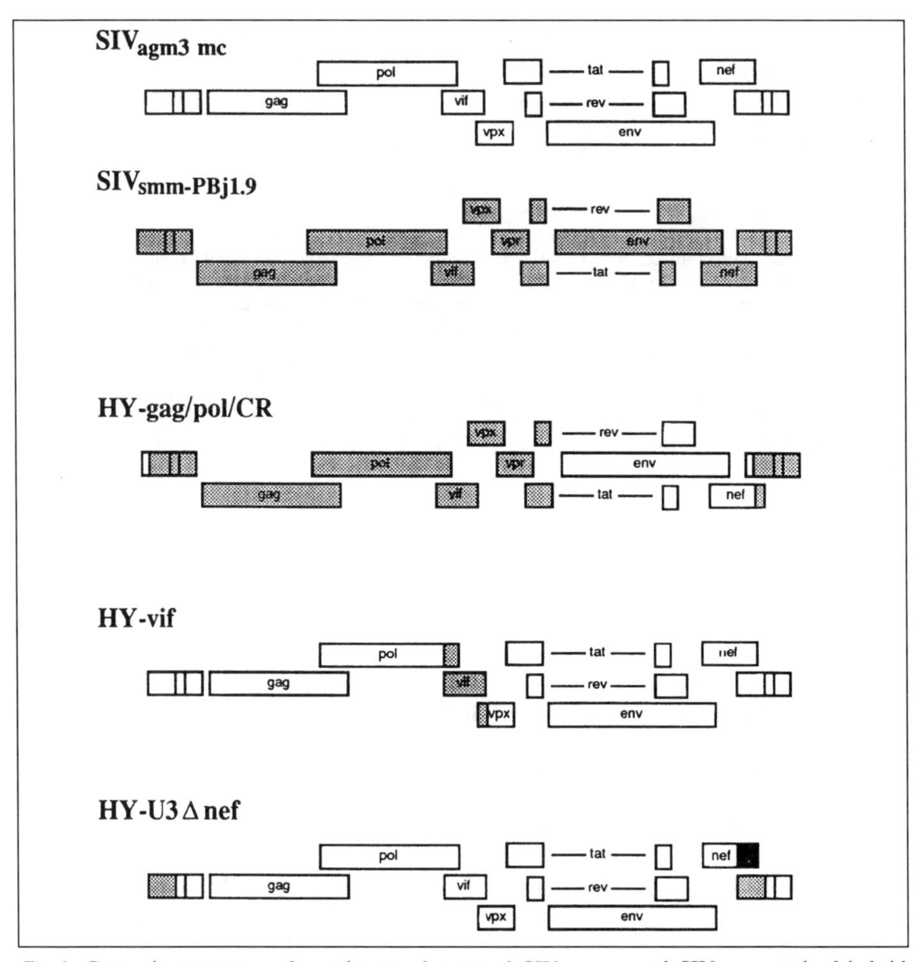

Fig. 1: Genomic structures of proviruses of parental $SIV_{smm\text{-}PBj1.9}$ and SIV_{agm3mc}, and of hybrid viruses HY-vif, HY-gag/polCR, and HY-U3Δnef.

Chimaeric SIV_{agm3mc}/$SIV_{smmPBj1.9}$ can be distinguished by its capacity to replicate in non-stimulated PBMC

As both cloned proviruses, $pSIV_{agm3}$ and $pSIV_{smmPBj1.9}$, comprise cistronic *gag*, *pol*, *env* as well as *vif*, *vpx* and *nef* genes, whereas only $pSIV_{smmPBj1.9}$ includes the *vpr* gene (Fig. 1), it was reasoned that this genetic similarity would allow the exchange of comparable subgenomic regions between both viruses without loss of replication competence. To identify subgenomic regions capable of conferring on SIV_{agm3mc} the ability to replicate in non-stimulated PBMC, chimaeric SIV_{agm3mc}/$SIV_{smmPBj1.9}$ proviruses were constructed (see Fig. 1) [14]. Briefly, provirus HY-vif comprises the *vif* gene of $pSIV_{smmPBj1.9}$ (nucleotides 5107 to 5754) in place of a comparable genomic region of $pSIV_{agm3}$ (nucleotides 4762 to 5459). Provirus

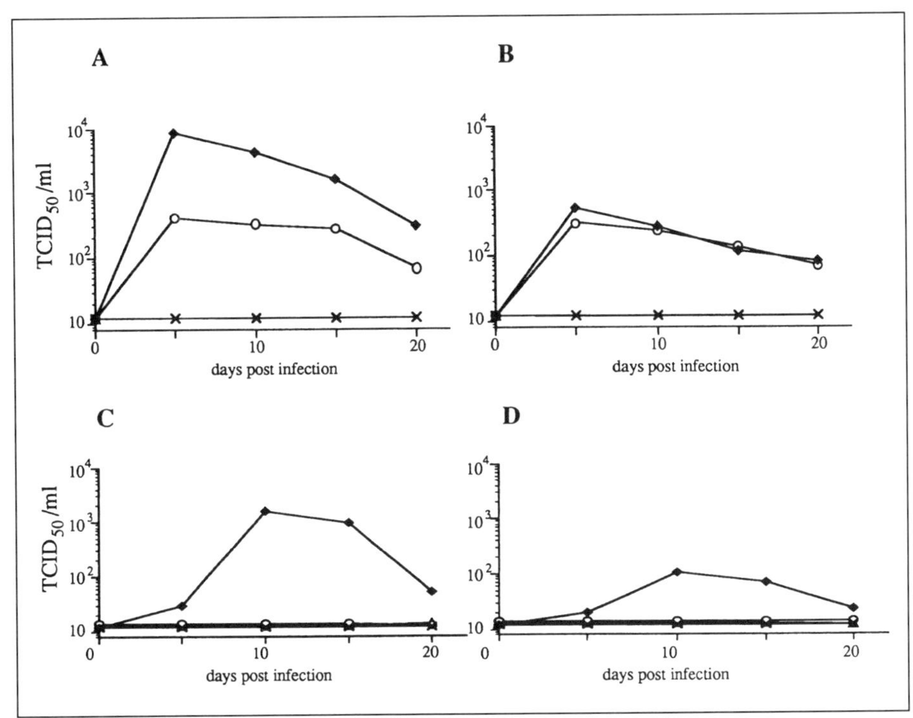

Fig. 2: Replication of $SIV_{smm\text{-}PBj1.9}$ (◆) and SIV_{agm3mc} (○) in stimulated PBMC (A and B) and in non-stimulated PBMC (C and D) from pig-tailed macaques (A and C) and from AGM (B and D); mock (x).

HY-gag/pol/CR contained the proviral region from the R region of the 5′ LTR up to the 3′-end of the first *tat* exon, which marks the end of the central region (CR) of $pSIV_{smmPBj1.9}$ in place of the respective proviral region of $pSIV_{agm3}$. The hybrid virus generated after transfection of provirus HY-gag/pol/CR was later shown to contain in addition the major part of the U3 region of $pSIV_{smmPBj1.9}$ within both LTRs (see below). Provirus HY-U3Δnef only comprised the U3 region of $pSIV_{smmPBj14}$ (nucleotides 9248 to 9786) instead of the comparable genomic region (nucleotides 8588 to 9096) comprising the U3 region of the $pSIV_{agm3}$ LTR. The *nef* gene of $pSIV_{agm3}$ was therefore truncated after codon 106.

Hybrid viruses were generated from all proviruses constructed as demonstrated by RT activity released into the supernatants of C8166 cells transfected with respective plasmids. Next, PBMC from pig-tailed macaques and AGM stimulated for two days in the presence of PHA and IL-2 were infected at a m.o.i. of 0.01 with the replication competent hybrid viruses. As shown in Figure 3 (A and B), all HY were able to replicate in PBMC from both monkey species. The infectious chimaeric viruses generated were then used for infection of non-stimulated PBMC from pig-tailed macaques and from AGM. As illustrated in Figure 3 (C and D), parental virus SIV_{agm3mc} and hybrid virus HY-vif were unable to grow in non-stimulated primary cells, whereas $SIV_{smmPBj1.9}$ and chimaeras HY-gag/pol/CR and HY-

U3Δnef replicated to high titres in PBMC from pig-tailed macaques. To our surprise, $SIV_{smmPBj1.9}$ and hybrid viruses HY-gag/pol/CR and HY-U3Δnef were also capable of replicating in non-stimulated AGM PBMC, although to about 15-fold lower titres than those achieved in non-stimulated pig-tailed macaque PBMC. Hybrid viruses HY-gag/pol/CR and HY-U3Δnef replicated to similar maximum titres and with similar kinetics as SIVsmmPBj1.9. Subgenomic chimaeric regions of virus HY-gag/pol/CR cloned before and after infection of PBMC showed a critical alteration of the viral LTR, which now comprised nucleotides 8588 to 8708 of $pSIV_{agm3}$ linked to nucleotides 9367 to 9787 of $pSIV_{smmPBj1.9}$. The putative viral promoter [6, 16] of this chimaeric virus was thus derived from $pSIV_{smmPBj1.9}$ and not from $pSIV_{agm3}$. Therefore, one genomic region of $SIV_{smmPBj1.9}$, namely the viral promoter within the U3 region of the viral LTR, was common to the infectious chimaeric viruses able to replicate in non-stimulated PBMC. In conclusion, a simple in vitro assay assessing the capacity of lentiviruses to replicate in non-stimulated PBMC isolated from pig-tailed macaques or AGM allowed us to select those candidate hybrid viruses putatively able to induce acute disease in infected monkeys.

In the light of these results we will be able to test in vivo the acute pathogenicity of the hybrid viruses in pig-tailed macaques using a very small number of animals. By constructing hybrid viruses encompassing all cistronic genes of SIV_{agm3mc} expressed by the viral promoter of $SIV_{smm\text{-}PBj1.9}$, the number of animal experiments may be reduced to three to six animals rather than twenty or more.

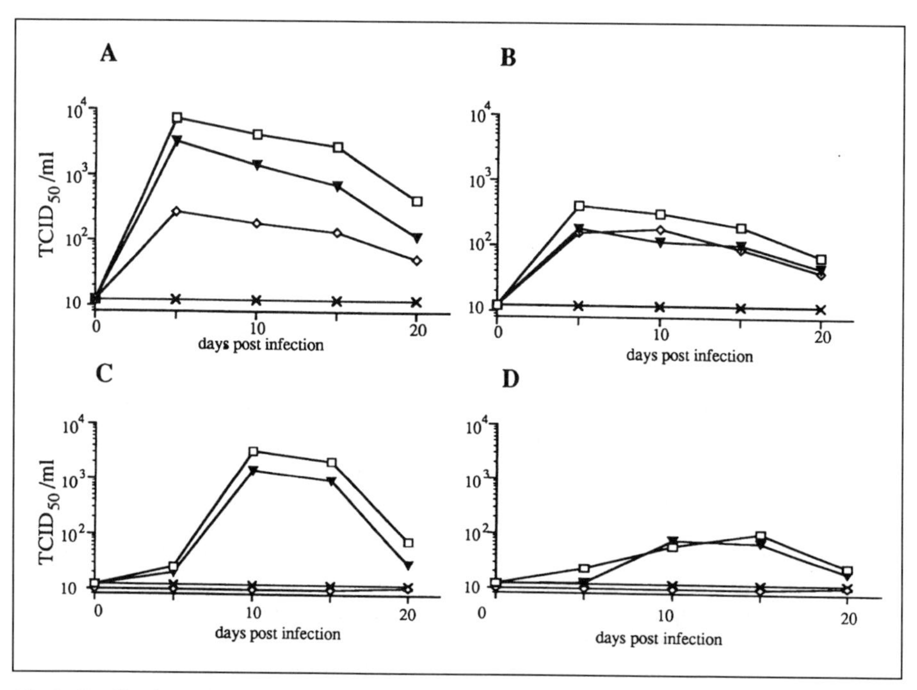

Fig. 3: Replication of hybrid viruses HY-vif (◇), HY-gag/pol/CR (▭) and HY-U3Δnef (▼) in stimulated PBMC (A and B) and in non-stimulated PBMC (C and D) from pig-tailed macaques (A and C) and from AGM (B and D); mock (x).

Acknowledgments

Portions of this article were adapted from Dittmar et al (1995) «The U3 promoter region of the acutely lethal simian immunodeficiency virus clone smmPBj1.9 confers related biological activity on the apathogenic clone agm3mc» [14]. We thank H. Merget-Millitzer and I. Treinies for constructive discussions. M. Selbert is acknowledged for expert automatic sequencing and S. Dewhurst (University of Rochester Medical Center, Rochester, NY, U.S.A.) for the kind donation of plasmid $pSIV_{smmPBj1.9}$.

Reference

1 Fultz PN, Anderson DC, McClure HM, Dewhurst S, Mullins JI: Dev Biol Stand 1990;72:253-258.

2 Fultz PN: J Virol 1991;65:4902-4909.

3 Novembre FJ, Johnson PR, Lewis MG, Anderson DC., Klumpp S, McClure HM, Hirsch VM: J Virol 1993;67:2466-2474.

4 McClure HM, Anderson DC, Fultz PN, Ansari AA, Lockwood E, Brodie A: Vet Immunol Immunpathol 1989;21:13-24.

5 Fultz PN, McClure HM, Anderson DC, Switzer WM: AIDS Res Hum Retroviruses 1989;5:397-409.

6 Dewhurst S, Embreston JE, Anderson DC, Mullins JI, Fultz PN: Nature (London) 1990;345:636-640.

7 Israel ZR, Dean GA, Maul DH, O'Nei SP, Dreitz MJ, Mullins JI, Fultz PN, Hoover EA: AIDS Res Hum Retroviruses 1993;9:277-286.

8 Yokoyama S: Mol Biol Evol 1988;5:645-659.

9 Myers G, McInnis K, Korber B: AIDS Res Hum Retroviruses 1992;8:373-386.

10 Johnson PR, Fomsgaard A, Allan J, Gravell M, London WT, Olmsted RA, Hirsch VM: J Virol 1990;64:1086-1092.

11 Baier M, Werner A, Cichutek K, Garber C, Müller C, Kraus G, Ferdinand FJ, Hartung S, Papas TS, Kurth R: J Virol 1989;63:5119-5123.

12 Baier M, Garber C, Müller C, Cichutek K, Kurth R: Virology 1990;176:216-221.

13 Baier M, Dittmar MT, Cichutek K, Kurth R: Proc Natl Acad Sci USA 1991;88:8126-8130.

14 Dittmar MT, Cichutek K, Fultz P, Kurth R: Proc Natl Acad Sci USA 1995;92:1362-1366.

15 Norley SG, Kraus G, Ennen J, Bonilla J, König H, Kurth R: Proc Natl Acad Sci USA 1990;87:9067-9071.

16 Dewhurst S, Embretson JE, Fultz PN, Mullins JI: AIDS Res Hum Retroviruses 1992;8:1179-1187.

Dr. K. Cichutek, Dept of Medical Biotechnology, Paul-Ehrlich-Institute, Paul-Ehrlich-Str. 51-59, D-63225 Langen, Germany

Brown F, Cussler K, Hendriksen C (eds): Replacement, Reduction and Refinement of Animal Experiments in the Development and Control of Biological Products.
Dev Biol Stand. Basel, Karger, 1996, vol 86, pp 175-182

Detection of Extraneous Agents in Vaccines Using the Polymerase Chain Reaction for Newcastle Disease Virus in Poultry Biologicals

L. Bruckner, N. Stäuber, K. Brechtbühl, M.A. Hofmann

Institute of Virology and Immunoprophylaxis, Mittelhäusern, Switzerland

Key words: Newcastle disease virus, poultry vaccines, RT-PCR.

Abstract: Reverse transcription-polymerase chain reaction (RT-PCR) in poultry vaccine was applied to the detection of Newcastle disease virus (NDV), using two primer pairs spanning the cleavage site of the F0 fusion protein coding sequence. Amplification of a specific cDNA segment was possible from live and inactivated, oil-adjuvanted NDV vaccines without previous treatment. The RT-PCR was able to detect between 5×10^2 EID_{50} (in live vaccine preparations) and 10^5 EID_{50} or 0.056 haemagglutinating units of NDV (in inactivated vaccine preparations). In addition, live vaccine preparations were inactivated with β-propiolactone (β-PL). Amplified cDNA was obtained after treatment with 0.1% β-PL, whereas at a concentration of 1% or 10% no specific bands were visible in the agarose gel. These results demonstrate the applicability of the method for the control of poultry vaccines by ensuring the absence of extraneous agents.

Introduction

Veterinary vaccines have to be tested by the manufacturer, and in several countries also by an independent analyst before they are approved for commercial use. According to the requirements prescribed in the European Pharmacopoeia [1] vaccines must be specifically tested for their identity, potency and efficacy, as well as for purity and safety. Whereas certain parameters can be checked by in vitro assays (e.g. bacterial sterility), animal experiments are required in many cases both for efficacy and viral contamination testing [2, 3]. The proof of absence of extraneous agents in poultry vaccines includes the vaccination of specific pathogen-free chicken with subsequent serological testing of the animals. The vaccine must not stimulate the induction of antibodies against several specified avian viruses.

To reduce the number of animals required for vaccine control there is a high demand for sensitive and reliable in vitro tests [2]. Furthermore, it has been shown that vaccines against avian diseases can be contaminated with NDV (Bruckner, Hoff-Jørgensen, unpublished).

NDV, or avian paramyxovirus-1 (PMV-1), belongs to the family *Paramyxoviridae*. The viral genome consists of an unsegmented single-stranded RNA of negative polarity which is coding for six gene products [4]. The polymerase chain reaction (PCR) allows the enzymatic amplification of minute amounts of DNA. It is widely used in diagnostic laboratories for virus detection [5]. If RNA is first reverse-transcribed into cDNA before amplification (RT-PCR), RNA viruses can be detected as well.

Materials and Methods

Vaccines and viruses

Two commercially available live vaccines containing either NDV strains La Sota or Hitchner B1, as well as a formalin-inactivated and mineral oil-adjuvanted PMV-1 (strain P_3G) vaccine for pigeons, were used for the establishment of the RT-PCR. The live vaccine preparations were inactivated with ß-propiolactone (ß-PL; Sigma) [6] at different concentrations. ß-PL is known to modify nucleic acids irreversibly [7] and hence could interfere with PCR. The inhibiting activity of ß-PL on PCR was determined by treating water with β-PL following the same methods as for vaccines, and then carrying out PCR with this water mixed 1:1 with live vaccines.

Seven live and seven inactivated vaccines against NDV were tested. One of the former preparations was polyvalent and additionally contained infectious bronchitis virus (IBV), whereas five of the latter contained inactivated Gumboro virus or egg drop syndrome virus or IBV, respectively. According to the manufacturers' outlines of production, inactivation was done either by formalin or by binary-ethyleneimine treatment.

Vaccines against other paramyxoviruses (live and inactivated canine parainfluenza-2 virus (strain SV5), live bovine parainfluenza-3 virus (strain RLB 103ts), inactivated parainfluenza-3 virus (strain Reissinger), live attenuated canine distemper virus (strains Snyder Hill and Onderstepoort), and vaccines against other poultry diseases known to be devoid of NDV served as negative controls in the PCR, as well as water treated identically to virus-containing samples.

Oligonucleotide primers for PCR

For the initial experiments two 18-mer primers, NCD1 and NCD2, were chosen (according to the cDNA sequence of NDV strain La Sota clone 30) [8]. Sequence comparison of primer NCD2 with several NDV strains revealed a second annealing site for NCD2 (nucleotides 274 to 290), leading to two amplified products of 274 and 316 bp with strains La Sota and Hitchner B1. A second primer pair was chosen within a conserved sequence in all available NDV strains [9, 10]. The sense primer, NCD3, was a 19-mer oligonucleotide with the sequence 5'-GGAGGATGTTGGCAGCATT-3', complementary to nucleotides 485 to 503 of the cDNA; the antisense primer, NCD4, was a 20-mer oligonucleotide with the sequence 5'-GTCAACATATACACCTCATC-3', identical to nucleotides 194 to 213 of the cDNA (Fig. 1).

Reverse transcription and polymerase chain reaction

2,5 µl of each dilution of a 1:10 dilution series of the vaccine were added without prior treatment to 15 µl 2x PCR buffer (50 mM Tris/HCl, pH 8.3, 150 mM KCl, 5 mM $MgCl_2$, 500 µM of each dNTP), 17 µl H_2O and 50 pmol of primer NCD2 or NCD4, respectively, and heated to 95°C for 5 min. The mixture was chilled on ice before the addition of 10 units RNase inhibitor (RNAsin, Promega) and 50 units reverse transcriptase (M-MLV Reverse Transcriptase, Gibco BRL), and incubated for 30 min. at 37°C. For hot start PCR [11] a piece of pastillated paraffin wax (congealing point 57-60°C; BDH Laboratory Supplies) was added into each tube, and the tubes were heated to melt the paraffin. After cooling, 0.5 units Taq DNA polymerase (SuperTaq™ DNA Polymerase, Stehelin, Basel) and 50 pmol of primer NCD1 or NCD3, respectively, were overlayed on the solidified paraffin layer. Amplification was carried out in a GeneAmp 9600 thermal cycler (Perkin-Elmer Cetus). The samples were subjected to one denaturation step at 95°C for 2 min. and then to 35 cycles of denaturation for 30 sec at 95°C, annealing for 1 min. at 51°C and extension for 1 min. at 72°C. The expected size of the amplified DNA band was 274 bp for primers NCD1/2 and 310 bp for primers NCD3/4 (Fig. 1).

Digestion with restriction enzymes Hae III and Nar I (New England Biolabs) was performed to identify PCR products, as described elsewhere [8]. Briefly, five units of the respective endonuclease were either added directly to 10 µl of the PCR reaction mix and incubated at 37°C for two hours, or the amplified DNA was first precipitated with ethanol and redissolved in the appropriate buffer before the enzyme was added.

Agarose gel electrophoresis

Aliquots of 10 µl PCR mix were analysed in 2% agarose gels (4% for restriction fragment patterns) in TBE buffer containing 0.5 µg/ml ethidium bromide. Amplified DNA and restriction fragments were compared with a DNA marker (plasmid pBlueScript (Stratagene), digested with Sau3A) to allow size determination.

Results

Amplification of NDV-specific cDNA

PCR amplification using primers NCD1 and NCD2 resulted in a specific DNA band of 274 bp as expected, when NDV strains Hitchner B1 (Fig. 2, lane 2), La Sota (Fig. 2, lane 5) or PMV-1 type P_3G (Fig. 2, lane 8) were used as template source, respectively. However, the two strains La Sota and Hitchner B1 (live virus vaccine preparations) yielded two additional DNA bands, one having a length of 316 bp (Fig. 2, lanes 2, 5), as predicted after sequence comparison of primer NCD2

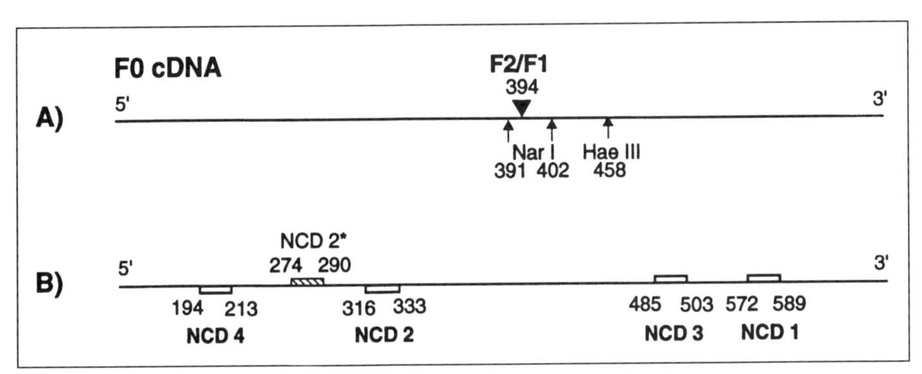

Fig. 1: Segment of cDNA of NDV F0 fusion protein gene used for RT-PCR, shown in positive (mRNA) sense, according to Toyoda et al [10]. A) Arrows indicate restriction sites of Nar I (strains D26/76, B1/47, BEA/45, TEX/48, QUE/66, ULS/67, [10]) and of Hae III (strains B1/47, BEA/45, TEX/48, LAS/46, [10]). The post-translational cleavage site of the fusion protein (F2/F1) is marked by an arrowhead.
B) Open boxes, annealing site of primers NCD 1, NCD 2, NCD 3, NCD 4; hatched box, second annealing site (NCD 2*) of primer NCD 2 (see text).

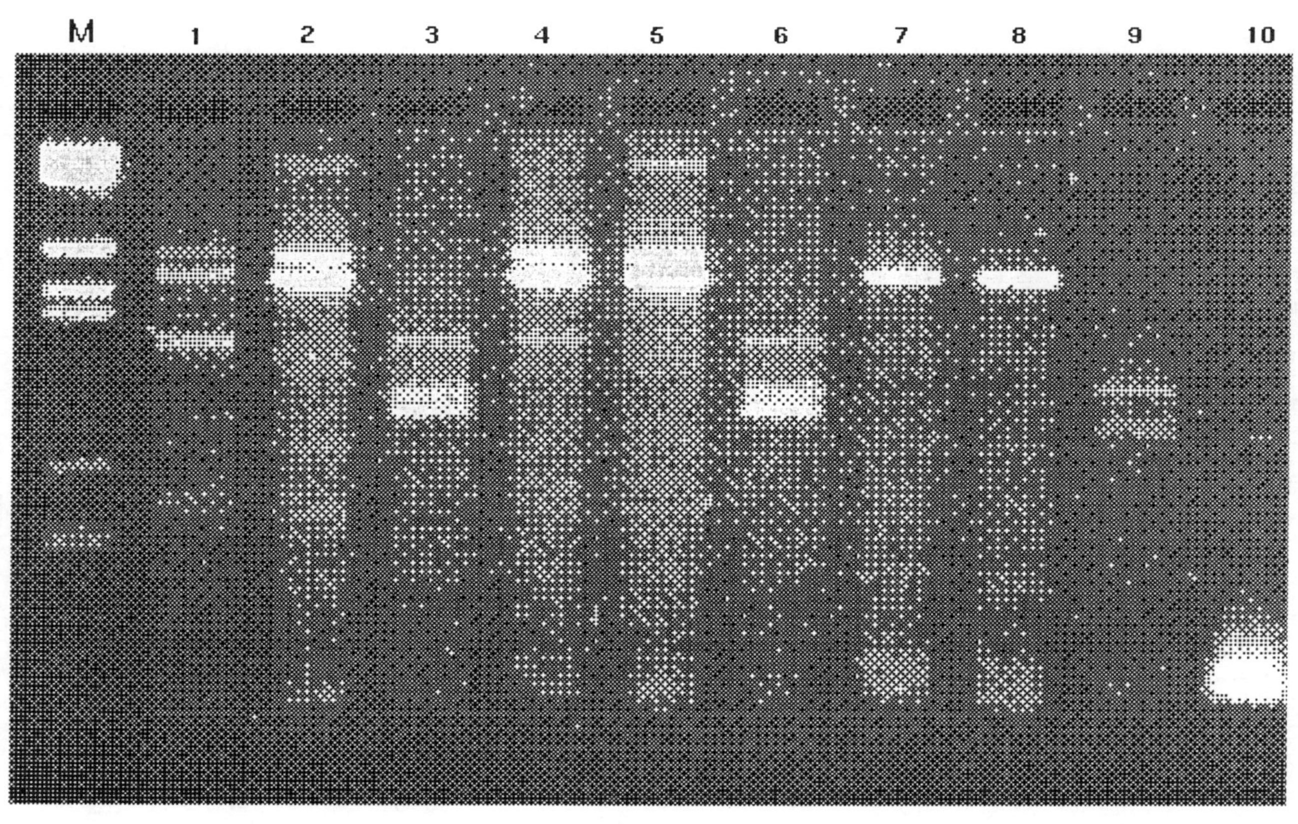

Fig. 2: Eletrophoresis in a 4% agarose gel of the PCR products with primer NCD2/NCD1. Live virus vaccine preparations NDV Hitchner B1 (lane 2) and NDV La Sota (lane 5), inactivated virus vaccine preparation strain P3G (lane 8). Digestion of amplified fragments was made with restriction endonuclease Nar I (HB1, lane 1; LAS, lane 4; P3G, lane 7) and Hae III (HB1, lane 3; LAS, lane 6; P3G, lane 9). M, DNA size marker; negative control (lane 10).

with NDV sequences, indicating the presence of a second (incomplete) annealing site (see Materials and Methods). The origin of the third DNA band of about 700 bp could not be determined. No amplified DNA was detected if other paramyxovirus vaccines or H_2O were used.

Digestion of PCR products from strains La Sota, Hitchner B1 as well as the inactivated PMV-1 type P_3G vaccine with Hae III resulted in two fragments of 131 and 144 bp, as predicted from the published sequences (Fig. 2, lanes 3, 6, 9). A third fragment of 185 bp was obtained for strains La Sota and Hitchner B1, originating from the cleavage of the 316 bp cDNA generated by the additional upstream annealing site of primer NCD2 (Fig. 2, lanes 3, 6). Digestion with Nar I did not cleave the amplified cDNA from strain PMV-1 type P_3G (Fig. 2, lane 7), whereas the fragments from strains La Sota and HB1 were cut only partially and three bands with sizes of 187, 129 and 87 bp were found (Fig. 2, lanes 1, 4).

As predicted, a 310 bp cDNA could be amplified by RT-PCR, using the primer pair NCD3 and NCD4 with the NDV strains Hitchner B1 (Fig. 3, lane 2), La Sota (Fig. 3, lane 5) and PMV-1 type P_3G (Fig. 3, lane 8). The cleavage of the PCR products by Hae III resulted in two fragments of 45 and 265 bp, respectively (Fig. 3, lanes 3, 6, 9). Digestion with Nar I partially cleaved the PCR-cDNA of strains La Sota and Hitchner B1 into two fragments of 101 and 209 bp, as predicted from the sequences established in the present study (Fig. 3, lanes 1, 4).

Live virus preparations La Sota and Hitchner B1 were inactivated with different concentrations of ß-PL. Amplified cDNA was obtained when ß-PL was diluted 1:1000 (0.1% (w/v) ß-PL, final concentration). At a 1:100 and 1:10 ß-PL dilution no specific bands were visible in the agarose gel. Following amplification of live vaccines mixed with hydrolysed β-PL in different dilutions, inhibition of the RT-PCR occurred only with a β-PL concentration of 10% (results not shown).

Sensitivity of the RT-PCR

The sensitivity was determined using the manufacturers specifications of the antigen or virus content of the vaccine. According to the manufacturers, live vaccine preparations contained at least 10^6 egg infectious doses (EID_{50}) NDV per dose. The highest dilution leading to a specific DNA band in the agarose gel was 1:1000, which corresponds to a virus titre of 5×10^2 EID_{50} per PCR assay (results not shown). The inactivated PMV-1 type P_3G vaccine showed a positive signal at a 1:10 dilution, being equivalent to 0.056 haemagglutinating units per PCR assay. For the other inactivated vaccines the detection limit of the PCR was at 10^5 EID_{50} of NDV.

No improvement of the sensitivity of the PCR was obtained after SDS-proteinase K and phenol-chloroform extraction and ethanol precipitation of viral RNA (results not shown).

Discussion

In an attempt to replace animal experiments for the control of live and inactivated poultry vaccines, nucleic acid amplification techniques were applied for the detection of NDV. Specific cDNA segments of the expected lengths could be amplified both with the published primers NCD1 and NCD2 [8] as well as with the

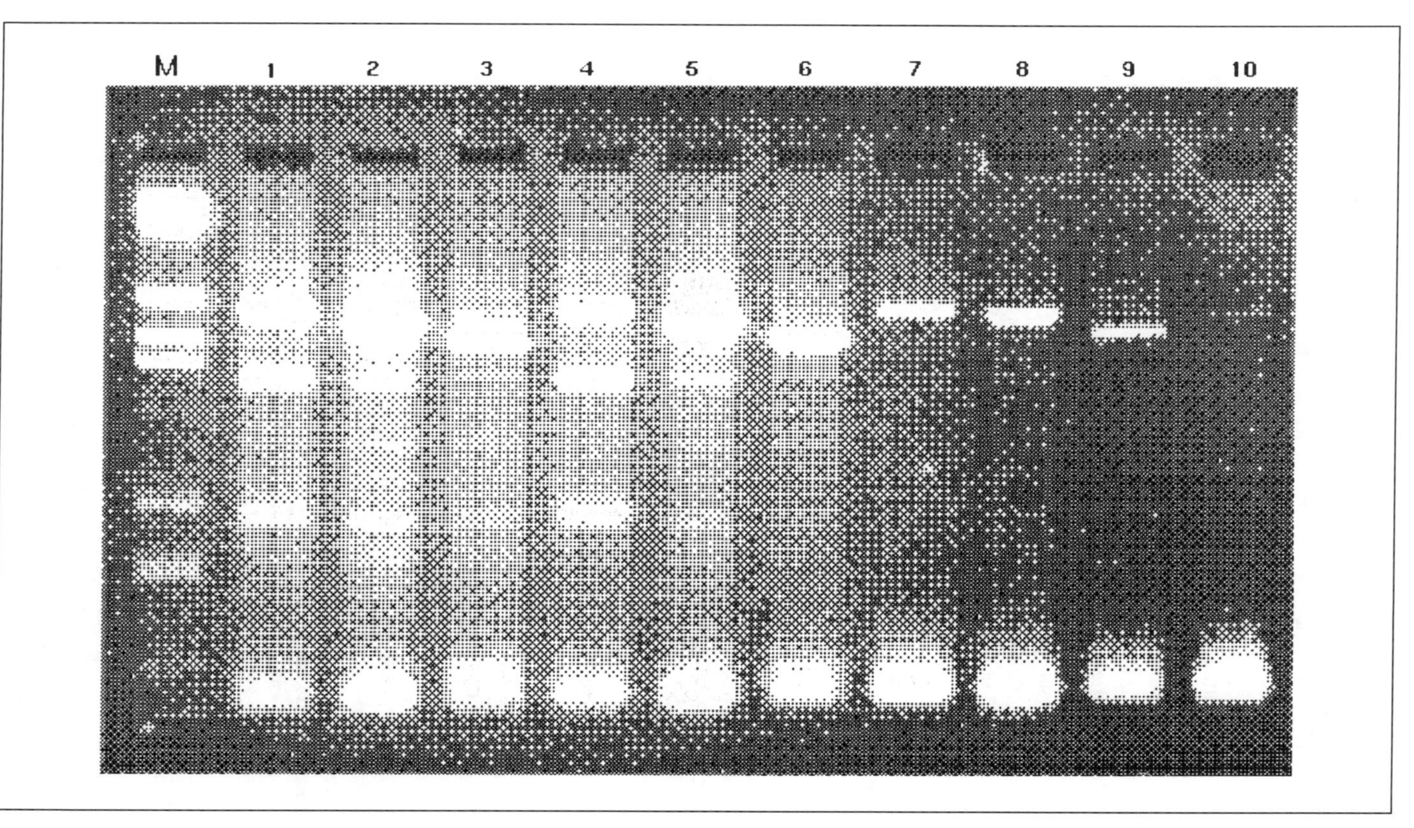

Fig. 3: Electrophoresis in a 4% agarose gel of the PCR products with primer NCD4/NCD3. Live virus vaccine preparations NDV Hitchner B1 (lane 2) and NDV La Sota (lane 5), inactivated virus vaccine preparation strain P3G (lane 8). Digestion of amplified fragments was done with restriction endonuclease Nar I (HB1, lane 1; LAS, lane 4; P3G, lane 7) and Hae III (HB1, lane 3; LAS, lane 6; P3G, lane 9). M, DNA size marker; negative control (lane 10).

primers NCD3 and NCD4 described in this study, and Hae III restriction enzyme digestion yielded cleavage products as predicted from the published sequences.

The PCR described in this work was able to detect between $5x10^2$ EID_{50} (in live vaccine preparations) and 10^5 EID_{50} or 0.056 haemagglutinating units of NCD virus (in inactivated vaccine preparations). In consideration of the antigen content of one vaccine dose, which is 10^6 EID_{50} for live virus vaccines and 10^8 EID_{50} for inactivated vaccines, the sensitivity of the PCR for the identity control of the vaccine antigen is sufficient. Furthermore, it can be expected that any possible contamination of poultry vaccines with NDV during the manufacturing process, particularly the antigen production, will result in a considerable NDV titre. This was confirmed in the case of an NDV-contaminated inactivated vaccine against infectious bronchitis (Bruckner, Hoff-Jørgensen, unpublished), where NDV in the vaccine was demonstrated by RT-PCR, after it had been shown that the vaccine stimulated the production of antibodies against NDV in chicken.

The RT-PCR described in this work allows direct detection of NDV in vaccines (live, inactivated, or oil-adjuvanted), whereas until now this was done by indirect in vivo assays. In vitro, a direct characterization has only been possible for live antigen. However, the interference of ß-PL-inactivated NDV with PCR had been shown, limiting its applicability, because ß-PL irreversibly degrades nucleic acids [7]. The results obtained in this study demonstrate that the alteration of nucleic acid molecules depends on the concentration of ß-PL and hydrolysed ß-PL only interferes with PCR at a concentration as high as 10%. Therefore, ß-PL concentrations between 0.025% and 0.1%, as used in the manufacturing process of vaccines, do not interfere with the detection of RNA in inactivated preparations. The fact that RNA can still be detected by RT-PCR after ß-PL-inactivation might be due to (i): nucleic acid modifications not always occurring in the short fragment to be amplified, or (ii): a few RNA molecules possibly not being affected by the inactivant [12]. In this context, it is worth mentioning that the efficiency of ß-PL inactivation should be re-examined in the light of the easily detectable RNA at ß-PL concentrations used during the vaccine manufacturing process.

Restrictions in the application of the PCR could be due to strain variations in the nucleic acid sequence, although this problem is limited to new NDV strains occurring in field isolates, and therefore is not applicable to the well-characterized virus strains used in vaccine preparations. However, we have used the RT-PCR described here for NDV detection in clinical samples of PMV-1-infected birds and preliminary results have confirmed the feasibility of the method (results not shown). Another limitation of the RT-PCR is the inability to allow direct quantification of the NDV titre, although serial dilutions can be used to estimate the amount of NDV in the vaccine. The method described here is therefore not applicable for vaccine potency testing, an application which would however be highly desirable.

References

1 Council of Europe. European Pharmacopoeia, Deuxième Edition 1989, V.2.1.3.5., Edt. Maisonneuve S.A. Sainte-Ruffine, France.

2 Levings RL, Henderson LM, Metz CA: In vitro potency assays for nonreplicating veterinary vaccines: comparison to in vivo assays and considerations in assay development. Vet Microbiol 1993;37:201-219.

3 Soulebot JP, Milward F, Prevost Ph: In vitro potency testing of inactivated biologics: current situation in the EEC. Vet Microbiol 1993;37:241-251.

4 Kingsbury DW: Paramyxoviridae and their replication, in Fields BN, Knipe DM (eds): Fields Virology, New York, Raven Press, 1990, Vol 1, pp 945-962.

5 Belàk S, Ballagi-Pordàny A: Review Article: Application of the polymerase chain reaction (PCR) in veterinary diagnostic virology. Vet Res Com 1993;17:55-72

6 King DJ. Evaluation of different methods of inactivation of Newcastle disease virus and avian influenza virus in egg fluids and serum. Avian Dis 1991;35:505-514.

7 Chen RF, Meiyal JJ, Goldwait DA: The reaction of ß-propiolactone with derivatives of adenine and with DNA. Carcinogenesis 1981;2:73-80.

8 Jestin V, Jestin A: Detection of Newcastle disease virus RNA in infected allantoic fluids by in vitro enzymatic amplification (PCR). Arch Virol 1991;118:151-161.

9 McGinnes LW, Morrison TG: Nucleotide sequence of the gene encoding the Newcastle disease virus fusion protein and comparisons of paramyxovirus fusion protein sequences. Virus Res 1986;5:343-356.

10 Toyoda T, Sakaguchi T, Hirota H, Gotoh B, Kuma K, Miyata T, Nagai Y: Newcastle disease virus evolution. II. Lack of gene recombination in generating virulent and avirulent strains. Virology 1989:169;273-282.

11 Cooke H: Inexpensive wax for PCR protocols. Trends in Genetics 1992;8:301.

12 Hofmann MA, Brian DA: Sequencing PCR DNA amplified directly from a bacterial colony. Biotechniques 1991;11:30-31.

Dr. L. Bruckner, Institute of Virology and Immunoprophylaxis, CH-3147 Mittelhäusern, Switzerland

Brown F, Cussler K, Hendriksen C (eds): Replacement, Reduction and Refinement of Animal Experiments in the Development and Control of Biological Products.
Dev Biol Stand. Basel, Karger, 1996, vol 86, p 183

SESSION VII

Toxoids Vaccines

Chairmen: J. Milstien (Geneva, Switzerland)
P. Castle (Strasbourg, France)

Brown F, Cussler K, Hendriksen C (eds): Replacement, Reduction and Refinement of Animal Experiments in the Development and Control of Biological Products.
Dev Biol Stand. Basel, Karger, 1996, vol 86, pp 185-197

Opportunities to Reduce the Use of Animals in the Potency and Toxicity Testing of Toxoid Vaccines

P.A. Knight

Welcome Research Laboratories, Beckenham, Kent, UK

Key words: Toxin neutralisation, toxoid vaccines, in vitro tests, cell culture, ELISA, TOBI.

Abstract: Alternative methods for titrating antitoxin are now available which should, in principle, permit a very large reduction in the numbers of animals required to test the potency of toxoid vaccines. More importantly they make it possible to eliminate the use of animals for the indication of excess toxicity almost completely. The full realisation of this potential is dependent upon the careful validation of all other methods and the introduction of more appropriate standard and reference preparations for titration of antisera and for the assay of DTP. It would be facilitated by a mechanism to facilitate the wider dissemination of relevant monoclonal antibodies and by a restructuring of veterinary vaccine potency tests to make full use of the additional serological information provided by the new methods.

INTRODUCTION

Vaccines prepared from bacterial toxoids are among the most efficacious of all inactivated preparations. Much of that efficacy derives from the use of tests based upon the toxin neutralisation (TN) principle which currently demand more animals than are required to test any other class of vaccine. Yet developments of alternative methods and the simplification of existing tests have now reached the point where there are no technical reasons why the use of animals for the control of toxoid vaccines should not be reduced to a fraction of the present level.

Toxoid vaccines have been prepared for almost every disease process in which an active bacterial exotoxin has been thought to play a part. These have included toxoid-derived Streptococcal and Staphylococcal haemolysins, enterotoxins and most recently pertussis toxin, but the most important toxoid vaccines in extensive current use are those listed in Table 1. They comprise diphtheria toxoid and a range of toxoids derived from clostridial toxins. Diphtheria and most of the clostridia are potential human pathogens but toxoid vaccines are only extensively used

Table 1: Application and test indicators of major toxoid vaccines.

Pathogen species	Application	Major toxin	Effects
C. diphtheriae	HV	diphtheria	lethal *dermonecrotic cytopathic
Clostridium tetani	HV & AV	tetanus	*lethal *paralytic
Cl. perfringens A	HP & AV	alpha	*lethal dermonecrotic haemolytic phospholipase C
Cl. perfringens B	AV	Beta/epsilon	*lethal dermonecrotic cytopathic
Cl. perfringens C	hv & AV	beta	*lethal dermonecrotic
Cl. perfringens D	AV	epsilon	*lethal dermonecrotic cytopathic
Cl. septicum	HP & AV	alpha	*lethal dermonecrotic cytopathic
Cl. novyi	HV & AV	alpha	*lethal dermonecrotic ? cytopathic
Cl. sordellii	HV & AV		*lethal cytopathic
Cl. botulinum A, B, E	hv		*lethal paralytic
Cl. botulinum C, D	AV		*lethal paralytic

* Official method under American or European requirements.
HV = Extensive use in human vaccines, hv = limited use in human vaccines.
HP = Human Pathogen; no vaccines, AV = extensive use in animal vaccines

in man for the control of diphtheria and tetanus with a much more limited use of vaccines against *Clostridium perfringens* C and *Cl. botulinum* A, B & E. Most toxoid vaccines are directed against clostridia which cause diseases in sheep, goats and cattle and are frequently combined in multivalent formulations. Finally, *Cl. botulinum* C & D toxoids are used to protect farmed mustellids against botulism.

Potency Tests

Potency tests for all three groups of vaccines are based on the TN principle. Veterinary botulinum vaccines and human vaccines are tested by in vivo neutralisation of a toxin challenge. Sera from vaccinated animals are titrated in vitro with in vivo indicators in the potency tests for vaccines for farm animals and human vaccines in the USA. In addition to the largely statutory tests on the final product, in process tests allow the quality and quantity of protective antigen and even protective epitopes within it to be monitored throughout the process.

TN tests were found to be critical to the development of efficacious antisera more than ninety years ago. They were equally essential in the development of the first vaccines in the 1930s and antitoxin standards were among the first to be produced.

The great disadvantage of TN tests was their use of animals as indicators and alternatives were sought from the beginning. The first of these was flocculation which was introduced by Ramon. It produced results which closely paralleled the TN test but occasionally produced deviant results. TN was therefore always used as the definitive test but flocculation was used for intermediate stages and enabled the final TN test to get on range straight away. The severity of the TN test could be reduced by the adoption of intradermal reaction instead of lethal indicators and these were introduced in some laboratories for diphtheria [1], *Cl. perfringens* [2, 3] and *Cl. septicum* [4] antitoxin titrations at least. In the 1950s passive haemagglutination [5] was introduced. As with flocculation it reflected TN most of the time but unlike flocculation it was suitable for sera of all species. Unfortunately, the correlation with TN broke down on rare occasions and it was not deemed good enough.

The first potency tests on vaccines depended on the measurement of absolute TN titres in vaccinated guinea pigs. In Britain the TN titres against diphtheria were measured by conventional TN test on serum or by intradermal challenge which effected the titration in vivo. For most other toxoid vaccines the intradermal method is not available.

During the 1960s it became apparent that absolute response levels were an unreliable index of vaccine potency. Animal strain, diet and even environment could alter the response to a vaccine profoundly. Table 2 shows that the response to a single dose of vaccine can be changed almost 10-fold by a change in the source of diet. Similarly, at least one strain of rabbit regularly produced titres to clostridial toxoids twice as high as other strains. To counter such problems the World Health Organisation (WHO) set up standard preparations of plain and adsorbed toxoids and encouraged the development of comparative assay methods to eliminate the effect of these variables. Despite the known variation in levels of immune response associated with animal strain and diet, non-comparative tests were retained for veterinary toxoid vaccines and for human toxoid vaccines in the Americas. Eventually WHO introduced methods based upon six-point assays to enable the potency of vaccines to be expressed in terms of those standards. These assays proved difficult to perform, with a high incidence of statistical invalidity and a use of animals that proved unacceptable in both developed and developing worlds.

These difficulties prompted a reappraisal of the methods by which vaccines might be compared with the International Standard Preparations [6]. The aim was to achieve the same precision and specificity that characterised the TN test and the same level of assurance that the potency of the test vaccine equalled or surpassed the standard, using fewer animals without abandonment of the comparative principle.

Table 2: Distribution of diphtheria and tetanus titres in guinea pigs following a single dose of 1 ml. of DTP.

Titres >	**0.1**	**0.2**	**0.5**	**1.0**	**2.0**	**5.0**	**10**	**20**	**Geom. Mean Titre**
Diets <	**0.2**	**0.5**	**1.0**	**2.0**	**5.0**	**10**	**20**	**50**	**(u/ml)**
a) Diphtheria									
Diet 18 + Greens				1	6	2	1		4.0
RGP + Greens	1	0	2	5	2				1.2
RGP alone		2	5	1	2				0.8
b) Tetanus									
Diet 18 + Greens							5	5	21.0
RGP + Greens				1	2	0	6	1	7.6
RGP alone				1	3	4	1	1	5.5

As a result of this reappraisal the concept of the single dose level test was developed [7, 8]. In this test the response of animals to a defined fraction of the minimum number of international units per human dose of standard vaccine is compared to the response of a similar group of animals to the same fraction of a human dose of the test vaccine. If the response to the test vaccine is significantly superior to that of the standard it is concluded that its potency exceeds the minimum requirement. If the true potency of the vaccine is only marginally greater than the minimum, the savings are relatively modest but many lots of DTP are substantially more potent than the minimum and these could be routinely released by tests which use fewer than 16 animals.

Simplified tests cannot be expected to replace full assays completely. Tests on new formulations, collaborative assays of standard preparations, stability tests and tests on batches of disputed potency will still require the performance of full assays. Furthermore, these tests are not available for batch release by manufacturers in Europe because they have not yet been accepted by the European Pharmacopoeia, conformity to which is specified in almost all product licences. The simplified test format is also applicable in principle to botulinum vaccines and also to some non-toxoid vaccines.

Although the simplified test format can deliver valuable reductions in the number of animals used, the greatest gains will come from the introduction of alternative means of titrating antitoxic sera. Savings in numbers will come from two sources – the direct elimination of animals from antitoxin titration and the substitution of serological response measurement for toxic challenge in comparative assays performed on diphtheria and tetanus vaccines under WHO requirements.

In Vitro Alternatives to TN Tests

The most objectionable feature of toxin neutralisation tests is their use of animals as indicators of excess toxin in toxin/antitoxin mixtures. This is as true of toxin challenge tests in which the toxin reacts with antitoxin in vivo as it is of test tube titrations. But in seeking to eliminate the unacceptable use of animals it is important to retain the unique specificity of these tests which recognise only the protective species of antibody elicited by the toxoid.

There are two main approaches to the development of alternatives: the replacement of the animal indicator with an in vitro indicator while retaining the TN principle and the design of serological tests which are more or less selective for neutralising antibody.

The first method, by retaining the toxin neutralisation principle, also retains its unique properties *provided that the agent which is active in the test is identical to the lethal toxin*. Table 1 shows that cytopathic, haemolytic or enzyme activities suitable for use as in vitro indicators are associated with the lethal toxins of *Corynebacterium diphtheriae* [9], *Clostridium perfringens* Types A, B & D [10], *Cl. septicum* [11], *Cl. sordellii* [12] & *Cl. novyi* [13]. A further cytopathic activity associated with *Clostridium perfringens* Types B & C filtrates has been shown to be unrelated to the beta toxin since it was not neutralised by a monoclonal antibody which neutralises the lethal and dermonecrotic activities of the beta toxin [10]. In contrast, the lethal, dermonecrotic and cytopathic effects of *Cl. perfringens* D filtrates could be neutralised with a single monoclonal, thus proving all three activities are attributable to the epsilon toxin. The identity of in vivo and in vitro indicators for excess diphtheria, *Cl. septicum*, *novyi* and *perfringens* toxins have not yet been proved with monoclonal antibodies but there is good evidence of co-purification of lethal and cytopathic activities from *C. diphtheriae*, *Cl. septicum* and *Cl. novyi* B. The evidence of correlation between titres of antisera raised against vaccines of different formulations in various species obtained using lethal and cytopathic indicators is consistent with identity.

Although the adoption of cell culture techniques may be novel for many serologists and bacteriologists, the basic methodology is the same as for classical in vivo TN tests. The total reaction volume is scaled down to 100 µl in microtitre wells, while retaining the same concentration of reagents as the in vivo test but using cell culture medium as diluent. Instead of injecting the reaction mixtures into animals, 50 µl of cell suspension – MDCK cells for *Cl. perfringens* epsilon [10] and VERO cells for the other toxins [11, 13] – are added to each well and the plates incubated at 37° C. Cytopathic effects can be determined microscopically by the use of pH indicators in sealed wells or by staining the fixed cell carpets at the end of the test. The latter method enables the end point to be easily determined by eye but a brief microscopic check is prudent. Cell culture indicators provide the same precision and specificity as in vivo indicators with each microtitre well providing the same information as one mouse.

The haemolytic activity of *Cl. perfringens* alpha toxin and its ability to coagulate ovolecithin have long been used as alternatives to the lethal indicator for the titration of antisera. The correlation of titres obtained by both in vitro methods with the lethal test has been shown to be excellent.

Most of the work on alternatives to the TN test has been directed to ELISA and agglutination tests. These tests cannot ever claim to equal the specificity of either in vivo or in vitro TN tests but they are seen as being easy to perform and to automate and in many cases they yield results that approximate sufficiently closely to TN titres to allow them to be substituted for them. It is particularly important that the relationship between TN test titres and their serological alternatives remains constant for sera raised against as wide a range of vaccine formulations as possible. Conformity to this requirement cannot be assumed and there have been instances where a serological test has failed to detect diphtheria antitoxin in sera raised against DPT while detecting antibodies raised against diphtheria vaccine [14]. Failure to meet this requirement indicates that the antibodies measured are wholly or largely irrelevant to protection.

The extent to which the results of a serological test can approximate to TN titres is dependent upon the purity of the antigens and the specificity of the antisera they use and the extent to which the test format is designed to recognise neutralising antibodies preferentially.

The specificity of direct ELISA and passive haemagglutination tests is completely dependent on the purity of the antigen used to sensitise the ELISA plate or the erythrocytes; the test will detect antibody to impurities shared by the sensitising antigen and the test or reference vaccine as well as to the toxin itself. Particular care needs to be taken to ensure that impurities present in the test vaccine but not in the reference vaccine are not present in the sensitising antigen, since such impurities may result in a systematic bias in favour of the test vaccine. The specificity of ELISA tests can be improved by the use of monoclonal antibodies (MAbs) to couple the sensitising antigen to the plate, particularly if the MAbs are unable to neutralise the toxicity of the toxin [15].

Work on serological alternatives to TN has been concentrated on tetanus, which is the only antigen extensively included in both human and animal vaccines and for which no cell culture or enzyme indicator is in prospect. Both direct ELISA and passive haemagglutination tests have been found to be unreliable for the detection of low levels of tetanus antitoxin in human sera. In response to this difficulty two test format strategies were developed. One strategy involved comparison of the ELISA response to sera alone and in the presence of a blocking dose of toxin [16]. The other method, the Toxin binding Inhibition (TOBI) assay [17] measures the extent to which the test antiserum blocks the binding of toxin to a plate sensitised with antitoxin. Both methods were effective in reducing the problem of false positives and although the TOBI method does not incorporate measures to improve specificity it has been successfully used to measure guinea pig and mouse sera in assays of diphtheria and tetanus vaccines. The TOBI test format is attractive because it simulates in part the prevention of binding of the toxin to its cellular receptor and because there is a potential to attain a high level of specificity by substituting a neutralising MAb for the polyclonal antitoxin used to sensitise the plate. TOBI tests are most effective at low levels of test and may give erroneous results with equine sera as a result of immune complex formation.

Measurement of very low titres is less critical to potency test requirements for veterinary tetanus vaccines and serological response assays and tests likely to be introduced under WHO. Recent studies have shown that satisfactory titre estimates can be obtained for tetanus antisera by passive haemagglutination and direct ELISA as well as more sophisticated methods such as TOBI or MAb-based capture ELISA. It is important to recognise that obtaining satisfactory results by one laboratory is no guarantee that another laboratory using similar methods will also achieve satisfactory results. For the simpler methods the purity of the sensitising antigen is critical and a partial re-validation is needed whenever a new lot of purified antigen is introduced. In this respect methods that depend upon a monoclonal antibody for their specificity enjoy a considerable advantage. Such antibodies have been produced against the majority of clostridial toxins by various laboratories. It would be extremely valuable if a central clearing house could be set up to facilitate the collection and maintenance of relevant hybridomas and the production and distribution of the antibodies.

In contrast to tetanus, both direct ELISA [15] and passive haemagglutination [14] have failed to reflect diphtheria TN titres of human and laboratory animal antisera, whereas both TOBI [18] and capture ELISA [15] yielded satisfactory results. However, in the case of diphtheria it is arguable that no serological test can match the assurance provided by the availability of an in vitro indicator [9] for the TN test.

Only limited data are available on other clostridial toxins and it is not clear how easy it will be to develop tests capable of reflecting TN titres satisfactorily. Direct ELISA for the titration of rabbit sera against most of the veterinary toxoids have been published [19] but more extensive validation data are needed to determine whether they remain reliable over a wider range of test sera. The only MAb-based test for veterinary clostridia was a competition assay for *Cl. perfringens* epsilon antitoxin [21] but there is a clear opportunity for monoclonal-based TOBI tests for *Cl. perfringens* beta and epsilon [10], *Cl. novyi* alpha [20] and *Cl. botulinum* C & D [22].

Direct replacement of the challenge tests for *Cl. botulinum* C & D vaccines is more difficult because of the extremely low titres associated with resistance to the challenge levels used. It may therefore be necessary to consider the use of larger dose levels or a two-dose vaccination schedule to achieve measurable titres. The use of a MAb-based TOBI test for the antitoxin would be worth investigating because of its potential for high specificity and sensitivity.

Table 3 shows the availability of in vitro indicators for TN tests for *Cl. perfringens* alpha, *Cl. perfringens* epsilon, *Cl. septicum* alpha, *Cl. novyi* and perhaps *Cl. sordellii* antitoxins. In view of this, the ease of application of direct ELISA and in the majority of cases the prospect of highly specific MAb-based TOBI and capture assays, it is difficult to see why the use of in vivo TN tests for release of human and veterinary toxoid vaccines should not be superseded over the next few years.

Laboratory Reference Preparations

In vitro tests for antitoxins should, like their in vivo counterparts, include a laboratory reference preparation which has been calibrated directly or indirectly against the relevant International Standard Antitoxin by the TN test. This is particularly important in absolute response tests such as those prescribed for almost all toxoid vaccines in the Americas and for veterinary clostridials in Europe. Laboratory reference preparations should, ideally, be representative pools of test sera. Unlike the reference preparations for in vivo tests, they should not be hyperimmune equine preparations for two reasons. The first is the obvious fact that most ELISA methods include an anti-species conjugate which cannot operate for both test and standard sera. The second is that even in those tests such as Passive Haemagglutination, in vitro TN and TOBI, which do not depend upon the recognition of bound antibody by its species of origin, the difference in avidity between equine reference preparations and test sera is large enough to distort the estimate of titre in the latter extensively. A similar distortion occurs when sera are titrated by the in vivo TN test, resulting in a progressive lowering of the estimate of titre as the level

Table 3: Availability of alternatives to in vivo TN and Challenge tests for antitoxins generated in vaccine potency tests.

Pathogen species	Major toxin	In vitro TN	Serology	Reported MAbs
C. diphtheriae	diphtheria	VERO cells	TOBI Capture ELISA	Yes Yes
Clostridium tetani	tetanus	No	TOBI Direct ELISA Competition Capture ELISA Haemagglutination	Yes N/A N/A Yes
Cl. perfringens A	alpha	Haemolytic Lecitho-vitellin		N/A N/A
Cl. perfringens C	beta	No	TOBI (comp)*	Yes
Cl. perfringens D	epsilon	MDCK cells	Direct ELISA Competing ELISA TOBI	N/A Yes Yes
Cl. septicum	alpha	VERO cells	Direct ELISA	No
Cl. novyi A & B	alpha	VERO cells	Direct ELISA (tobi, comp)*	Yes
Cl. sordellii	Two toxins	cytopathic	None	No
Cl. botulinum A, B, E		No	?	Yes
Cl. botulinum C, D		No	(TOBI, comp)*	Yes

* Opportunities to develop monoclonal antibody based tests are bracketed.

of test is decreased. In compendial tests the extent of this distortion is kept sensibly constant by controlling the concentrations of toxin and antitoxin and the number of indicating doses present in the test mixture. Cell culture tests tend to exaggerate the depression of titre seen in the animal test whereas ELISA tends to overestimate titres while passive haemagglutination may overestimate them by up to 10 times. Use of a reference preparation which is typical of the test samples but which has been carefully calibrated by TN ensures that the in vitro test results obtained will be directly comparable with TN results that would have been obtained on the same test sera.

Reference preparations are less crucial for the titration of sera derived from comparative assays. They should nevertheless be included to ensure that the results of titrations carried out on different plates or on different days are comparable and can be included safely in a single vaccine assay.

Although some laboratories rely on a single dilution of test antiserum to determine its titre by ELISA, it is generally advisable to use a range of dilutions of both test and reference antisera covering the linear portion of the regression of log. optical density on log. serum dilution. The potency of each serum can then be determined by the same method as is used to estimate the relative potency of the test vaccine from the serological response data.

Although there is no difference in principle between the neutralisation of injected toxin by circulating antitoxin and the measurement of the antitoxin content of a serum in vitro, the levels of response that are optimal for the two methods of enumeration are different. Estimates of potency for combined vaccines such as DPT, in which the pertussis component is an adjuvant for toxoids, relative to monovalent standards, are higher when large immunisation and challenge doses are used [23]. This may result in apparent differences in estimates of potency when the toxoid components of DTP are assayed against their monovalent standard preparations using toxin challenge and serum titration methods. These apparent discrepancies are not peculiar to the alternative methods but provide further support for the introduction of a standard DTP preparation which would resolve this problem as well as that of the dual criteria for tetanus in DTP.

Replacement of toxin challenge by in vitro measures of the antitoxin response in comparative assays and tests, as well as avoidance of the use of animals for titration purposes, increases the amount of useful information available from each test animal. This increase makes it possible to achieve a given level of precision with only half the animals required for a toxin challenge assay. In non-comparative potency tests on veterinary clostridial vaccines, the availability of cheap in vitro alternative methods of titration could facilitate a restructuring of the whole test method. In the current test eight or ten rabbits are vaccinated and bled but because of the high cost of TN tests on the several antitoxin species elicited by multicomponent vaccines, the sera are pooled and only the single pool is titrated. A high proportion of the information available from the rabbits is lost as a result of this procedure. Almost all the in vitro methods of titration – in vitro TN, ELISAs, TOBI and passive haemagglutination – are performed in 96 well plates on which ten sera can be titrated almost as cheaply as one pool. The feasibility of titrating individual rabbit sera offers a prospect of making the test comparative by treating half the rabbits with a reference preparation, thereby eliminating the effects of animal strain and diet on the outcome of the potency test.

Prospects for Complete Elimination of Potency Tests

The foregoing measures offer an opportunity of reducing the numbers of animals used in potency tests of toxoid vaccines to a fraction of those used at present. Furthermore the severity of the procedures to which those animals are exposed can be greatly reduced. The establishment of valid tests for potency which avoid the use of animals entirely is likely to prove more difficult.

Bacterial toxoids, unlike polysaccharides, the potency of which is assessed in terms of antigenic mass, vary in the extent to which the epitopes have been damaged during toxoiding, in the efficiency with which they have been combined with adjuvant and in the positive or negative effects of other components in combined vaccines [23]. Any adequate method of potency prediction would have to measure these factors by in-process controls and take the results into account, preferably by a mathematical formula leading to a single numerical potency estimate.

The replacement of in-process animal tests has received little attention because of the small numbers of animals required and their absence from most compendia.

Although the LF test is used as a convenient measure of antigen content for most toxins and toxoids, it is often unreliable, particularly for some of the veterinary clostridia. It is therefore good practice to measure L + titres of toxins and total combining power (TCP) of toxoids as a more direct measure of the epitopes which induce protection. The L + test is a direct TN test titration, the results of which can be reproduced readily by in vitro TN methods for toxins which are cytopathogenic or haemolytic. ELISAs based on binding of neutralising monoclonal antibodies have been found to reflect the L + titres of tetanus toxins satisfactorily.

The TCP test is performed as a TN back-titration in which varying volumes of the test toxoid are allowed partially to neutralise a fixed dose (2 units) of antitoxin. The unreacted antitoxin is then exposed to a fixed dose of toxin (1 unit) and inoculated into animals to determine whether the original volume of toxoid neutralised more or less than 1 unit of the antitoxin. TCP tests are important not only as an index of quantity of effective antigen but also, in conjunction with the LF test, as an index of the immunogenic integrity of the toxoid. In vitro TN TCP tests work well for toxoids derived from haemolytic toxins but cell culture indicators suffer interference from the high non-specific cytopathogenicity of many clostridial cultures and veterinary vaccines. Since the concentration of reagents in TCP tests has to be kept high, it is necessary to transfer volumes of 10 µl or less of the reaction mixtures to wells containing 100 µl of cell culture. Development of ELISA tests to reflect the results of TCP tests has also proved difficult. However, simple ELISA methods based upon the binding of neutralising monoclonal antibodies referred to above, although not reflecting TCP values directly, would also be sensitive to the quality and quantity of unimpaired protective epitopes and might provide an alternative index of antigen integrity.

Charge and polymerisation of the toxoid molecule are other qualities of toxoids which has been found to relate to immunogenicity. Both of these factors affect the electrophoretic mobility of toxoid in gels under alkaline pH and it was shown by Lingood et al [25] that slow rates of migration under these conditions are associated with high immunogenicity.

The nature and quantity of aluminium adjuvant present in toxoid vaccines profoundly affects their potency but since each manufacturer should conform to an exact specification, a demonstration that the aluminium content is exactly equal to that of the original proving batches for the formulation should be sufficient. Achievement of satisfactory binding between antigen and adjuvant is also important and much less automatic. Holt [26] showed more than fifty years ago that the resistance of adsorbed diphtheria toxoid to elution with serum proteins and its potency increased during the months following the blending of adsorbed vaccine. Resistance to elution can be conveniently measured by incubation of the residue from centrifuged vaccine with 10% BSA overnight and determining the antigen content of the supernatant. It is convenient to capture the antigen with hyperimmune equine antitoxin, to eliminate interference from BSA. Bound antigen can then be recognised by appropriate monoclonal antibodies. Liberation of more than normal quantities of the bound antigen under these conditions is associated with suboptimal potency.

The other major determinants of potency in combined toxoid vaccines are the adjuvant or competitive effects of other vaccine components. In DTP both toxoids are potentiated by whole cell pertussis which contains two adjuvants, incompletely inactivated pertussis toxin and endotoxin. Van der Gun (this volume) may shed some light on the impact of pertussis components on estimates of potency.

The interactions of other components in the more complex blends prepared against the veterinary clostridia result in reduced responses relative to monovalent vaccines and are probably too complex to quantify.

In conclusion, it is probably possible to quantify all the factors which govern the potency of monovalent toxoid vaccines. It may be possible to measure the factors which determine the potency of the toxoid components of DTP by in vitro methods but determination of a means of linking all those elements to derive a single estimate of potency would require considerable effort. It is doubtful whether the problem could be solved for the multicomponent veterinary clostridia.

Specific Toxicity

Opportunities to replace specific toxicity tests are limited to toxoids derived from toxins with defined in vitro activities, i.e. diphtheria, *Clostridium perfringens* alpha and epsilon, *Cl. novyi* alpha, *Cl. septicum* alpha and perhaps *Cl. sordellii.* The main difficulty with the cell culture indicators relates to the high levels of non-specific cytopathogenicity in many crude toxoid and vaccine preparations.

These difficulties can be resolved in two ways, by dialysis [28] or by dilution. Dialysis has the advantage that it is applicable to all preparations irrespective of the relative levels of toxoid and cytopathogenic activity. The difficulty is the possibility that specific toxicity might be sequestered during the dialysis or diafiltration process. Because safety is at stake very careful validation with marginally toxic material will be needed to show that such toxicity is quantitatively retained during processing. The possible use of dilution is dependent on the extreme sensitivity of VERO cells to diphtheria, *Cl. septicum* alpha and *Cl. novyi* alpha toxins. The sensitivity of MDCK cells for *Cl. perfringens* epsilon toxin is not sufficient for toxicity testing. For diphtheria toxoids at vaccine strength it is possible to dilute the toxoid to the point where non-specific cytopathogenicity is eliminated and still retain sensitivity equal to that of the in vivo test. Cell culture toxicity tests on diluted toxoids are only applicable to bulk toxoids and not to final vaccines for *Cl. septicum* and novyi alpha toxoids. Any toxicity tests performed on diluted samples would need to be performed in duplicate in the presence or absence of blocking antitoxins to avoid confusion with abnormal levels of non-specific toxicity.

Conclusions

Although the major obstacles to the replacement of TN tests with valid alternatives and a reduction in the numbers of animals needed to test the potency of toxoid vaccines can be resolved, many difficulties remain. Some are problems inherent in the in vivo test methods which have come to light when in vitro tests yielded different results. Many arise from the use of standard and reference preparations which differ too much from the test materials, e.g. the use of hyperimmune equine sera to standardise tests on non-avid sera from laboratory animals and the use of monovalent toxoid preparations for assays of DTP. The introduction of a DTP standard vaccine would do more than anything to facilitate the successful introduction of valid alternative methods in the testing of DTP.

To maximise the effectiveness of the change from in vivo to in vitro methods for measuring the response of animals to vaccines, the following measures are proposed:

1. Provision of an International Standard DTP preparation.
2. Production of homologous laboratory reference antitoxin preparations which have been calibrated against existing standards by TN. These would not need to be provided centrally.
3. Provision of an international facility to collect, maintain and distribute monoclonal antibodies that are of use in vaccine control tests.
4. A restructuring of veterinary clostridial potency tests to make full use of the additional data available as a result of the use of alternative titration methods.

Complete elimination of animal testing for toxoid vaccines is far from imminent although some progress may be possible with monovalent vaccines. Replacement of specific toxicity tests by in vitro methods is possible in some cases but would require extremely careful validation and control.

References

1 Glenny AT, Llewellyn-Jones M: The intracutaneous method of testing diphtheria toxin and antitoxin. J Path Bact 1931;34:143-156.

2 Glenny AT, Barr M, Llewellyn-Jones M, Dallin T, Ross HE: Multiple toxins produced by some organisms of the *B. welchii* group. J Path Bact 1933;37:53-74.

3 Batty I, Glenny AT: Titration of *Cl welchii* toxins and antitoxins. J Exp Path 1947;28:110-126.

4 Glenny AT, Llewellyn-Jones M, Mason JH: The intracutaneous method of testing the toxins and antitoxins of the «Gas Gangrene» organisms. J Path Bact 1931;34:201-211.

5 Boyden SV: Adsorption of proteins on erythrocytes treated with tannic acid and subsequent haemagglutination by anti protein sera. J Exp Med 1951:93;107-120.

6 Manclark CR (ed): Proceedings of an Informal Consultation on the World Health Organisation Requirements for Diphtheria, Tetanus, Pertussis and Combined Vaccines. Department of Health and Human Services, United States Public Health Service, Bethesda Maryland. DHHS publication, n° (FDA) 91-1174, 1991.

7 Kreeftenberg JG: Report of an informal meeting about alternative methods for the potency control of the diphtheria and tetanus components in vaccines. Develop Biol Stand 1986;65:261-267.

8 Knight PA: Guidelines for performing one-dilution tests for ensuring that potencies of Diphtheria and Tetanus containing vaccines are above the minimum required by WHO. World Health Organisation 1989;WHO BS/89.1618.

9 Kreeftenberg JG, van der Gun J, Marsman FR, Sekhuis VM, Bhandari SK, Maheshwari SC: An investigation of a mouse model to estimate the potency of the diphtheria component in vaccines. J Biol Stand 1985;13:229-234

10 Knight PA, Queminet J, Blanchard JH, Tilleray JH: In vitro tests for the measurement of clostridial toxins, toxoids and antisera. II. Titration of *Clostridium perfringens* toxins and antitoxins in cell culture. Biologicals 1990;18:263-270.

11 Knight PA, Tilleray JH, Queminet JH: In vitro tests for the measurement of veterinary clostridial toxins, toxoids and antisera. I. Titration of *Clostridium septicum* toxins and antitoxins in cell culture. Biologicals 1990;18:181-189.

12 Batty I: Unpublished data.

13 Knight PA, Burnett C, Whitaker AM, Queminet J: The titration of clostridial toxoids and antisera in cell culture. Dev Biol Stand 1986:64;129-136.

14 Sheffield F, Perkins FT: The responses of guinea-pigs to the diphtheria toxoid component of combined antigens of different formulations. Symposium Series Immunobiological Stand 1967;7:121-128.

15 Knight PA, Tilleray JH, Queminet J: Studies on the correlation of a range of immunoassays for diphtheria antitoxin with the guinea-pig intradermal test. Dev Biol Stand 1986;64:25-32.

16 Simonsen O, Schou C, Heron I: Modification of the ELISA for the estimation of tetanus antitoxin in human sera. J Biol Stand 1987;15:143-157.

17 Hendriksen CFM, van der Gun J, Nagel J, Kreeftenberg JG: The toxin binding immunoassay as a reliable in vitro alternative to the toxin neutralisation test in mice for the estimation of tetanus antitoxin in human sera. J Biol Stand 1988;16:287-297.

18 Hendriksen CFM, van der Gun J, Kreeftenberg JG: The use of the toxin binding inhibition (TOBI) test for the estimation of the potency of the diphtheria component of vaccines. J Biol Stand 1989;17:241-247.

19 Wood KR: An alternative to the toxin neutralisation assay for the potency testing of the *Clostridium tetani, Clostridium septicum, Clostridium novyi* Type B and *Clostridium perfringens* Type D epsilon components of multivalent sheep vaccines. Biologicals 1991;19: 281-286.

20 Pietrzykowski E, Cox J, Zachariou M, MacGregor A: Development of an enzyme immunoassay for the detection of *Clostridium novyi* Type B antitoxin. Biologicals 1991;19 :293-298.

21 Sojka MG, White VJ, Thorns CJ, Roeder PL: The detection of *Clostridium perfringens* epsilon antitoxin in rabbit serum by monoclonal antibody based competition ELISA. J Biol Stand 1989;17:117-124.

22 Oguma K, Syuto B, Kubo S, Iida H: Analysis of antigenic structure of *Clostridium botulinum* Type C1 and D toxins by monoclonal antibodies. In Monoclonal Antibodies against Bacteria. Vol II. Academic Press 1985.

23 Knight PA: Dose response relationships of diphtheria and tetanus components in combined and single vaccines. Symp Series Immunobiol Stand 1967;7:235-242.

24 Knight PA: Are potency tests for tetanus vaccines really necessary? Dev Biol Stand 1986;64:39-45.

25 Lingood FV, Stevens MF, Fulthorpe AJ, Woiwod AJ, Pope CG: The toxoiding of purified diphtheria toxin. Brit J Exp Path 1963;44: 177-188.

26 Holt LB: Developments in diphtheria prophylaxis. Heinemann. London, 1950.

27 Abreo CB, Stainer DW: A tissue culture assay for diphtheria toxicity testing. Dev Biol Stand 1986;64:33-37.

Dr. P.A.Knight, c/o Mrs W. Leyshon, Biologicals QA, Bld 114, Wellcome Research Laboratories, Langley Court, Beckenham, Kent BR3 3BS, UK

Brown F, Cussler K, Hendriksen C (eds): Replacement, Reduction and Refinement of Animal Experiments in the Development and Control of Biological Products.
Dev Biol Stand. Basel, Karger, 1996, vol 86, pp 199-206

Validation of the Toxin-Binding Inhibition (ToBI) Test for the Estimation of the Potency of the Tetanus Toxoïd Component in Vaccines

J. van der Gun, A. Akkermans, C. Hendriksen, H. van de Donk

National Institute of Public Health and Environmental Protection, Bilthoven, The Nertherlands

Abstract: A validation study has been performed to determine the suitability of the toxin binding inhibition (ToBI) test for the serological estimation of the potency of the tetanus toxoïd component in vaccines. 37 Murine serum pools over a wide range of antibody levels were titrated in both toxin neutralization (TN) and ToBI test. A good correlation was found between both assays. Sixteen DPT-polio, twelve DT-polio and seven T vaccines were tested in the mouse lethal challenge test and the in vitro serological test, using the ToBI test for determining vaccine-induced tetanus antibodies. For all three types of vaccine a statistically valid correlation between both assays was found. However, for two batches of DPT-polio vaccine an «overestimation» of the tetanus potency was observed in the serological assay compared to the challenge assay. This phenomenon could not be explained by the difference in immunization period nor by misinterpretation of the ToBI test of DPT-polio-induced antibodies. In the LPF test high LPF activity was observed for the deviating DPT-polio vaccines. Therefore, the effect of pertussis toxin (PT) on the potency of the tetanus component in the serological assay was examined. The addition of 2 μg of PT to a «normal» DPT-polio vaccine resulted in a nearly twofold increase of the tetanus potency. It was concluded that pertussis toxin has a vaccine dose-dependent adjuvant effect on the potency of tetanus toxoïd resulting in high potency values when determined by ToBI procedure. It is unclear how these findings should be interpreted with respect to the behaviour of such vaccines in man.

Introduction

For the potency estimation of the tetanus toxoid component in monovalent or combined vaccines the mouse challenge test has been in use for many years. The test procedure is described in the European Pharmacopoeia [1] and by WHO [2]. For scientific and animal welfare reasons, efforts have been made to develop methods in which the lethal challenge procedure can be replaced by a serological method. Huet et al [3] and Nyerges et al [4] have described a haemagglutination test to determine vaccine-induced antibody levels. Several authors [5-7] have described an ELISA for the determination of tetanus antibodies. In our laboratory Hendriksen et al [8] developed the in vitro toxin-binding inhibition (ToBI) test as an alternative to the in vivo toxin neutralization assay. To estimate the validity of the ToBI test as an alternative to the lethal challenge procedure, two points were considered to be of particular importance. First the ability of this test to correlate with the in vivo toxin neutralization test and second the correlation of potency values and fiducial limits of vaccines obtained with the lethal challenge procedure and the serological procedure using the ToBI test. This paper describes the validation of the ToBI test as a serological procedure for the determination of the potency of the tetanus toxoid component in vaccines as well as the experiences gained with two deviating batches of DPT-polio vaccine.

Materials and Methods

Vaccines

The DPT-polio, DT-polio and tetanus vaccines used in this validation study were routinely produced aluminium-adsorbed lots from the National Institute of Public Health and Environmental Protection (RIVM) in the Netherlands. The Dutch reference vaccine Ta 75/1 is a lyophilised aluminium adsorbed product.

Animals

All animals used were male and female, outbred RIVM: NIH mice, from the RIVM breeding colony. The animals weighed 10-14 g and were randomly distributed among the cages at the start of the experiment. Within an experiment animals of only one sex were used.

Toxin Neutralization test

The toxin neutralization test was performed according to the procedure described by Hendriksen et al [8].

Lethal challenge test in mice

Tests were made according to the method described in the WHO Technical Report Series TRS 800.

Serological test using the ToBI procedure

Groups of eight mice were immunized subcutaneously with two-fold dilutions of the various vaccines and the reference preparation. Five weeks after vaccination all animals were bled individually by bleeding from the orbital plexus under halothane narcosis. Serum was separated from the clot and stored at -20°C until titrated for tetanus antibodies. Serum samples were randomly distributed over the plates. Ten 2-fold dilution series of each serum were made in phosphate buffered saline (PBS, pH 7.2) in round-bottomed plates in 100 µl volumes. Thereafter 20 µl (0.1 LF/ml) of purified tetanus toxin (Statens Serum Institute, Denmark) were added to each well of the microtitre plate except the wells of column 12 which served as a negative control (PBS). Column 11 was used as a positive control containing toxin only. Plates were gently shaken and incubated overnight at 37°C in a humid atmosphere. Next day, 100 µl of the pre-incubation mixtures were transferred to the corresponding wells of an immunoassay plate coated overnight at 37°C with equine anti-tetanus antitoxin (1 AU/ml). Plates were incubated for two hours at 37°C. After washing with tapwater containing 0.05% Tween 80, plates were incubated with 100 µl per well peroxidase-labelled equine anti-tetanus IgG (lot 32-33, RIVM) for 1.5 hours at 37°C. Finally the binding of toxin was visualized by the addition of 100 µl of Tetramethyl benzidine substrate. After 10 minutes the reaction was stopped by the addition of 100 µl of 2M H_2SO_4. Absorbance values were measured at 450 nm using an automatic micro-plate reader (Biotek EL312). By using a computer programme (TOBISCOR, RIVM) for each serum the antibody titre was expressed as a score, being the number of the last well with an optical density below the value representing 50% (OD_{50}) of the maximum optical density. The maximum optical density is defined as the difference between the mean OD of the positive control (PBS + tetanus toxin) and the negative control (PBS only).

Results

To establish the correlation between the TN test and ToBI test, a total number of 37 murine serum pools ranging in antibody titre from 0.01-1.28 AU/ml were titrated in both models. The results in Table 1 confirm our previous conclusion based on human serum samples that there is good agreement between the two methods. In a subsequent experiment the effect of difference in the time lapse between vaccination and bleeding was evaluated. Groups of eight mice were vaccinated with various dilutions of the Dutch reference tetanus toxoid. At three, four and five weeks respectively, groups of eight animals were bled and tetanus antibody levels were assessed by ToBI test. A relatively high antibody response was obtained five weeks post vaccination. The subcutaneous route proved to be satisfactory and was also selected for practical reasons. The suitability of the ToBI test to estimate potency values has been investigated for sixteen DPT-polio, twelve DT-polio and seven T vaccines by comparing ToBI results with those obtained in the lethal challenge procedure (Tables 2, 3, 4). A statistically significant correlation was demonstrated between the results obtained in the ToBI model and in the challenge test. The correlation coefficient based on all vaccines was 0.89. The ratio between ToBI and lethal challenge calculated for each vaccine type is 1.025 for DPT-polio

Table 1: Tetanus antitoxin scores of pooled serum samples (n=37) as determined in the in vitro ToBI test and the in vivo TN Test.

ToBI test	**TN test antitoxin scores**								
antitoxin scores	**0**	**1**	**2**	**3**	**4**	**5**	**6**	**7**	**8**
0	2	1							
1	1	2							
2		1							
3				1	1				
4				1	4	2			
5						5	3		
6							4	3	
7							1	1	2
8									2

Table 2: Tetanus potencies of DPT-polio vaccines performed with the ToBI test and lethal challenge test.

Vaccine	ToBI-test		Lethal challenge test	
	IU/ml	95% interval	IU/ml	95% interval
DPT-polio A	225	(174-297)	359	(247-544)
DPT-polio B	163	(125-215)	203	(157-265)
DPT-polio C	371	(275-530)	246	(187-335)
DPT-polio D	320	(232-474)	381	(258-573)
DPT-polio E	547	(400-809)	309	(193-483)
	381	(273-575)	249	(165-369)
DPT-polio F	493	(346-777)	255	(189-344)
	540	(343-1019)	309	(207-459)
DPT-polio G	497	(359-747)	311	(213-451)
	528	(362-868)	328	(236-454)
DPT-polio H	222	(161-317)	231	(138-359)
DPT-polio I	247	(173-374)	289	(165-438)
DPT-polio J	249	(179-366)	210	(145-302)
DPT-polio K	354	(241-556)	359	(238-568)
DPT-polio L	222	(171-292)	248	(165-359)
	332	(213-529)	279	(184-423)
DPT-polio M	131	(96-176)	240	(176-324)
DPT-polio N	333	(253-439)	294	(198-433)
DPT-polio O	237	(171-342)	197	(130-285)
DPT-polio P	307	(210-492)	204	(147-293)
Mean rel. intv.	100%	(72%-209%)	100%	(68%-217%)

p = 0.09 ratio ToBI/challenge = 1.025 (0.996-1.055)

Table 3: Tetanus potencies of DT-polio vaccines performed with the ToBI test and lethal challenge test.

Vaccine	ToBI-test		Lethal challenge test	
	IU/ml	95% interval	IU/ml	95% interval
DT-polio A	73.4	(58.2-92.7)	78.5	(51.0-122)
DT-polio B	118	(91.9-155)	80.8	(53.2-120)
DT-polio C	94.3	(73.1-122)	105	(69.3-158)
DT-polio D	109	(81.0-148)	90.4	(59.1-138)
DT-polio E	102	(76.8-143)	140	(96.7-201)
DT-polio F	132	(99.1-179)	99.3	(70.0-141)
DT-polio G	151	(120-195)	87.7	(56.1-122)
DT-polio H	154	(122-198)	81.8	(60.6-111)
DT-polio I	92.4	(66.3-129)	109	(80.8-148)
DT-polio J	102	(73.0-142)	99.0	(68.8-142)
DT-polio K	90.8	(68.0-125)	113	(71.1-172)
DT-polio L	92.0	(69.0-127)	114	(79.2-169)
Mean rel. intv.	100%	(76%-177%)	100%	(68%-215%)

p = 0.40 ratio ToBI/challenge = 1.019 (0.977-1.062)

vaccine, 1.019 for DT-polio vaccine and 0.997 for T vaccine. Despite the good overall correlation for two batches of DPT-polio vaccine (lots F and G), a relatively high potency was found in the ToBI model. Vaccine DPT-polio G is produced according to a modified procedure. As the deviating results are exclusive for these two batches, both products were investigated. In the first experiment the effect of

Table 4: Tetanus potencies of T vaccines performed with the ToBI test and lethal challenge test.

Vaccine	ToBI-test		Lethal challenge test	
	IU/ml	95% interval	IU/ml	95% interval
TA	265	(206-337)	438	(323-595)
TB	401	(308-525)	234	(157-349)
TC	313	(235-427)	378	(118-1059)
TD	190	(137-252)	231	(145-367)
TE	312	(226-428)	405	(276-603)
TF	313	(236-413)	303	(205-368)
TG	330	(262-414)	228	(174-298)
Mean rel. intv.	100%	(76%-175%)	100%	(64%-304%)

p = 0.84 ratio ToBI/challenge = 0.997 (0.941-1.053)

Table 5: Relevant information on DPT-polio vaccines.

Vaccine	Potency	LPF[1]				
		MWG	16 OU	8 OU	4 OU	Bulk product
DPT-polio B	4.6 (2.1-9.7)	105%	21.200	24.300	14.800	87/89-HC-2
DPT-polio C	2.1 (1.2-3.5)	125%	16.800	14.900	15.400	87/89-HC-2
DPT-polio D	4.1 (2.0-8.6)	104%	22.900	22.500	22.100	89-HC-4
DPT-polio E	5.4 (3.7-7.8)	96%	23.800	26.300	19.400	89-HC-4
DPT-polio F	7.5 (4.1-14.1)	102%	30.500	21.700	22.900	89-HC-4
DPT-polio G	13.5 (4.0-49.5)	108%	48.900	27.700	24.200	PU 909/912
DPT-polio H	5.7 (3.2-10.4)	116%	21.600	16.600	15.200	89-HC-4
DPT-polio I	7.2 (4.2-12.2)	107%	23.600	18.300	15.000	89-HC-4
DPT-polio J	4.3 (2.4-7.5)	92%	21.500	16.400	20.200	89-HC-4
DPT-polio K	7.6 (5.2-11.2)	138%	18.400	14.900	14.400	89-HC-4
DPT-polio L	6.9 (3.2-14.9)	101%	24.800	20.800	16.200	90-HC-1

[1] = number of circulating leucocytes per mm^3.

the immunization period on the potency result was investigated. Five batches of DPT-polio vaccine were tested in the ToBI test and lethal challenge test, both with an immunization period of three and five weeks. No clear difference was observed although the ToBI results after an immunization period of three weeks tend to be higher than after five weeks. To investigate the neutralising capacity of serum samples obtained after immunization with DPT-polio G, a toxin neutralization test on seven serum pools was performed in parallel with a ToBI test. The results, not shown, indicate that for the serum samples tested, ToBI does not overestimate the neutralising activity. Further investigations were directed at the composition of the deviating vaccines. It is known that the pertussis component present in the combined vaccine has a potentiating effect on the T component. Relevant information was therefore collected for all DPT-polio vaccines examined in the validation study and is summarized in Table 5. With respect to both DPT-polio vaccines, deviating results were obtained in the LPF test performed according to v. Straaten-v.d.Kappelle [9]. Both vaccines showed a relatively high LPF activity. Based on the LPF activity of the reference vaccine it was concluded that DPT-polio G had an «extra» LPF activity of approximately 2 µg of pertussis toxin (PT) with respect to

Table 6: Effect of PT on Tetanus potency estimated in the ToBI test.

Vaccine	Potency	(95% interval)	LPF 16 OU[1]
DPT-polio G	300	(181-488)	48.900
DPT-polio L	162	(92.2-264)	18.000
DPT-polio L+2 μg/ml PT	317	(193-513)	36.900

[1] = number of circulating leucocytes per mm³.

Table 7: Effect of PT on Tetanus potency at different stages of the immunisation period estimated in the ToBI test.

Vaccine	Potency (DPT-polio L=100%)			
	2 weeks	3 weeks	4 weeks	6 weeks
DPT-polio G	112	158	127	185
DPT-polio L	100	100	100	100
DPT-polio L+PT	103	128	165	195

other «normal» vaccines. To determine the effect of surplus PT on the tetanus potency as estimated in the ToBI model, 2 μg of PT (JNIH-5) were added to DPT-polio L. The PT-enriched vaccine (DPT-polio L+PT) was tested for T potency and LPF activity together with DPT-polio G and L. The results, summarized in Table 6, show that both potency and LPF activity of DPT-polio L increased twofold after the addition of 2μg of PT up to a level equal to DPT-polio G. In the same experiment the potency of DPT-polio G, L and L+PT was investigated with an immunization period of two, three, four and five weeks respectively. The results (Table 7) show that during the course of the immunization period the relative difference increases between DPT-polio L on the one hand and DPT-polio G and L+PT on the other hand. In Figures 1 and 2 the antibody response three and five weeks after immunization with DPT-polio G, L, L+PT and the Dutch reference toxoid are plotted. An immunization period-related difference between DPT-polio L and DPT-polio G and L+PT is observed due to presence of PT.

Discussion

The aim of this study was to evaluate the suitability of the ToBI test as a serological method to estimate the potency of the tetanus component in vaccines. The ToBI test proved to correlate well with the in vivo toxin neutralization test. The results show that a five week immunization period results in relatively high antibody titre. A significant correlation with the results obtained in the lethal challenge test was demonstrated. An overestimation of the potency of two batches of vaccine in the ToBI test has been observed and is most probably due to the relatively high LPF activity of both products. These findings raise the question of which result is relevant for the situation in man. Crucial in this respect is whether the protection in humans is higher for both deviating vaccines. No such information has been found in the public domain.

Provided consistency is established and/or the LPF value is shown not to exceed a set level the use of the ToBI test is indicated. When consistency in pro-

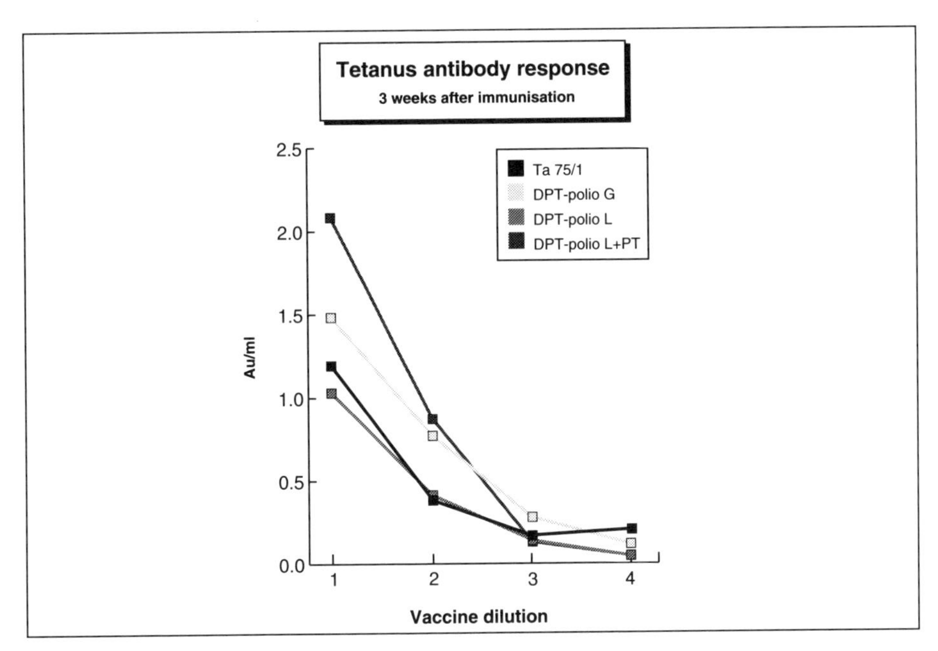
Tetanus antibody response
3 weeks after immunisation
Ta 75/1
DPT-polio G
DPT-polio L
DPT-polio L+PT
Au/ml
2.5
2.0
1.5
1.0
0.5
0.0
1
2
3
4
Vaccine dilution

Fig. 1.

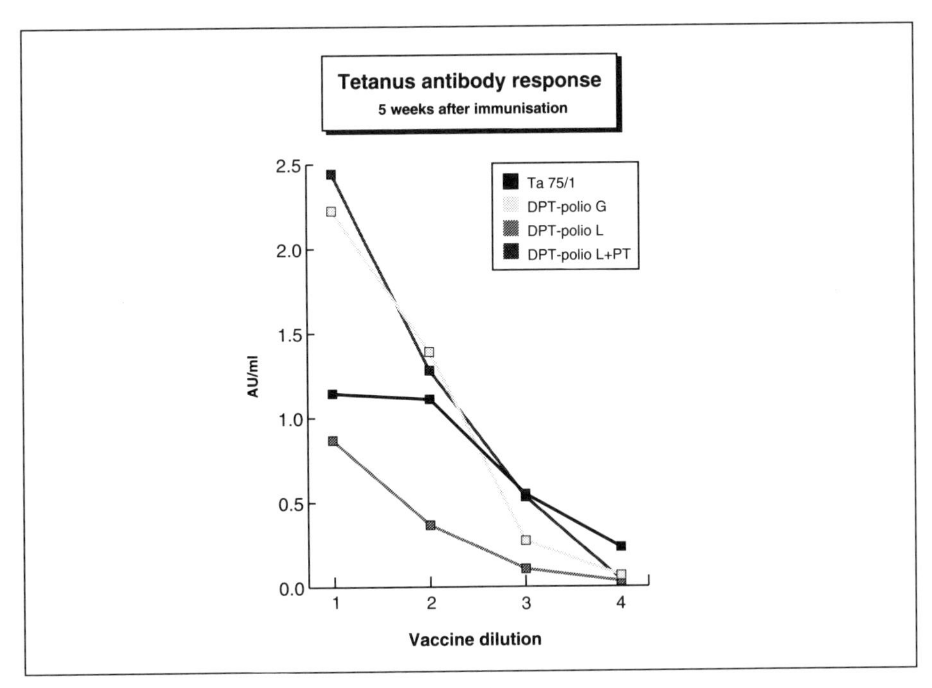
Tetanus antibody response
5 weeks after immunisation
Ta 75/1
DPT-polio G
DPT-polio L
DPT-polio L+PT
AU/ml
2.5
2.0
1.5
1.0
0.5
0.0
1
2
3
4
Vaccine dilution

Fig. 2.

duction and quality control is well established, the ToBI method is superior to the lethal challenge test both in precision and in reducing the number of animals.

References

1 European Pharmacopoeia: Methods for the assay of Tetanus vaccine (adsorbed) V 2.2.9. Document PA/PH/Ex. 3/T (92) 66, DEF, January 1993.

2 World Health Organization: Requirements for Diphtheria, Tetanus, Pertussis and combined vaccines, TRS 800, 1990.

3 Huet M, Relyveld E, Camps S: Methode simple de controle de l'activité des anatoxines tétaniques adsorbées. Biologicals 1990;18:61-67.

4 Nyerges G, Virag E, Lutter: The potency testing of diphtheria and tetanus toxoids as determined by the induction of antibody in mice. J Biol Stand 1983;11:99-103.

5 Manghi M, Pasetti M, Brero M, Deluchi S, di Paola G, Mathet V, Eriksson P:Development of an ELISA for measuring the activity of Tetanus toxoid in Vaccines and comparison with the Toxin Neutralization test in mice. J of Immunological Methods 1994;168:17-24.

6 Hagenaars A, van Delft R, Nagel J: Comparison of ELISA and toxin neutralization test for the determination of tetanus antibodies. J of Immunoassay 1984;5:1-11.

7 Simonsen O, Schou C, Heron I: Modification of the ELISA for the estimation of tetanus antitoxin in human sera. J Biol Stand 1987;15:143-157

8 Hendriksen C, van der Gun J, Nagel J, Kreeftenberg J: The Toxin Binding Inhibition test as a reliable in vitro alternative to the Toxin Neutralization test in mice for the estimation of tetanus antitoxin in human sera. 1988;16:287-297.

9 Van Straaten-Van De Kappelle I, Wiertz J, MArsman F, Borsboom D, van de Donk H, Kreeftenberg J: The Modified Leukocytosis Promoting Factor (LPF)-test: A valuable Supplement to the Mouse Weight Gain (MWG)-test in Toxicity Control of Whole Cell Pertussis Vaccine. 1992;20:277-282.

Dr. J.W. van der Gun, Laboratory for Control of Biological Products, National Institute of Public Health and Environmental Protection, P.O.Box 1, NL-3720 BA Bilthoven, The Netherlands

Brown F, Cussler K, Hendriksen C (eds): Replacement, Reduction and Refinement of Animal Experiments in the Development and Control of Biological Products.
Dev Biol Stand. Basel, Karger, 1996, vol 86, pp 207-215

Use of In Vitro Vero Cell Assay and ELISA in the United States Potency Test of Vaccines containing Adsorbed Diphtheria and Tetanus Toxoids

R.K. Gupta, G.R. Siber

Massachusetts Public Health Biologic Laboratories, Boston, MA, USA

Key words: Vero cell assay, ELISA, toxin neutralization, potency test, tetanus toxoid, diphtheria toxoid, avidity.

Abstract: Current United States (US) regulations for potency testing of vaccines containing adsorbed diphtheria and tetanus toxoids require in vivo toxin neutralization (TN) tests on the pooled sera of immunized guinea pigs. To reduce the number of animals required for testing, two in vitro tests have been evaluated, the Vero cell assay for diphtheria antitoxin and ELISA for IgG antibody to tetanus toxin; these have been correlated with in vivo TN tests. In the Vero cell method, diphtheria antitoxin titres of the guinea pig sera, obtained four weeks after immunization as per US potency requirements, were markedly dependent on the toxin dose level used in the assay. A toxin dose level termed the Lcd/1 dose (*limit* of cytopathic dose at 1 IU/ml) for Vero cells gave comparable estimates of antitoxin activity to the in vivo TN test performed at L+/1 dose of toxin. When lower dose levels of toxin were used in the Vero cells (Lcd/10 to Lcd/1000), diphtheria antitoxin levels in four weeks guinea pig sera were two to 11.7 times lower than with the Lcd/1 dose level. The most likely reason for these differences is that guinea pig sera at 4 weeks are of lower avidity than the equine antitoxin standard. Antibodies of low avidity bind antigen less well at low reactant concentrations. Therefore, to obtain similar estimates of diphtheria antitoxin in the Vero cell method and in vivo TN test, the use of a toxin dose for the Vero cell method similar to that for the in vivo TN test is suggested. Another alternative, in which any dose of toxin may be used for the Vero cell method, is the use of a reference guinea pig serum (calibrated in IU/ml by the in vivo TN test at L+/1 level of toxin) that has similar avidity or similar immunization status as the test sera (i.e. 4 week serum).
IgG antibodies to tetanus toxin in guinea pig sera were found early in the course of immunization when tetanus antitoxin could not be detected by TN test. Tetanus toxin IgG antibody levels of guinea pig sera calculated in IU/ml against an ELISA guinea pig reference serum (calibrated in IU/ml by TN test) depended upon the immunization status of the animals. To obtain similar estimates of tetanus antibodies in IU/ml by TN and ELISA, the ELISA reference guinea pig serum should have similar immunization status (and presumably similar avidity) as the test serum (i.e. six week serum). We propose that the Vero cell method and ELISA deserve further evaluation to determine whether they can replace in vivo TN tests for titration of diphtheria and tetanus antitoxins in the US potency test.

Introduction

The United States (US) potency test for vaccines containing adsorbed diphtheria and tetanus toxoids involves immunization of guinea pigs with half the total human dose of the vaccine and titrating pooled sera of at least four guinea pigs by the in vivo toxin neutralization (TN) tests for diphtheria antitoxin at four weeks and for tetanus antitoxin at six weeks [1-3]. The test requires a large number of animals, particularly for titration of antitoxins in the guinea pig sera. An in vitro method for titration of diphtheria antitoxin on Vero cells has been used for many years and has been found highly specific, reproducible and sensitive [4-7]. However, this method did not give results comparable to the in vivo TN method on the sera from immunized guinea pigs for the US potency test on diphtheria toxoid [8]. Several in vitro tests for measuring antibodies to tetanus toxoid (TT), including indirect haemagglutination [9-13], several versions of ELISA [14-21], and the toxin-binding inhibition test [22] have been reported to correlate well with the TN test, but are not used in the potency testing of vaccines. Recently, the World Health Organization proposed these in vitro methods for potency testing of tetanus and diphtheria toxoids [23].

In this study, we evaluated the Vero cell assay for the titration of diphtheria antitoxin and ELISA for measuring IgG antibodies to tetanus toxin in the sera of immunized guinea pigs as a replacement for the in vivo TN tests.

Materials and Methods

Animals

Female outbred mice (CD-1), weighing 17-20 g and guinea pigs (Hartley) weighing 250-350 g and 450-550 g, were purchased from Charles River, Wilmington, MA.

Reagents

Chromatographically-purified tetanus toxin for coating ELISA plates, crude diphtheria toxin containing 100 Limes flocculation Units (Lf)/ml for TN assays both in guinea pigs and Vero cells and partially purified tetanus toxin for L+/1000 dose level of TN test were from our laboratories. US Standards for tetanus and diphtheria antitoxins, and tetanus toxin for L+/100 level were from the Center for Biologics Evaluation and Research, FDA, Bethesda, MD. A reference anti-TT guinea pig serum for ELISA was prepared by pooling sera from six guinea pigs injected with 7.5 Lf of aluminium phosphate adsorbed TT and bled at six weeks. The pooled serum was calibrated in IU/ml by the TN test. Alkaline phosphatase-conjugated goat anti-guinea pig IgG, bovine serum albumin (BSA), Brij-35 and phosphatase substrate were from Sigma Chemical Co., St. Louis, MO. Vero cells originally obtained from the American Type Culture Collection, Rockville, MD (CCL#81), were preserved at passage level 7 in our Laboratories as cell stock. Usually cells from 15-20 further passages of our stock were used for the assay. Medium 199 (with Hanks' balanced salt solution and glutamine), foetal bovine serum (FBS) and other cell culture reagents were from Biowhittaker, Walkersville, MD.

Vaccines

Several licensed lots of adsorbed diphtheria-tetanus-pertussis (DTP) vaccine, containing 20 Lf/ml of diphtheria toxoid (DT), 10 Lf/ml of TT and eight protective units/ml of whole cell pertussis vaccine adsorbed to 1.36 mg/ml of aluminium phosphate, and several investigational lots of adsorbed diphtheria-tetanus-acellular pertussis (DTaP) vaccine, containing the same concentrations of tetanus and diphtheria toxoids and aluminium phosphate as in our licensed DTP vaccine, were used for immunizing guinea pigs.

Investigational preparations of TT of varying purity [24] were adsorbed on to various adjuvants including aluminium phosphate, calcium phosphate and stearyl tyrosine as described elsewhere [24] at a final concentration of 10 Lf/ml.

Immunization of guinea pigs

For diphtheria potency experiments, groups of eight guinea pigs were injected subcutaneously with 0.75 ml of undiluted DTP or DTaP vaccine (1/2 of total human dose) containing 15 Lf of DT according to the US potency method [1]. Four weeks after injection the guinea pigs were bled by cardiac puncture after anaesthesia and serum collected. Sera from four or more guinea pigs were pooled in equal volumes. Sera used for the Vero cell method were inactivated at 56°C for 30 minutes.

A group of six guinea pigs was injected subcutaneously with 0.75 ml of undiluted aluminium phosphate-adsorbed TT (7.5 Lf, 1/2 of total human dose) and the guinea pigs were bled at 9, 28, 42, 91, 182, 273 and 365 days after injection. The guinea pigs were boosted with 0.05 ml of undiluted vaccine (0.5 Lf) at 365 days and bled two weeks later. For tetanus potency experiments, groups of five guinea pigs were injected subcutaneously with 7.5 Lf TT on different adjuvants and were bled after six weeks [2, 24]. The sera from four or more guinea pigs were pooled in equal volumes.

In vivo Toxin Neutralization test

Determination of tetanus antitoxin

The TN test for tetanus antitoxin was performed at L+/100 and L+/1000 levels as described elsewhere [24, 25]. Briefly, tetanus toxin was diluted in 1% peptone water to 5-10 L+/100 or L+/1000 doses per ml. Then various dilutions of US standard tetanus antitoxin and serum samples were mixed with three doses of L+/100 or L+/1000 doses of toxin. The volume was made to 1.5 ml with normal saline. The toxin antitoxin or serum mixtures were incubated at room temperature (RT) for one hour and two mice from each mixture were injected subcutaneously with 0.5 ml. Mice were observed for five days for tetanic symptoms and deaths. The level of antitoxin was calculated against the standard and expressed in IU/ml.

Determination of diphtheria antitoxin

The in vivo TN test on pooled sera was performed at the L+/1 dose level of diphtheria toxin according to the US potency method [1]. The diphtheria antitoxin levels of most serum pools were tested at one level only (2 IU/ml); end points were not determined because of the high cost of testing serum samples at several dilutions in guinea pigs.

Enzyme linked immunosorbent assay for tetanus toxin IgG antibodies

Anti-TT IgG in guinea pig sera was determined by ELISA [19, 24] by coating microtitre plates (Corning Glass Works, Corning, NY) with purified tetanus toxin (5 µg/ml) at RT overnight. In column 2 reference guinea pig serum for ELISA was diluted serially and columns 3 to 12 were used for the two-fold dilution of serum samples. After two hours incubation and washing of plates, alkaline phosphatase labelled goat anti-guinea pig IgG was added. The plates were incubated at RT for two hours and washed. Finally, substrate was added and the plates were read on an ELISA reader at 405 nm. The content of TT IgG antibodies of serum samples was expressed in IU/ml against the reference serum calibrated in IU/ml by TN test.

Vero cell method for titration of diphtheria antitoxin

Guinea pig serum pools were assayed for diphtheria antitoxin by the Vero cell method at Lcd/1 dose level of diphtheria toxin as described [26]. Briefly, the unknown serum samples were serially diluted 1.5-fold to improve precision. Each serum sample and standard diphtheria antitoxin were tested in duplicate. For standard diphtheria antitoxin, five dilutions at 0.8, 0.9, 1.0, 1.1 and 1.2 IU/ml were made. A volume of 50 µl of these dilutions was transferred in appropriate wells. Appropriate cell and toxin controls were included on each plate. Then 50 µl of diphtheria toxin diluted in Medium 199 with 5% FBS containing Lcd/1 toxin dose were added to wells with serum samples, standard antitoxin and toxin control. The plate was incubated at room temperature for one hour and 100 µl of Vero cell suspension was added to each well. The plates were read after four days for cytopathic effect and the content of unknown samples was determined against standard diphtheria antitoxin.

Results and Discussion

Vero Cell Method for Titration of Diphtheria Antitoxin

We have described an in vitro Vero cell method for the titration of diphtheria antitoxin in the sera of guinea pigs immunized according to US potency method which gives estimates of diphtheria antitoxin similar to those obtained by the in vivo TN test [26]. The method is based on a *limit* test using a concentration of diphtheria toxin analogous to that used in the in vivo L+/1 TN test in guinea pigs. By testing the potency of diphtheria toxoid in 30 lots of licensed and investigational vaccines, we found that the Vero cell method at the Lcd/1 level of toxin and the in vivo TN test correlated well as judged by their ability to discriminate lots inducing ≥ 2 IU/ml (pass potency) versus <2 IU/ml (fail potency) (Table 1). However, when the Vero cell method was performed at lower concentrations of toxin, the diphtheria antitoxin concentrations of sera were lower [26]. At Lcd/10, Lcd/100 and Lcd/1000 dose levels of diphtheria toxin the mean diphtheria antitoxin concentrations of 12 serum pools was 2-, 4.7- and 11.3-fold, respectively lower than the mean concentration of diphtheria antitoxin determined at the Lcd/1 level. These discrepancies were attributed to the lower avidity of guinea pig sera at four weeks compared to the avidity of equine antitoxin used as a standard because sera from guinea pigs immunized twice did not show these discrepancies [26]. These observations are consistent with earlier findings on the avidity of antitoxins where the declining antitoxin activity of sera with decreasing concentration (termed the dilution ratio) has been used as a measure of the avidity of antibody responses [27-32].

Recently, we determined diphtheria antitoxin levels of individual sera from guinea pigs immunized with vaccines containing DT by the Vero cell method and found that the geometric means (GM) of the titres were lower than the titres measured by the in vivo TN test on pooled sera (Table 2). Nevertheless, titration of individual guinea pig sera would provide statistical analysis of the data which may be very useful in studying differences in the potency of DT while developing formulations for combined vaccines. Therefore, it may be desirable and feasible to titrate individual guinea pig serum samples in the US potency test of adsorbed diphtheria toxoid by the in vitro Vero cell method which was not practical with the in vivo TN test due to the use of large numbers of guinea pigs.

ELISA for Measuring IgG Antibodies to Tetanus Toxin

We have described an ELISA for measuring IgG antibodies to tetanus toxin in the sera of guinea pigs immunized according to the US potency method [19]. The

Table 1: Diphtheria antitoxin concentrations of pooled sera of immunized guinea pigs determined by the in vivo TN method at the L+/1 dose level and the in vitro Vero cell method at Lcd/1 dose level. Groups of eight guinea pigs were injected with 0.75 ml of undiluted vaccine containing aluminium phosphate adsorbed diphtheria toxoid and bled at four weeks.

Vaccine*	Diphtheria antitoxin levels (IU/ml) determined by	
	in vivo TN method	Vero cell method
DTP-1	4	4.0
DTP-2	<2	1.0
DTP-3	2.5	3.0
DTP-4	2.7	3.0
DTP-5	>2	2.5
DTP-6	>2	2.5
DTP-7	>2	2.5
DTP-8	>2	4.4
DTP-9	>2	2.5
DTP-10	>2	2.5
DTP-11	>2	4.4
DTP-12	>2	3.4
DTP-13	>2	3.4
DTP-14	>2	2.3
DTP-15	>2	2.3
DTaP-1	>2	3.4
DTaP-2	>2	5.1
DTaP-3	>2	4.2
DTaP-4	>2	3.4
DTaP-5	>2	3.4
DTaP-6	<2	1.5
DTaP-7	>2	3.4
DTaP-8	>2	3.4
DTaP-9	2	1.8
DTaP-10	>2	2.0
DTaP-11	>2	2.4
DTaP-12	>2	4.4
DTaP-13	>2	2.5
DTaP-14	>2	3.6
DTaP-15	>2	3.4

From Gupta et al: Biologicals 1994;22: 65-72.

* DTP, Diphtheria-tetanus-pertussis vaccine with whole cell pertussis vaccine.
DTaP, Diphtheria-tetanus-pertussis with acellular pertussis vaccine.

assay gives estimates of tetanus antibodies comparable to those measured by the TN test at six weeks after immunization as required by the US potency test (Table 3), showing a correlation coefficient of 0.88 ($P< 0.01$) between TN and ELISA titres. This was accomplished by using an ELISA reference serum obtained six weeks after immunization, similar to the test sera. The GM titres of individual serum samples were similar to the titres on the pooled sera. There were no statistically significant differences between TN and ELISA titres of 18 guinea pig serum pools ($P> 0.4$). Comparable estimates of TN and ELISA titres were obtained with sera from guinea pigs injected with different types of TT (standard TT, purity approximately 1500 Lf per mg protein nitrogen and chromatographically purified TT, purity approximately 2500 Lf per mg protein nitrogen) adsorbed to different adjuvants (aluminium phosphate, calcium phosphate and stearyl tyrosine) or injected as fluid. Therefore the vaccine formulation did not influence the correla-

Table 2: Diphtheria antitoxin concentrations of sera of immunized guinea pigs determined by the in vivo TN method at the L+/1 dose level and the in vitro Vero cell method at the Lcd/1 dose level. Groups of eight guinea pigs were injected with 0.75 ml of undiluted vaccine containing aluminium phosphate adsorbed diphtheria toxoid and bled at four weeks.

Vaccine	Diphtheria antitoxin levels (IU/ml) determined by	
Lot	in vivo TN method Pooled sera	Vero cell method GM (95% CI)*
1	3	1.90 (1.53-2.36)
2	3	1.64 (1.11-2.42)
3	≥4	2.33 (1.76-3.06)
4	≥4	2.44 (2.02-2.94)
5	≥4	2.10 (1.38-3.19)

*Geometric mean (95% Confidence Intervals).

Table 3: Tetanus antibody concentrations in sera of immunized guinea pigs determined by the in vivo TN method at the L+/100 dose level and ELISA. Groups of five guinea pigs were injected with 0.75 ml of various preparations of adsorbed tetanus toxoid and bled at six weeks.

Tetanus toxoid	Tetanus antibody levels (IU/ml) determined by		
	TN method	ELISA (IgG)	
Lot		Geometric mean	Pooled sera
1	16.0	8.8	10.8
2	10.0	6.1	7.0
3	20.0	11.8	16.0
4	5.0	3.3	4.3
5	5.0	3.0	3.3
6	7.0	5.0	7.3
7	4.0	3.2	2.8
8	4.0	3.3	2.7
9	8.0	4.9	4.6
10	13.2	16.7	15.0
11	13.0	5.2	15.0
12	4.0	2.3	3.6
13	4.0	4.1	3.6
14	5.0	5.2	7.0
15	4.5	7.4	6.8
16	7.0	9.3	10.4
17	2.0	2.3	2.6
18	4.0	2.8	3.9

tion between TN and ELISA titres of immunized guinea pig sera. Similar results were obtained with sera from immunized mice [19]. Other workers have also reported good correlation between TN and ELISA titres of human sera [14, 15]. Simonsen et al [17] found good correlation between ELISA and TN titres of sera from vaccinated human populations but no correlation with sera from persons having an incomplete vaccination history.

Tetanus-specific IgG antibodies estimated by ELISA were detected in the sera of guinea pigs on the ninth day post inoculation when there were no neutralizing antibodies (below the lowest detection level of 0.0025 IU/ml) (Table 4). Four weeks after inoculation the ELISA IgG titre (determined against six weeks ELISA reference serum) was still approximately eight times higher than the TN titre (Table 4). IgG antibodies which appeared during the early course of immunization may be of low avidity. These antibodies bound in ELISA but did not neutralize tetanus toxin. At nine and 28 days, the ratios between ELISA and TN titres of the sera were greater than one because the ELISA reference serum had higher avidity than the test sera before six weeks, whereas at 42 days (six weeks) the ratio was close to unity (Table 4) due to similar avidities of the ELISA reference serum and test sera. After six weeks, the ratio was less than unity showing lower avidity of the ELISA reference serum than the test sera. It is therefore proposed that to obtain similar TN and ELISA titres, the ELISA reference serum must be obtained at the same time after immunization as the test serum.

Conclusions

Two in vitro assays, the Vero cell method for titration of diphtheria antitoxin and the ELISA for the measurement of IgG antibodies to tetanus toxin gave estimates of antitoxins comparable to those measured by the in vivo TN tests in the sera of guinea pigs immunized according to the US potency method. The discrepancies between in vitro assays and in vivo TN tests result from the differences between avidity of the reference serum and unknown sera. To obtain similar estimates of antitoxins by the in vitro assays and the in vivo TN tests, the use of a reference serum with avidity or immunization status similar to the test serum is proposed. For the potency test of vaccines containing adsorbed diphtheria toxoid, an analogous dose of toxin used in the in vivo TN test is suggested for use in the Vero cell method. In addition to saving on animals and cost, the in vitro assays have several advantages over the in vivo tests: 1) the response of individual animals can be evaluated by the in vitro assays, permitting statistical analyses to assess differences between vaccine formulations, 2) sera can be titrated to end point which is not usually done for the in vivo TN tests, thus providing information on the precise

Table 4: Tetanus toxin antibodies in the pooled sera of guinea pigs injected with 0.75 ml (7.5 Lf) of aluminium phosphate adsorbed tetanus toxoid. Guinea pigs were bled at different intervals and tetanus toxin antibodies were measured by toxin neutralization test and ELISA.

Days after injection	Tetanus toxin antibodies (IU/ml) by		Ratio ELISA: TN titres
	TN test	ELISA	
9	<0.0025	0.38	>152
28	0.6	4.92	8.20
42	7.0	5.87	0.84
91	16.5	6.94	0.42
182	18.8	5.73	0.30
273	8.0	5.17	0.65
365	8.0	3.83	0.48
379*	20.0	7.05	0.35

* Two weeks after a boost of 0.5 Lf of aluminium phosphate adsorbed TT at 365 days.

potency of vaccine from lot to lot which is useful for trend analysis and 3) the variations in the in vitro assays can be controlled better than in the in vivo assays.

Acknowledgements

The authors thank Roger Anderson and Doug Cecchini for developing several vaccine formulations used in this study; Bradford Rost, William Latham, Lee Campbell and Dalida Yeroshalmi for tetanus and diphtheria toxin preparations; Paul Griffin, Jr., Kimberley Benscoter and Jin Xu for technical assistance with Vero cell assay and Lizette Montanez-Ortiz, Anthony Pelligrino and Louis Nunnally for the in vivo TN test for diphtheria antitoxin.

References

1 United States Minimum Requirements: Diphtheria Toxoid. U.S. Department of Health, Education and Welfare, National Institutes of Health, Bethesda, MD, USA, 4th Revision, 1947.

2 United States Minimum Requirements: Tetanus toxoid. U.S. Department of Health, Education and Welfare, National Institutes of Health, Bethesda, MD, USA, 1952.

3 Fitzgerald EA: Overview of the methods for potency testing of diphtheria and tetanus toxoids in the United States, in Manclark CR (ed): Proceedings of an Informal Consultation on the World Health Organization Requirements for Diphtheria, Tetanus, Pertussis and Combined Vaccines. Department of Health and Human Services, United States Public Health Service, Bethesda, MD, DHHS Publication No. [FDA] 91-1174, 1991, pp 61-64.

4 Miyamura K, Tajiri E, Ito A, Murata R, Kono R: Micro cell culture method for determination of diphtheria toxin and antitoxin titres using VERO cells II. Comparison with the rabbit skin method and practical application for sero-epidemiological studies. J Biol Stand 1974;2:203-209.

5 Kriz B, Sladky K, Burianova-Vyoska B, Mottlova O, Roth Z: Determination of diphtheria antitoxin in guinea pig sera by the Jensen and tissue culture methods. J Biol Stand 1974;2: 289-296.

6 Sharma SB, Sharma K, Maheshwari SC, Gupta RK, Bhandari SK, Ahuja S, Saxena SN: An in vitro method for titration of diphtheria antitoxin on Vero cells. J Com Dis 1985;17:177-180.

7 Aggerbeck H, Heron I: Improvement of Vero cell assay to determine diphtheria antitoxin content in sera. Biologicals 1991;19:71-76.

8 Dular U: Comparative studies on the in vivo toxin neutralization and the in vitro Vero cell assay methods for use in potency testing of diphtheria component in combined vaccines/toxoids. 1: Standardization of a modified Vero cell assay for toxin-antitoxin titration of immunized guinea-pig sera. Biologicals 1993;21:53-59.

9 Peel MM: Measurement of tetanus antitoxin I. Indirect haemagglutination. J Biol Stand 1980;8:177-189.

10 Gupta RK, Maheshwari SC, Singh H: The titration of tetanus antitoxin II. A comparative evaluation of the indirect haemagglutination and toxin neutralization tests. J Biol Stand 1984;12:137-143.

11 Gupta RK, Maheshwari SC, Singh H: The titration of tetanus antitoxin III. A comparative evaluation of indirect haemagglutination and toxin neutralization titres of human sera. J Biol Stand 1984;12:145-149.

12 Relyveld EH, Saliou P, Le Cam N: Dosage d'anticorps tétaniques à l'aide du vacci-test T Pasteur. Eurobiologiste 1990;26:219-228.

13 Huet M, Relyveld E, Camps S: Simplified activity evaluation of several tetanus vaccines. Biologicals 1992;20:35-43.

14 Melville-Smith ME, Seagroatt VA, Watkins JT: A comparison of enzyme-linked immunosorbent assay (ELISA) with the toxin neutralization test in mice as a method for the estimation of tetanus antitoxin in human sera. J Biol Stand 1983;11:137-144.

15 Gentili G, Pini C, Collotti C: The use of an immunoenzymatic assay for the estimation of tetanus antitoxin in human sera: a comparison with seroneutralization and indirect haemagglutination. J Biol Stand 1985;13:53-59.

16 German-Fattal M, Bizzini B, German, A: Immunity to tetanus: tetanus antitoxin and anti-BIIb in human sera. J Biol Stand 1987;15:223-230.

17 Simonsen O, Bentzon MW, Heron I: ELISA for the routine determination of antitoxic immunity to tetanus. J Biol Stand 1986;14:231-239.

18 Simonsen O, Schou C, Heron I: Modification of the ELISA for the estimation of tetanus antitoxin in human sera. J Biol Stand 1987;15:143-157.

19 Gupta RK, Siber GR: Comparative analysis of tetanus antitoxin titers of sera from immunized mice and guinea pigs determined by toxin neutralization test and enzyme-linked immunosorbent assay. Biologicals 1994;22:215-219.

20 Manghi MA, Pasetti MF, Brero ML, Deluchi S, di Paola G, Mathet V, Eriksson PV: Development of an ELISA for measuring the activity of tetanus toxoid in vaccines and comparison with the toxin neutralization test in mice. J Immunol Methods 1994;168:17-24.

21 Knight PA: The use of ELISA for the assay of antitoxins from human and laboratory animals, in Manclark CR (ed): Proceedings of an Informal Consultation on the World Health Organization Requirements for Diphtheria, Tetanus, Pertussis and Combined Vaccines held at Geneva on May 30 – June 1, 1988. Department of Health and Human Services, United States Public Health Service, Bethesda (DHHS Publication No. (FDA) 91-1174), 1991, pp 123-126.

22 Hendriksen CFM, Gun JW, Nagel J, Kreeftenberg JG: The toxin binding inhibition test as a reliable in vitro alternative to the toxin neutralization test in mice for the estimation of tetanus antitoxin in human sera. J Biol Stand 1988;16:287-297.

23 World Health Organization: Laboratory methods for the testing for potency of diphtheria (D), tetanus (T), pertussis (P) and combined vaccines. World Health Organization, Geneva, Unpublished working document, BLG/92.1.

24 Gupta RK, Siber GR: Comparison of adjuvant activities of aluminium phosphate, calcium phosphate and stearyl tyrosine for tetanus toxoid. Biologicals 1994;22:53-63.

25 Gupta RK, Maheshwari SC, Singh H: The titration of tetanus antitoxin IV. Studies on the sensitivity and reproducibility of the toxin neutralization test. J Biol Stand 1985;13:143-149.

26 Gupta RK, Higham S, Gupta CK, Rost B, Siber GR: Suitability of the Vero cell method for titration of diphtheria antitoxin in the United States potency test for diphtheria toxoid. Biologicals 1994;22:65-72.

27 Turpin A, Bizzini B, Raynaud M: Titrage des anticorps antitétaniques. Son intérêt en pratique médicale. Med Mal Infect 1973;3:65-70.

28 Bizzini B: Tetanus, in Germanier R (ed): Bacterial Vaccines. Academic Press, Orlando. 1984, pp 37-68.

29 Glenny AT, Barr M: The dilution ratio of diphtheria antitoxin as a measure of avidity. J Pathol Bacteriol 1932;35:91-96.

30 Cinader B, Weitz B: Interaction of tetanus toxin and antitoxin. J Hyg 1953;51:293-310.

31 Cinader B, Weitz B: ß and γ globulin tetanus antitoxin of the hyper immune horse. Nature 1950;166:785.

32 Raynaud M, Relyveld EH: Avidity Determinations. Dilution ratio as an indirect measure of avidity, in Williams CA, Chase MW (eds): Methods in Immunology and Immunochemistry, vol. IV, New York, Academic Press, 1977, pp 302-303.

Dr. R.K. Gupta, Massachusetts Public Health Biologic Laboratories, State Laboratory Institute, 305 South St., Boston, MA, 02130, USA

Brown F, Cussler K, Hendriksen C (eds): Replacement, Reduction and Refinement of Animal Experiments in the Development and Control of Biological Products.
Dev Biol Stand. Basel, Karger, 1996, vol 86, pp 217-224

Vero Cell Assay Validation of an Alternative to the Ph. Eur. Diphtheria Potency Tests

A.M. Gommer

National Institute of Public Health and Environmental Protection (RIVM), Bilthoven, The Netherlands

Abstract: In the framework of the Biological Standardisation Programme of the European Pharmacopoeia Commission, in 1993 a collaborative study was organised for the validation of an alternative to the diphtheria in vivo challenge tests required by the Ph.Eur. monograph V.2.2.7. The alternative assay is based on the detection of neutralising antibodies in the sera from mice immunised with the vaccines to be tested (Vero cell assay). In the study this assay method was validated against intradermal and lethal challenge in guinea-pigs, performed in conformity with Ph.Eur. Therefore the potency of the diphtheria component in a number of different types of vaccine, representative for those currently on the European market, was assayed in parallel by the different assay methods.
Seventeen laboratories, from eleven different countries, participated in the study. Three laboratories performed the intradermal challenge assay, while three other laboratories performed the lethal challenge assay. All seventeen laboratories performed the Vero cell assay.
The results of the study suggest that the potency of the diphtheria component of both monovalent diphtheria vaccines and combined diphtheria-tetanus vaccines can be estimated adequately by means of the Vero cell assay. It does not yet seem possible for all combined diphtheria-tetanus-pertussis vaccines to replace a potency assay based on the protective capacity of a vaccine in guinea-pigs by the Vero cell assay. This may be due to an adjuvant effect of the pertussis component of the vaccine, in combination with the adsorbent used, which may be more pronounced in mice than in guinea-pigs and may also differ between different strains of mice.

Introduction

The main objective of the Biological Standardisation Programme of the European Pharmacopoeia Commission, a collaboration between the European Community and the Council of Europe, is to improve the control of biological products and to facilitate its harmonization. In addition to the establishment of Biological Reference Preparations (BRPs) for defined biological products, this is achieved by validating new test methods.

In 1993 a collaborative study was organised for the validation of an alternative to the diphtheria potency tests required by Ph.Eur. monograph V.2.2.7. In general, the relevance and reliability of alternative potency assays is evaluated by comparing the results with those obtained by means of accepted bioassays. Therefore the following three test methods were included in the collaborative study: intradermal challenge in guinea-pigs (in conformity with Ph.Eur. V.2.2.7.; page 2), lethal challenge in guinea-pigs (in conformity with Ph.Eur. V.2.2.7.; page 3) and the alternative test based on the detection of neutralising antibodies in the sera from mice immunised with the vaccines to be tested (Vero cell assay). In the study the Ph.Eur. working standard was used as reference. Different types of diphtheria vaccines, representative of the vaccines currently on the European market, were included in the study as samples to be tested. The vaccine samples were obtained and distributed by the European Pharmacopoeia Secretariat. The study was coordinated by the RIVM's Laboratory for Medicines and Medical Devices.

Material and Methods

Vaccines

The vaccine used as a reference in the study was the Ph.Eur. working standard: diphtheria vaccine (adsorbed) BRP, ref.no. D2700 (freeze-dried). This reference was manufactured by Behringwerke (Germany), each ampoule containing 75 Lf diphtheria-toxoid (based on the manufacturing data of the manufacturer). The reference has been assigned a potency of 140 IU per ampoule [1].

The following vaccines were included as test samples:

- DT-vaccine, batch no. J0069, provided by Pasteur Mérieux
- DTP-vaccine, batch no. J5050, provided by Pasteur Mérieux
- D-vaccine, batch no. 12136-03, provided by Berna
- DT-vaccine of inferior quality, batch no. 510 A, provided by the RIVM

The vaccine of inferior quality was prepared at the RIVM by incubating the (non-commercial) batch of DT-vaccine for four hours at 56°C.

In Table 1 the manufacturers' information on the potencics (IU/ml), the diphtheria-toxoid Lf-content and the aluminium content of the different vaccine-samples is given. In the study the vaccines were coded as indicated in the table.

The potencies claimed for the samples A and B are based on the results of one intradermal challenge assay, performed according to the method of the European Pharmacopoeia. The potency claimed for sample C is based on the results of one lethal challenge assay, performed according to the method of the European Pharmacopoeia. The potency claimed for sample D is based on the results of one Vero cell assay («Kreeftenberg method» [2]), before heat-treatment.

Diphtheria toxin

In the study ampoules containing 1000 Lf of freeze-dried diphtheria toxin (Dt 79/1) provided by the RIVM were used. The participants were instructed to reconstitute the diphtheria toxin with 20 ml saline to obtain a concentration of 50 Lf/ml. After reconstitution, the toxin was stored at -20°C.

Diphtheria antitoxin

In the study the First International Standard for Diphtheria Antitoxin (equine), provided by the World Health Organisation (WHO), was used. The antitoxin concentration of this solution is 10 IAU/ml.

Control antiserum

Each participant received 1.5 ml control antiserum to be used in the alternative potency test as an assay-control. This control serum was provided by the RIVM.

At the RIVM a hundred mice were immunised subcutaneously with 0.5 ml of a dilution of the Ph.Eur. Diphtheria working standard corresponding to 4 Lf/ml. Five weeks after this immunisation

Table 1: Claimed potencies, Lf-contents and aluminium content of the vaccine samples.

Sample	Code	Claimed potency and 95% Confidence Limits (IU/ml)	Lf-content (per ml)	Aluminium content (per ml)
Reference (after reconstitution in 3 ml saline)	original label (uncoded)	46.67	25	1.5 mg $Al(OH)_3$
DT-vaccine (Pasteur Mérieux)	A	94 (68-132)	50	3.4 mg $Al(OH)_3$
DTP-vaccine (Pasteur Mérieux)	B	140 (100-196)	50	3.6 mg $Al(OH)_3$
D-vaccine (Berna)	C	113 (84-151)	50	3.2 mg $Al(PO_4)_3$
DT-vaccine of inferior quality (RIVM)	D	25 (14-43)	5	3.0 mg $Al(PO_4)_3$

the mice were bled. The blood was pooled and successively incubated for two hours at 37°C and two hours at +4°C. Following these incubations the blood was centrifuged for 20 minutes at 800 g. The serum was siphoned off, inactivated for 30 minutes at 56°C and stored at -20°C, until it was sent to the participants on dry-ice.

Vero cells

Each participant of the study received one vial, containing one ml of a frozen 4.10^6 cells/ml Vero cell suspension. The cells were sent to the participants on dry-ice. Most laboratories did not succeed in culturing these Vero cells and used their own Vero cells for this study.

Experimental animals

- guinea-pigs: participants performing a lethal challenge were asked to use guinea-pigs with a body-weight of 250 to 350 g.
- mice: participants performing the alternative test method were asked to use mice, male or female with a body-weight of 10 to 14 g.

Participants

Seventeen laboratories from eleven different countries participated in the study and sent in their results.

Study design

The study included three test methods:
- intradermal challenge in guinea-pigs (in conformity with Ph.Eur. V.2.2.7.; page 2)
- lethal challenge in guinea-pigs (in conformity with Ph.Eur. V.2.2.7.; page 3)
- the alternative potency test based on the detection of neutralising antibodies in the sera from mice immunised with the vaccines to be tested (Vero cell assay).

The participants were asked to perform each assay method twice on two separate occasions.

As far as the Vero cell assay is concerned, the participants were asked to perform the test in conformity with the «Kreeftenberg method» [2].

The toxic effect of diphtheria toxin on Vero cell cultures is inhibited and/or neutralised by diphtheria antibodies, present in the sera of mice immunised with vaccine. By culturing Vero cells in the presence of a series of dilutions of the serum samples and a fixed amount of diphtheria toxin, the survival of the Vero cells can be taken as a measure of neutralising antibodies in the sera.

The reading of the Vero cell assay is based on visual determination of colour change in wells of microtitre plates from red to yellow caused by metabolic formation of acid.

In addition to the «Kreeftenberg method» two variant versions of the Vero cell assay were included in the study:

- A Vero cell assay in which the visual assessment of the end point is replaced by a spectrophotometric measurement [3].
- A Vero cell assay in which the end point is assessed after adding a thiazolyl blue solution to the tissue culture plates [4].

Three participating laboratories performed the lethal challenge assay. Three other laboratories performed the intradermal challenge assay. All seventeen participating laboratories performed the Vero cell assay in conformity with the «Kreeftenberg method». The scores of one laboratory were omitted from further calculations, because of an inconsistency in the test protocol.

Results

Estimated potencies

The potencies of the samples A, B, C and D were calculated on the basis of the data obtained by means of the various methods.

In general, immunisation with the various dilutions of sample D did not result in a dose-response curve as a consequence of the immunising potency of this vaccine being too low.

The potencies based on the lethal challenge data were calculated by means of probit analysis. The potencies based on the individual guinea-pig scores in the intradermal challenge test were calculated by parallel line analysis, after the frequency distribution of each treatment group was checked for deviations from normality. The potencies based on the Vero cell assay data («Kreeftenberg method») were also calculated by parallel line analysis. The percentage of valid potency results was rather low, which could either be ascribed to deviations from parallelism or linearity.

Statistical evaluation

In view of the very few valid potency results available for the different types of assay, all valid potencies are taken together, to compare the Vero cell assay results with the results of both the lethal and intradermal challenge assay results. No attempt is made to determine an intra-laboratory variation.

A one-way analysis of variance was performed on the three groups of data. As far as samples A, B and C are concerned, no significant difference was found between the potencies calculated by the three types of assay. However for sample B the potencies calculated on the basis of the Vero cell assay scores showed a tendency to be somewhat higher than the potencies calculated on the basis of the challenge data. The p-values resulting from the analysis of variance were 0.599 for sample A, 0.063 for sample B and 0.421 for sample C respectively. The mean potencies and the corresponding values of standard deviation (SD) are presented in Table 2.

Table 2: Means and standard deviations (SD) of estimated potencies.

Sample	Assay method	Number of data	Mean potency (IU/ml)	SD
A (DT-vaccine)	Vero cell assay	17	82.5	48.2
	Intradermal challenge	3	105.4	15.3
	Lethal challenge	3	69.2	28.3
B (DTP-vaccine)	Vero cell assay	7	443.1	256.2
	Intradermal challenge	6	191.2	58.8
	Lethal challenge	2	178.9	59.8
C (D-vaccine)	Vero cell assay	14	89.3	30.4
	Intradermal challenge	4	67.5	32.3
	Lethal challenge	4	83.0	12.3

Only three laboratories were able to assign a potency to the inferior sample D (DT-vaccine) included in the study. The potency estimated by Vero cell assay was 1.3 IU/ml with a 95% confidence interval of 0.0-4.6 IU/ml. This potency correlated well with the potency estimated by intradermal challenge assay, which was 3.0 IU/ml with a 95% confidence interval of 0.8-6.0 IU/ml. However the potency estimated by lethal challenge assay was significantly higher: 32 IU/ml with a 95% confidence interval of 16-57 IU/ml.

Spectrophotometric measurement of colour formation (Aggerbeck method)

Three participating laboratories performed the Vero cell assay in conformity with the «Aggerbeck method» (spectrophotometric measurement of colour change). From the data of one of these laboratories it was not possible to calculate valid potencies due to poor dose-response curves. There is no significant difference between the potencies of the samples estimated by means of the «Kreeftenberg method» and those estimated by means of the «Aggerbeck method».

MTT cytotoxicity assay

The results obtained by means of the MTT cytotoxicity assay, performed by one of the participants, did not differ from the results of the Kreeftenberg method. This means that the end point assessed by adding MTT solution to the microtitreplates in order to detect viable cells did not essentially differ from the end point assessed on the basis of colour change of the culture medium.

Using the optical density data the relative activity of each mouse serum was estimated, i.e. relative to within plate reference serum (First International Standard for Diphtheria Antitoxin). With these relative activity estimates of the mouse sera the potencies of the coded samples A, B, C and D were calculated relative to the reference vaccine, the potency of which was taken as 1. Calculated this way, no valid estimates were obtained for sample B. However, it was possible to calculate one valid estimate of relative potency for samples A and D and two for sample C. It appears that the potency estimates of samples A, C and D, calculated on the basis of relative activity of mouse sera, are lower than those based on quantal analysis. However, the ranking order of the potency estimates is consistent with the claimed potencies.

Control antiserum

The toxin neutralising activity of the control antiserum, as determined in the Vero cell assay by all participating laboratories, indicates that the variation in test results between laboratories can be ascribed mainly to differences in the immune response of the mice, whereas the serum titration part of the assay does not affect the inter-laboratory variation of the potency assay.

Discussion

For all three types of assay only a small number of valid results have been obtained. Most laboratories did not succeed in producing valid replicate results. It is therefore not possible to determine variation within laboratories, nor to draw conclusions about the significance of differences between laboratories.

For the DT-vaccine (sample A) and the D-vaccine (sample C) most laboratories have succeeded in producing at least one valid estimate of the potency by means of the different types of assay performed. For the DTP-vaccine (sample B) only a few laboratories were able to produce a valid potency estimate by means of the Vero cell assay. For the samples A, B and C no significant difference was found between the potencies determined by either the Vero cell assay, the intradermal challenge assay, or the lethal challenge assay. Only three laboratories were able to assign a potency to the inferior sample D (DT-vaccine) included in the study.

The aim of the study was to investigate the suitability of an assay, based on the detection of neutralizing antibodies in the sera from mice immunised with the vaccines to be tested, to determine the potency of diphtheria vaccines. Both for the Ph.Eur. challenge assays and the Vero cell assay the percentage of valid results was rather low. The most important reason seems to be that, in general, the doses used to immunise the animals (mice or guinea-pigs) were too low. It appears that choosing an immunisation dose merely based on Lf-content is not always convenient. The immunising potency is also dependent on the composition of the vaccine.

It is expected that most laboratories will experience much fewer invalid assays when dealing with vaccines and reference preparations with which they have become familiar.

As far as the statistically valid Vero cell assays are concerned, no significant difference was found between the Vero cell assay and the challenge assays. The mean potencies estimated for sample A (DT-vaccine) and sample C (D-vaccine) by Vero cell assay are between the mean potencies estimated by intradermal and lethal challenge assay. Only a few laboratories have been able to estimate a potency for sample B (DTP-vaccine) by means of the Vero cell assay. The most important reason for this was the absence of a clear dose-response curve, which appears not to be related to the immunisation dose. In earlier studies [1, 5] a lack of homogeneity of assays in mice was found with respect to potency estimates of the tetanus toxoid component of adsorbed diphtheria-tetanus-pertussis vaccine. In these studies the homogeneity in mice was very much less than that in guinea-pigs.

In the present study the estimated potencies for the combined diphtheria toxoid (DTP-vaccine) differed in mice and guinea-pigs. The mouse assay clearly gave higher values than the guinea-pig assays. This phenomenon was also found in the earlier studies concerning the tetanus toxoid component of DTP-vaccine. It was ascribed to a possible adjuvant effect of the pertussis component, being more pronounced in mice than in guinea-pigs.

To ascertain whether the deviant results found for the DTP-vaccine can be ascribed to an inappropriate range of the immunisation dose, the Vero cell assay was repeated for sample B in a follow-up study. In this follow-up study the Ph.Eur working standard was again included as reference. A commercial batch of DTP-Polio-vaccine, produced by RIVM, was included in the follow-up study, acting as an assay-control. Both the Laboratory for the Control of Biological Products and the Laboratory for Medicines and Medical Devices of the RIVM participated in the study. It appeared that the problems encountered in producing a clear dose-response curve for the DTP-vaccine is not related to the immunisation dose range, but is likely to be inherent in the composition of the vaccine. Furthermore it became clear from the follow-up study that, in addition to the presence of the pertussis component, other properties of the DTP-vaccine are responsible for the deviant results in the Vero cell assay. This conclusion is based on the fact that the DTP-Polio-vaccine produced by the RIVM can be estimated by means of the Vero cell assay, despite the pertussis component present in this vaccine.

Conclusions

In view of the small number of valid potency estimates it seems necessary for a laboratory to be familiar with the vaccines tested in the Vero cell assay, to select immunisation doses resulting in adequate dose-response curves. Provided the appropriate immunisation doses are chosen, the potencies of the diphtheria vaccine and combined diphtheria-tetanus vaccine, estimated by Vero cell assay, correlate well with the potencies estimated by the Ph.Eur. challenge methods. In all three types of assay the inferior DT-vaccine fails to show any significant immunising potency.

It appeared to be nearly impossible to assign a potency to a combined diphtheria-tetanus-pertussis vaccine by means of the Vero cell assay. Therefore, caution is recommended regarding the implementation of a potency assay using mice as an alternative for assays using guinea-pigs as far as combined diphtheria-tetanus-pertussis are concerned, in agreement with similar observations reported by other workers.

Acknowledgements

During the preparation of this study and the processing of the data I was advised by Mr. J. van der Gun, Dr. C.F.M. Hendriksen and Mr. F.R. Marsman.

I thank them for their helpful advice. I thank Dr. Aggerbeck for drafting the protocol of the Aggerbeck-method and for processing the data sent by some of the participating laboratories. Finally I would like to thank all participants for their co-operation.

References

1 Mussett MV, Sheffield F: A collaborative investigation of methods proposed for the potency assay of adsorbed diphtheria and tetanus toxoids in the European Pharmacopoeia. Journal of Biological Standardization 1973;1:259-283.

2 Kreeftenberg JG, van der Gun J, Marsman FR, Sekhuis VM, Bhandari SK, Maheshwari SC: An investigation of a mouse model to estimate the potency of the diphtheria component in vaccines. Journal of Biological Standardization 1985:13:229-234.

3 Aggerbeck H, Heron I: Improvement of a Vero cell assay to determine diphtheria antitoxin content in sera. Biologicals 1991;19:71-76.

4 Hoy CS, Sesardic D: In vitro assays for detection of diphtheria toxin. Toxicology in Vitro 1994;8(4):693-695.

5 Van Ramshorst JD, Sundaresan TK, Outschoorn AS: International collaborative studies on the potency assays of diphtheria and tetanus toxoids. Bulletin of the World Health Organization 1972:46:263-276.

Dr. A.M. Gommer, National Institute of Public Health and Environmental Protection (RIVM), Antonie van Leeuwenhoeklaan 9, P.O. Box 1, 3720 BA Bilthoven, The Netherlands

Brown F, Cussler K, Hendriksen C (eds): Replacement, Reduction and Refinement of Animal Experiments in the Development and Control of Biological Products.
Dev Biol Stand. Basel, Karger, 1996, vol 86, pp 225-241

Passive Haemagglutination Tests Using Purified Antigens Covalently Coupled to Turkey Erythrocytes

E.H. Relyveld[1], M. Huet[2], L. Lery[3]

[1] Institut Pasteur, 92430 Marnes la Coquette, France
[2] Institut Bouisson-Bertrand, 34090 Montpellier, France
[3] Institut Pasteur de Lyon, 69365 France

Keywords: Passive haemagglutination, tetanus, diphtheria, antibody, toxin, immunodeficiency, HBs Ag, HIV-1 gp 160 antigen.

Abstract: Passive haemagglutination tests have been developed by covalent coupling purified antigens to turkey red blood cells. Circulating antibodies can be assessed in 20 minutes using one drop of blood. False positive reactions are avoided by using highly purified antigens; sensitized erythrocytes are stable in the absence of freeze-drying and blood samples can be preserved on paper discs. This method, applied to the determination of circulating tetanus (T) and diphtheria(D) antibodies and titres compared to other in vivo or in vitro methods, gave good correlation. The titration of circulating antibodies can be applied in emergency care units and field trials to establish whether the individuals are adequately protected. Results of surveys by several health care centres have shown that tetanus immune coverage was insufficient in France. The decrease of both T and D immune coverage with age has been established. The antibody response of pregnant women, vaccinated with two different adsorbed T toxoids exhibiting a low and a high titre as expressed in international immunizing units (I.I.U.), was studied. No significant difference in circulating antibody titres was obtained after the first injection of either vaccine, but titres after the second injection were much higher for the vaccine having a low value expressed in I.I.U.The activity of commercial and reference T vaccines can be evaluated in mice after immunization and titration of the antitoxin levels. This simple method is much easier than the official challenge tests. Antibody levels determined after vaccination also made it possible to assess the evolution of immunodeficiency in certain diseases.The passive haemagglutination test has also been used to measure anti-HBs and anti-gp 160 antibodies.

INTRODUCTION

The in vivo titration of circulating antitoxins which needs a large number of laboratory animals, is expensive and time-consuming [1]. Several in vitro assays have been developed such as: flocculation tests; enzyme-linked immunoabsorbent assays (ELISA); toxin binding inhibition (ToBI) tests; radio-immuno-assays

(RIA); immunodiffusion techniques (simple and double diffusion, rocket immunoelectrophoresis, single radial immunodiffusion, etc...); passive agglutination techniques with toxoids attached to latex particles and passive haemagglutination (HA) tests with toxoids adsorbed on tanned and formalized human, sheep or horse erythrocytes. However some of the techniques have low precision, are cumbersome and time-consuming and even need labelled toxins or toxoids and large volumes of sera of high titres (as for flocculation tests). Valid results are only obtained by the use of highly purified antigens or monovalent antibodies. Passive haemagglutination with the use of turkey cells and adsorbed toxoids give quick results because the nucleated cells sediment rapidly, but are not very stable.

The use of passive haemagglutination tests by covalent coupling of purified antigens and especially tetanus toxin to turkey red blood cells is described. Circulating antibodies can be titrated by using only one drop of blood or serum and be evaluated 20 minutes after adding the sensitized red blood cells to the wells. False positive reactions are avoided by using highly purified toxins, blood samples can be preserved on paper discs and the sensitized red blood cells are stable.

The test can be applied as a guide for prophylaxis management at the time of exposure to tetanus:

- to evaluate the protection of civilians and to assess the results of field trials;
- for the selection of plasmas for the preparation of specific human immune serums;
- to titrate the potency of vaccines;
- to estimate the evolution of immunodeficiency in certain diseases.

Material and Methods

1. Preparation of sensitized erythrocytes

Passive haemagglutination tests have been developed since 1951 [2] and used for titrating circulating antibodies by the use of animal or human red blood cells coated with antigens after first modifying the surface with tannic acid, chromium chloride or bisdiazobenzidine.

The use of erythrocytes of birds and especially turkey cells was emphasized in 1958 [3], because these erythrocytes are nucleated and settle rapidly. Turkey erythrocytes have been used by Pitzurra (4), after formalization, tanning and adsorption of tetanus toxoid to titrate tetanus antibodies.

Covalent coupling of purified tetanus toxin by the use of glutaraldehyde, a bifunctional cross-linking agent is carried out by the following steps: blood is collected in Alsever's solution and after centrifugation and washing with PBS, resuspended with the same buffer.

A 0.025M solution of glutaraldehyde is added by stirring and after one minute of contact a solution of tetanus toxin is added and the mixture is stirred for one hour. The sensitized red cells are separated by centrifugation and washed with the PBS buffer. A 0,1M solution of glycine is then poured onto the centrifugation pellet and the suspension is left stirring for 30 minutes.

The red cells thus treated are again isolated by centrifugation and after washing resuspended in a 5% concentration in the PBS buffer containing 2:10.000 (w/v) of sodium azide. Details of the method illustrated in Figure 1 are given elsewhere [5].

Crystalline diphtheria toxin can be coupled by the same method, but in this case the period of reaction with cells and glutaraldehyde is two hours. It should be noted that both toxins are also toxoided by the reaction with glutaraldehyde.

Titration of tetanus antibodies is made as shown in Table 1 and Figures 2 and 3. The test can either be performed on whole blood or serum. After pricking the finger with a disposable microlance, 10 μl blood was collected with a micropipette and diluted with 190 ul of distilled water or collected on a disk of blotting paper that absorbs the amount of blood necessary to release 10 μl for titration. After drying, the disc is stored and suspended at the time of assay in 200 μl of distilled water and eluted for one minute for titration.

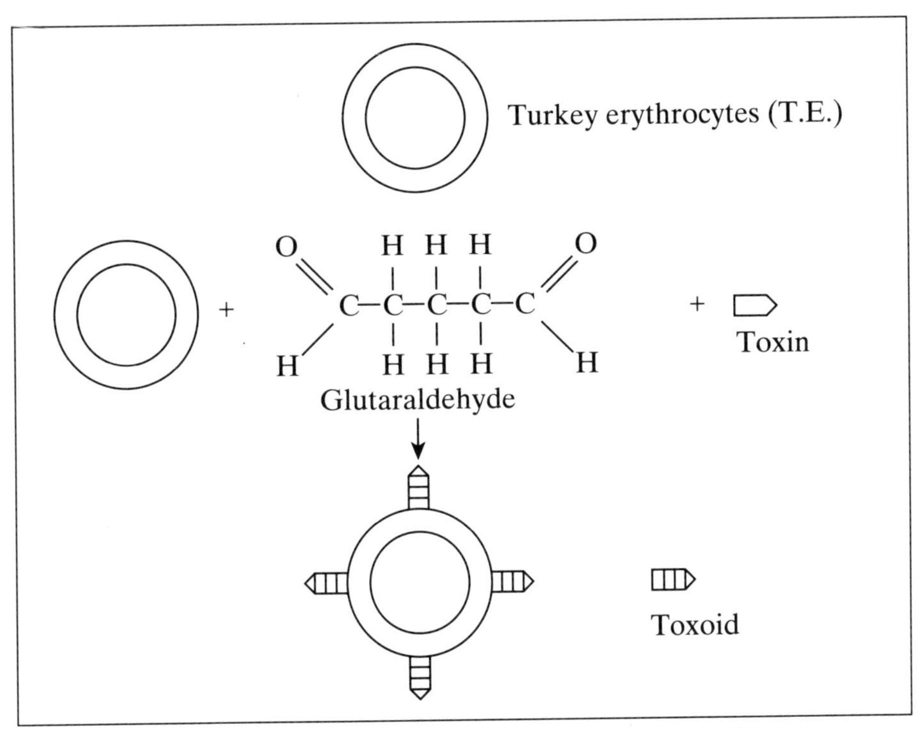

Fig. 1: Covalent coupling of toxins to turkey erythrocytes (T.E).

Table 1: Titration of tetanus antibodies.

1. Use plastic microplates with V-wells.
2. Dilute blood or serum 1/20 (10 µl in 190 µl of distilled water).
3. Distribute 50 µl of PBS buffer in wells 1 to 12.
4. Add 50 µl of 1/20 diluted samples into wells 1.
5. Prepare serial two-fold dilutions by transfer of 50 µl from well 1 to 2 and repeat after mixing up to well 11; well 12 is a negative control (N.C.) (Well 1 = 1/40 dilution).
6. Add 50 µl sensitized cells to wells 1 to 12.
7. Agitate the plate for one minute.
8. Cover and incubate the plate for 20 minutes at room temperature.
9. Determine the last dilution which gives a positive agglutination for:
   - the reference serum = 1 IAU/ml
   - the samples
10. Calculate antibody titres in HU/ml of serum:

$$\frac{\text{positive limit dilution of the reference serum}}{\text{positive limit dilution of the sample}}$$

Titres must be multiplied by two for whole blood samples, which contain 50% cells.

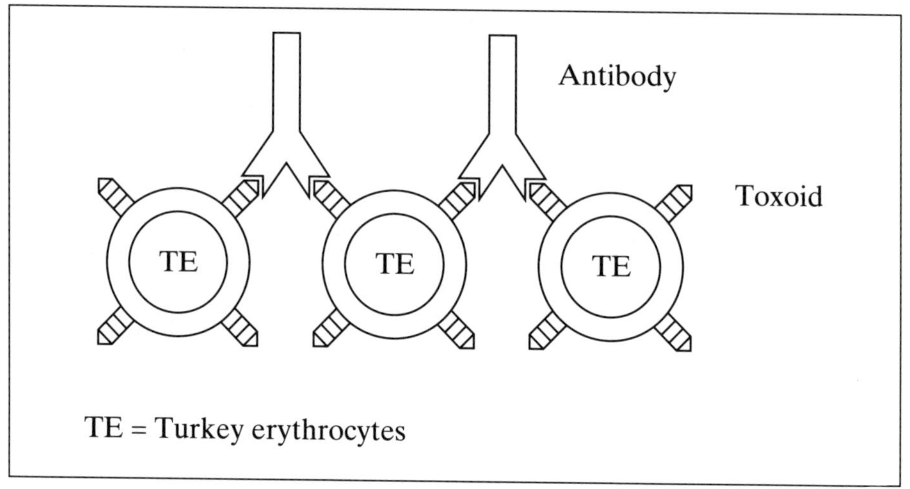

Fig. 2: Passive haemagglutination using highly purified toxoids covalently coupled to turkey erythrocytes (T.E.).

Antibody titres of the samples are determined by comparing the last dilution which induces haemagglutination, with the threshold value of the reference human serum. Titres are expressed in haemagglutination Units/ml (HU/ml) corresponding to International Antitoxic Units which are units of the reference serum (IAU):

$$\text{Titres in HU} = \frac{\text{positive limit dilution of reference serum}}{\text{positive limit dilution of sample}}$$

The titres in HU units are multiplied by two to obtain antibody levels per ml of serum when the test is performed on a blood sample containing 50% of cells.

Comparative Evaluation of the Tests

Several studies have been made [6-16] to compare the passive haemagglutination (HA), toxin neutralization (TN) and enzyme-linked immunoabsorbent assay (ELISA) titres for the determination of circulating tetanus antibodies in humans. Only part of the results are presented here:

1. Correlation between in vivo toxin neutralization and haemagglutination titres. Titrations have been carried out at the Pasteur Institute. The toxin neutralization test was performed at the L +/10 - - - - L +/ 1000 levels as described elsewhere [17]. Serum samples from 194 pregnant women from Cameroon were titrated before and after vaccination and classified in four groups (Table 2). Good correlation was obtained in 155 of 194 titrations.

   Serum samples of 21 persons classified as not protected by the HA test were shown as having either low protection [19] or protection [2] when tested by the TN method. Their HA titres were lower than estimated but this had no serious consequence for prophylactic management. Results of the passive haemagglutination test for titration of antibodies to tetanus toxoid in human and guinea pig

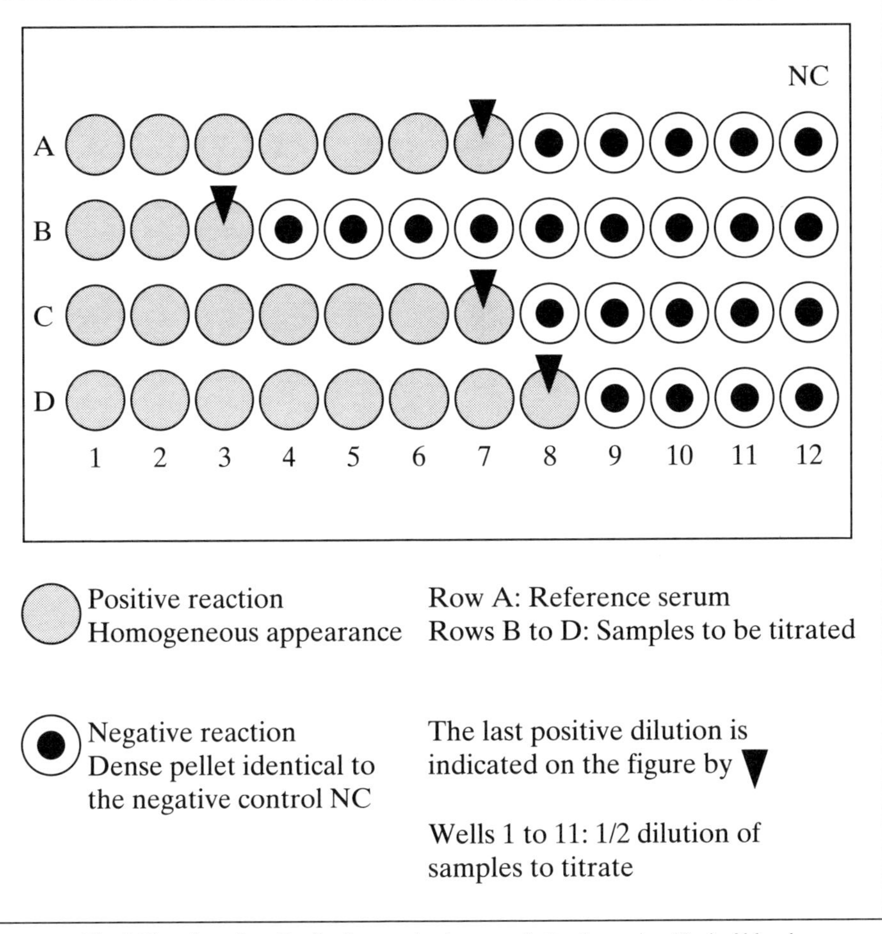

Fig. 3: Titration of antibodies by passive haemagglutination using 10 µl of blood.

sera have been published and show a very good correlation with the toxin neutralization test [18-20].

2. Correlation between haemagglutination and ELISA titres. Titrations at the Val de Grâce Hospital in Paris (Table 3) showed statistically significant correlation (student's t-test) for 103 titrations, of which 78 were the same and 19 were in good agreement. This means that practical results were the same for 94% of titrations. For six results, protection was low for ELISA and negative for HA, again without practical consequence.
3. Titrations at the Pasteur Institute: correlation between titres for protection and no protection was obtained during a field trial in Senegal for 211 sera to evaluate a combined DTC + polio + Hepatitis B vaccination. Good correlation was found for 92,4% of titrations (Table 4). Overestimation of ELISA titres

Table 2: In vivo and in vitro titration of circulating tetanus antibody levels in 194 pregnant women before and after vaccination (Cameroon).

		In vitro haemagglutination Vaccitest – T Pasteur – HU/ml			
		NP < 0.06	**LP ⩾ 0.06 – < 0.125**	**P ⩾ 0.125 – < 0.5**	**VGP ⩾ 0.5**
In vivo neutralization of toxin	**NP < 0.01**	94			
	LP ⩾ 0.01 – < 0.1	19	15	1	
	P ⩾ 0.1 – < 0.5	2	10	10	
	VGP ⩾ 0.5			7	36

NP: Not Protected LP: Low Protection P: Protected VGP: Very Good Protection

Table 3: Correlation between ELISA and HA for 103 sera, Hôpital Val de Grâce, Paris.

HA \ ELISA	Number of sera		
	NP (<0.01 IAU/ml)	**LP (0.01-0.06 IAU/ml)**	**GP (>0.06 IAU/ml)**
NP (<0.06 HU/ml)	3	6	0
LP (0.06 HU/ml)	0	9	1
GP (>0.06 HU/ml)	0	18	66

Student's t-test: significant correlation between the two methods.
NP: No Protection LP: Low Protection GP: Good Protection

[21-24] was again demonstrated for six TN titrations, giving negative antitoxin values < 0,01 IAU/ml.

4. Titrations at the University Hospital of Clermont-Ferrand [12, 15]:
   Statistically significant correlation was again established for 83 titrations and 95% correlation for titres > 0,016 and < 1 units. Several other studies have been made to correlate the results of geographically different laboratories to compare TN and HA titres of human tetanus sera in field trials or to compare HA, ELISA and RIA values obtained by the three different methods, a statistically significant correlation was nearly always found.

Table 4: Correlation between ELISA and HA for 211 sera during a field trial in Senegal.

HA \ ELISA	<0.06 IAU/ml	⩾ 0.06 IAU/ml
<0.06 HU/ml	80	15*
⩾ 0.06 HU/ml	1	115

Correlation: 92.4%

* 6 sera titrated «in vivo»: resluts <0.01 IAU/ml

Tetanus Immune Coverage in France

Even though tetanus vaccination is compulsory in France and the protection efficacy of vaccines is well established, results of several surveys have shown that the immune coverage is insufficient. In 1976, 288 cases were found and even if the number fell there were still 61 cases in 1993 (Fig. 4) most of them in persons over 60 years old.

The costs of tetanus treatment are very high, including hospitalization in intensive care units and afterwards in medical units amounting to a total of 44 days for a sum of about U.S. $ 22,080 per case (Table 5). The situation is worse in developing countries, where over a million adult lives could be spared every year if vaccination was carried out correctly and where about 600,000 newborn babies die every year because pregnant women are not highly protected [25].

Our results of circulating antibody titrations by HA allowed four groups of protection to be defined as a guide to tetanus prophylaxis treatments to be given in each of them (Table 6). The test can be applied in emergency care units, in cases of injury to establish whether the individual is protected or not [11, 12, 15]. This would allow a saving on the annual wasteful administration of about 1,3 million doses of human tetanus immunoglobulin preparations at a price of 16$/dose.

The haemagglutination test enabled the study of the protection by the French Protection Health Service (C.E.S.) in about 5,000 adults and 1,000 children half of whom were male and half female in towns and in urban or regional cities such as Paris, Nice, Chartres, Vandœuvre [26]. Titrations were carried out in 3,485 adults again in equal numbers of males and females. The results presented in Figure 5 show that 19% were not protected and only 67.6% had very good protection. It was also found that 30% of the Paris population and 42% of those in the suburbs were not protected at all. Vaccination rates are higher for males and about 75% have very good protection compared to only 60% in women who are not protected at all. The younger people have higher vaccination rates compared to older people and also higher immunity since 90% under 30 years of age have very good protection, whereas 35% of subjects over 60 years old are not protected.

The distribution of protection for men and women according to age, which is presented in Figure 6, shows that men are always better protected compared to women whatever the age, which could be the result of a booster injection at the time of their military service. People living in the country have higher booster rates

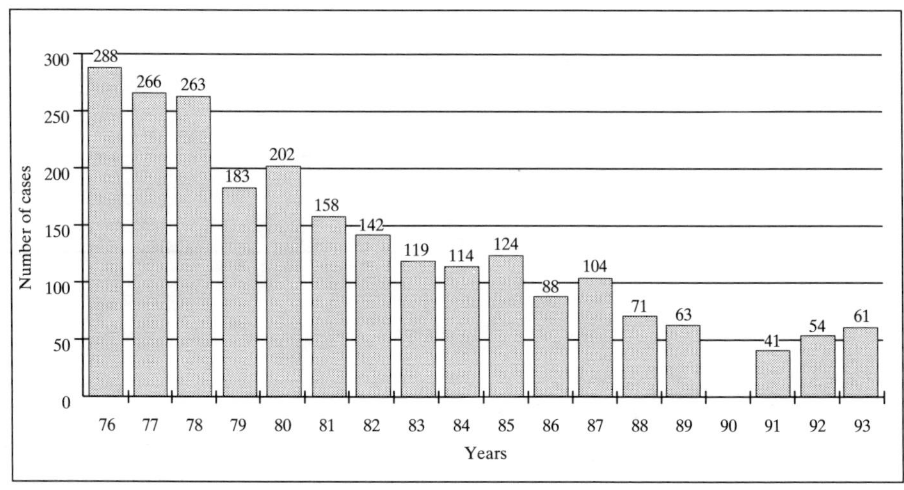

Fig. 4: Numbers of reported tetanus cases in France by years (data D.G.S.).

Table 5: Costs of treating tetanus in France*.

	Days of hospitalization	Daily costs	Costs/case
Hospitalization days: Intensive care units	32	600$	19.200$
Hospitalization days: Medical units	12	240$	2.880$
Total	44		22.080$

* In U.S. dollars (1$ = 5 FF).

Table 6: Guide to tetanus prophylaxis management.

Titres HU/ml	Immune status	Immune serum	Toxoid
<0.06	NP	1 injection	2 injections
⩾0.06 – <0.125	LP	0	1 injection
⩾0.125 – ⩽0.5	P	0	1 injection
> 0.5	VGP	0	0

NP: No Protection LP: Low Protection P: Protection VGP: Very Good Protection

and are better protected compared to those living in towns or suburbs with rates of 80.7; 67.9 and 58.7%; this also depends on the regions with e.g. 86.9% in Vandoeuvre compared to 53.9% in Paris.

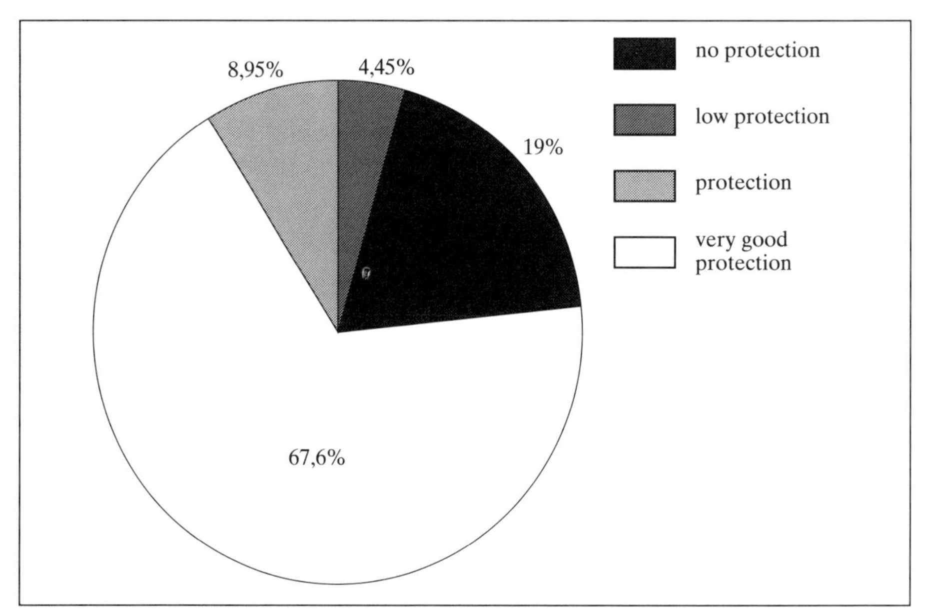

Fig. 5: Results of the tetanus immune coverage in France for eight cities: Paris, Nice, Chartres, etc. and 3,485 titrations in adults (French Preventive Health Services).

It is also interesting that when the test was commercialized in April 1988, among people visiting the yearly medical exhibition (MEDEC) only 41% had very good protection and 20% were not protected at all [16]. Results were worse when those attending the Medical Olympic Games in Lyon were tested in July of the same year, because 35% of the young athletics were not protected and only 49% had very good protection [27]. Immunodeficiency was also noted in the population, among the 3,485 tested, a hundred who had written proof of vaccination and booster injections (55 women and 45 men) were not protected. The explanation of this group of «non-responders» is under study. The results of these studies show that the protection of the population can be easily and rapidly checked in an economic way and should encourage subjects to have booster injections.

Activity Evaluation of Tetanus Vaccines

Potency of toxoid vaccines has been assessed for many years by the measurement of circulating antibody levels or challenge with toxin in immunized animals, which is still applied in some countries. A different method has been introduced by the WHO and adopted by the European Pharmacopoeia [28] in which a large number of laboratory animals (mice or guinea-pigs) are immunized with three dilutions of a reference vaccine and the vaccines to be tested and titres calculated after a subsequent toxin challenge on the basis of lethal or paralytic effects. However the validity of the latter method has not been unanimously accepted mainly because even if the vaccine batches satisfy the minimum requirements, titres do not corre-

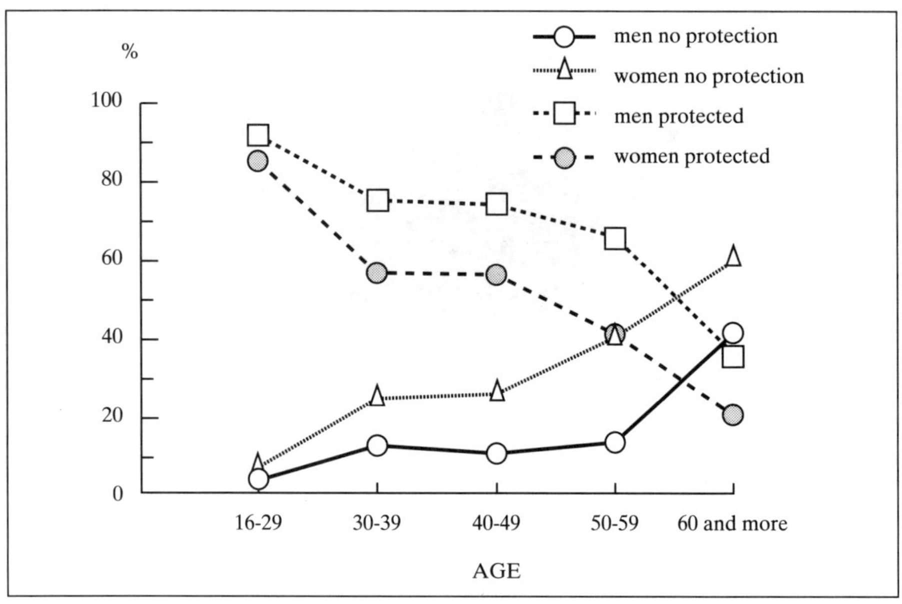

Fig. 6: Tetanus protection related to age and sex.

spond to immune response in humans [29, 30]. Variations of potency values have been observed depending on the species of animals used in the potency test, by the use of different challenge toxins and for toxoids of comparable quality containing similar adjuvants [31]. High antibody titres have been found in children immunized with vaccines having low protection values. Non-parallelism of dose-responses due to dissimilarities between reference and test vaccines and difficulties in estimating ED_{50} values made titrations of vaccines and even new reference preparations impractical [32]. The method is certainly not reliable because the vaccines are diluted in saline for immunization of the laboratory animals and the adjuvant effect diluted out. In fact highly purified toxin eg toxoided purified (crystalline) diphtheria toxin gives a low antibody response in humans in the absence of adjuvants.

This means that immunogenicity of vaccines has to be evaluated after the injection of undiluted preparations in man or even in small quantities in animals. In this way the equilibrium between the antigen and the adjuvant or other factors such as pH, ionic strength, etc. is not modified [33, 34]. The three-dilution method of titration is also very expensive, time-consuming and needs a large number of animals of good quality which are sometimes unavailable in many countries, especially guinea-pigs.

Antigenicity of toxoid vaccines can easily be measured in humans by an evaluation of the antibody response with the turkey cell passive haemagglutination test [30]. This has been done in pregnant women vaccinated with two different adsorbed toxoids. Vaccine A was prepared with toxoided purified tetanus toxin (TPTT) and vaccine B with purified toxoided tetanus toxin (PTTT). Titres expressed in International Immunizing Units (I.I.U.) were 69 per 0,5 ml for vaccine A adsorbed on to calcium phosphate and 212 per 0,5 ml for vaccine B adsorbed aluminium phosphate. As shown in Table 7 no significant difference

Table 7: Vaccination of pregnant women by two different adsorbed tetanus toxoids. Number of pregnant women immunized and antibody titres.

	Vaccine A 69 IIU/0.5 ml		**Vaccine B 212 IIU/0.5 ml**	
Period	**Total**	**Mean titre**	**Total**	**Mean titre**
1. First dose, month five of pregnancy	60		66	
2. Two months after first dose, second injection	39[a]	0.26	38[a]	0.21
3. Two months after second dose at birth	29[b]	1.12	29[b]	0.36
4. Six weeks after birth	17[b]	0.66	24[b]	0.29

Vaccine A: Prepared by toxoiding purified tetanus toxin and adsorbed on to calcium phosphate.
Vaccine B: Prepared by purifying formalized crude tetanus toxin and adsorbed on to aluminium phosphate.
a: Some women were absent or excluded due to the presence of a protective antibody level before immunization.
b: Some women were absent.

between circulating antibodies was obtained after the first injection but titres were much higher after the second injection for vaccine A having a low titre as compared to vaccine B having a high titre. These results show again that potency of vaccines cannot be evaluated by the injection of diluted doses in animals (regardless of the method), but have to be measured in man according to the practical immunization schedules.

The results also allow us to consider again that the calcium phosphate is a very efficient adjuvant for establishing basic immunity, and followed by high antibody response levels after a second or booster injection as shown before [35-39] and recently again in the Danish Army [40].

A simplified activity test for tetanus toxoids based on a comparison of circulating antitoxin titres in mice four weeks after injection of a reference toxoid and the vaccines to be tested has been developed [41, 42]. Antibody titrations were carried out by the HA test. This method needs a small number of animals, the results are reproducible and within confidence limits, and can replace the official potency test. Results of the measurement of circulating tetanus antitoxin levels in 15-20 mice after injection of 1/20 of the officially required human minimal dose (40 IIU/ml) for 16 different experiments are presented in Figure 7. The overall mean value is represented as a horizontal line and has been used to measure and compare the activity of 22 commercialized vaccines of which some have been titrated one to three times. Results are presented in Figure 8 giving antibody levels for each vaccine after separate titrations and mean values for repeated evaluations. It should be noted that several vaccines have high immunogenicity when compared to the European Reference Preparation but some preparations have levels that are below the official requirement.

In conclusion the results show that commercialized vaccines can be titrated in an easy and cheap manner by national control of countries that have no or insufficient local production and in this way can assist in making the optimal choice between available preparations.

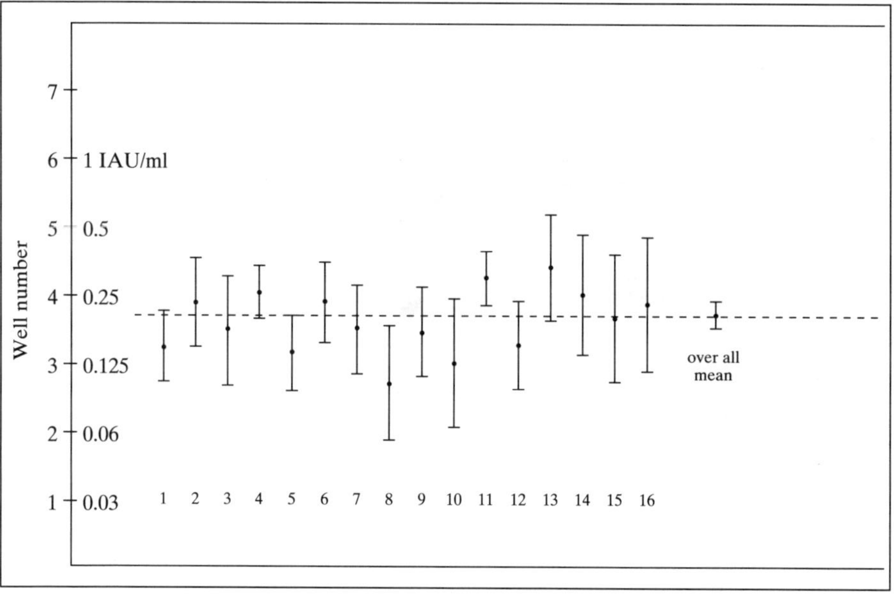

Fig. 7: Circulating tetanus antitoxin levels in mice after injection of two IIU of the European Reference Preparation. Mean values and 95% confidence limits for each assay in 15-20 mice, repeated 16 times.

Evaluation of Diphtheria Immunity

More than 25 cases of diphtheria epidemics have been reported during the last five years mostly in Russia but also in Algeria [43-47].

In Ukraine diphtheria was a common disease at the beginning of the century and its annual incidence rate reached a level of nearly 1% in 1910. Since 1960 a steady decrease has been noted, with only eight cases in 1975. However, from 1980 the annual number of reported cases increased, reaching about 3,000 in 1993. The number of diphtheria cases declined in 1994, suggesting that the epidemic had been brought under control [48]). Reasons for the resurgence of diphtheria in Russia include low immunization coverage. Control measures are urgently needed to help achieve high immunization coverage.The diphtheria HA test is again a rapid screening method for people about to be vaccinated. In this way it is easy to detect unprotected persons and also to avoid giving booster injections to protected persons and eliminate Arthus-type or allergic reactions especially if the vaccines are not prepared with a highly purified toxoided diphtheria toxin. In fact, we have shown before that reactions are not due to diphtheria toxoid itself but to impurities present in crude toxoids, which can not be eliminated by purification after interaction with formaldehyde. For this reason the diphtheria HA test was prepared by coupling crystalline diphtheria toxin to the turkey red blood cells allowing us to detect only antitoxic antibodies. Overall results in the French population for diphtheria immunity were about the same as for tetanus and it was found again that males were better protected compared to females, again because of DT boosters given in the army or for professionnal reasons.

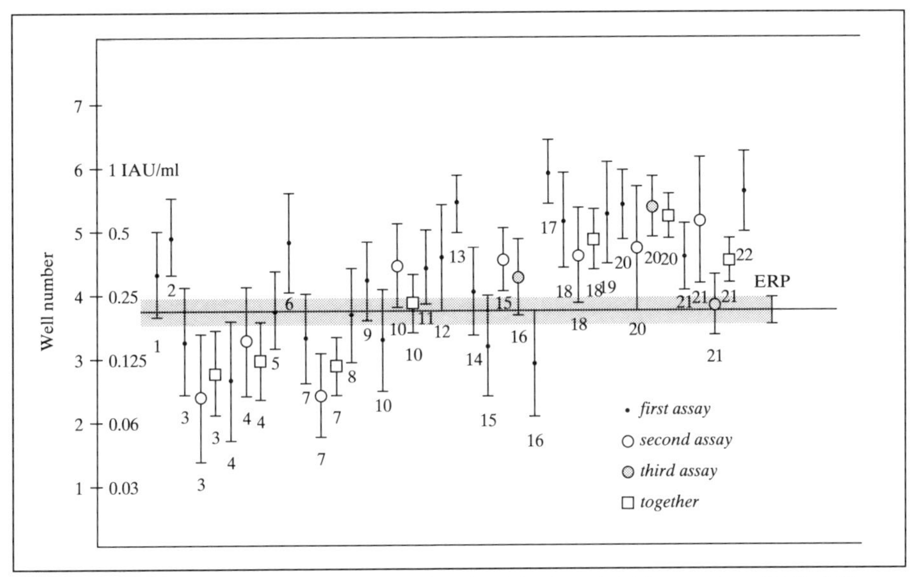

Fig. 8: Circulating tetanus antitoxin levels in mice after injection of 1/20 human dose (HD) of 22 commercialized vaccines as compared to two IIU of the European Reference Preparation (ERP) illustrated by a horizontal gray strip. Mean values and 95% confidence limits for each vaccine.

A study of protection in the elderly was recently published by Lery et al [49], which found that 53% of males had circulating antibody titres higher than 0,1 HU/ml compared to only 23% in females. It was again stressed that non-response to a booster injection was not related to the toxoid concentration but to vaccine immunogenicity. It is also more beneficial to boost with high doses of toxoided purified toxins and avoid adverse reactions instead of using the outdated low dose adult-type vaccines prepared from reactive crude toxoids.

It should be reiterated that diphtheria immunity can be easily detected by the use of a Schick toxin prepared using a crystalline preparation [50]. In this way the injection of a heated reference preparation to detect allergic reactions can be avoided. Comparative results of Schick tests and HA test should be very constructive.

Passive Haemaglutination Tests with Coated Viral Antigens

Turkey red blood cells were covalently coupled with purified recombinant HBs Ag and HIV-1 gp 160 antigen. Studies carried out in different laboratories to measure circulating anti-HBs antibodies showed good correlation between ELISA and HA titres. Strong correlation was also found between the ELISA and the HA test for the measurement of anti gp-160 antibodies. However the viral-coated turkey red blood cell tests have not yet been used in field trials or for large scale applications.

Evaluation of Immunodeficiency

One major application of the HA test for the future may be the evaluation of immunodeficiency in several diseases. Circulating tetanus antibodies have been measured in babies born from HIV-1 positive and negative mothers and vaccinated with BCG, DTP, polio and measles vaccines in Kigali (Rwanda), in a study of the University of Bordeaux II (INSERM U 330) in collaboration with the «Aids Task Force» of the EEC, the Belgian Medical Cooperation, the WHO and UNICEF and several laboratories and Public Health Authorities. Antibody levels were measured for tetanus at 6, 12 and 18 months and it was found that HIV+ babies born from HIV+ mothers were not protected, as opposed to those having HIV- mothers.

The tetanus immune response was studied in chronic lymphocytic (CLL) and myelocytic (CML) leukaemias in correlation with the occurrence of infections in non-protected subjects [52-53]. Immune response measured by HA is given in Table 8. After one, two or three injections, circulating antibody titres were higher in the control group compared to CML patients; CLL patients at stage C were devoid of antibodies or had very low titres which means that protection decreased according to the progression of CLL. The annual incidence of infection was the same in the control and CML groups but significantly increased in the CLL groups, susceptibility to infections even being correlated with the immune response.

Conclusion

The passive haemagglutination test with antigens covalently coupled to turkey erythrocytes is an economic and rapid method, which can easily be applied in countries with poor medical facilities in case of epidemics.

This method does not need sophisticated laboratory equipment or even electricity. No coating, washing nor long incubation of the plates at 37°C is necessary.

Table 8: Circulating tetanus antibody titres after vaccination.

	Controls	CML	CLL A	CLL B	CLL C
Before vaccination	0	0	0	0	0
After 1 injection	1.4 (2.37)	0.92 (1.29)	1.45 (2.25)	0.47 (0.95)	0
After 2 injections	3.03 (4.85)	1.78 (3.82)	2.68 (4.73)	0.63 (0.97)	0.02 (0.05)
After 3 injections	3.07 (4.82)	1.79 (3.82)	2.70 (2.32)	0.65 (0.96)	0.12 (0.13)

CML: Chronic myelocytic leukaemia
CLL: Chronic lymphocytic leukaemia
A, B, C: Stages of CLL.
Antibody titres are expressed in haemagglutination units: mean values and standard deviation.

Results are known after dilution of the blood samples in the wells, mixing with the sensitized red blood cells, and incubation for 20 minutes at room temperature. The coated turkey red blood cells have to be prepared in specialized laboratories using highly purified antigens and all the reference sera have to be titrated in vivo. Results of different research and hospital laboratories have been reported, mainly for the titration of tetanus antibodies to measure immune coverage in field trials. The technique is also valuable for screening blood samples to prepare immune sera and to evaluate the potency of vaccines in a simple way. Some applications to measure diphtheria immunity have also been reported. It is important to evaluate protection in regions with epidemics and to avoid reactions after useless boosting. No other available method can be used to evaluate protection rapidly and simply in isolated regions. Other toxins such as pertussis, staphylococcal, streptococcal etc. can also be coupled by glutaraldehyde [54] to the turkey cells to measure circulating antibodies. Applications to assess immunodeficiency in diseases can easily be studied, by measuring circulating antibodies after daily vaccination in one drop of blood and even for months. It should be stressed that results are estimated by reading of the last positive reaction of serial two-fold dilutions by comparison with a reference serum; this can be less quantitative compared to dose-response curves based on optical densities used in ELISA, capture ELISA and ToBI assays [55]. Guides to tetanus prophylaxis management have been given for four groups of protection.

References

1 Huet M: Rapid diagnostic methods, in Nisticò G, Bizzini B, Bytchenko B, Triau R (eds): Eight International Conference on Tetanus. Roma, Milan, Phytagora Press, 1989, pp 457-464.

2 Boyden SV: The adsorption of proteins on erythrocytes treated with tannic acid and subsequent hemagglutination by antiprotein sera. J Exp Med 1951;93:107-120.

3 Fauconnier B: Utilisation des hématies hyper-formolées en virologie. I. Préparation et propriétés générales de ces hématies. Ann Inst Pasteur 1958;95:777-780.

4 Pitzurra M, Bistoni F, Pitzurra L, Marconi P: Use of turkey red blood cells in the passive haemagglutination test for studying tetanus immunity. Bull WHO 1983;61 331-338.

5 Relyveld EH: Reagent for the determination by haemagglutination of antibodies to bacterial toxins, method of preparation and application thereof. United States Patent 1994;Jun.7;N° 5, 318, 913.

6 Saliou P, Le Guellec A, Buisson Y, Antoine HM, Relyveld EH: Intérêt du dosage extemporané des anticorps tétaniques pour une prophylaxie adaptée. Bull Mém Soc Méd Paris 1986;14:19-22.

7 Relyveld EH, Saliou P, Coursaget P, Marié FN: Evaluation of antitoxic antibody levels using passive haemagglutination with toxins covalently coupled to turkey erythrocytes: Intern.Conf. on Covalently Modified Antigens and Antibodies in Diagnosis and Therapy, Lyon, Inserm-Cytogen 1987, p 64.

8 Saliou P, Relyveld EH, Marié FN, Le Cam N, Coursaget P: Comparison of the results of field trials using various assessment methods for estimation of circulating tetanus antibody levels, in Nistico G, Bizzini B, Bytchenko B, Triau R (eds): Eight International Conference on Tetanus.Roma, Milan, Phytagora Press, 1989, pp 267-273.

9 Saliou P, Le Guellec A, Relyveld EH: Evaluation d'une technique d'hémagglutination passive pour le dosage rapide des anticorps tétaniques. Médecine et Armées 1987;15:595-597.

10 Saliou P, Marié FN, Relyveld EH: Evaluation de la fiabilité et de l'intérêt d'un test d'hémagglutination passive pour le dosage rapide des anticorps tétaniques. In Fondation Marcel

Mérieux 3e Séminaire International sur les Vaccinations en Afrique, Niamey 29-31 janvier 1987, pp. 663-668.

11 Beytout D, Cailleba A, Dolci A, Nguyen TT, Mamouret A, Lauras H: Evaluation du test rapide de mesure de l'immunité antitétanique dans la pratique d'un service d'urgences. La Presse Méd 1988;17:2091.

12 Mamouret-Beytout A, Nguyen Trung T, Laveran H, Dolci A, Lauras H, Beytoud D: Utilisation de la réaction d'hémagglutination passive (Vacci-test Pasteur) pour évaluer l'immunité contre le tétanos. Méd Mal Infect 1988;11:802-806.

13 Beytout D, Nguyen Trung T, Laveran H, Mamouret-Beytout A: Contrôle de l'état d'immunité antitétanique dans la population du Puy-de-Dôme. Méd Mal Infect 1988;12:897-899.

14 Le Cam N, Relyveld EH: Intérêt du dosage extemporané des anticorps tétaniques par le Vacci-test T Pasteur. Le Médecin du Midi 1989;93: 1 & 4.

15 Beytout J, Lauras H, Cailleba A, Nguyen TT: Rapid evaluation of tetanus immunity by a haemagglutination test in the injured at a hospital emergency unit. Biomed & Pharmacother 1989; 43:621-625.

16 Relyveld EH, Saliou P, Le Cam N: Dosage d'anticorps tétaniques à l'aide du Vacci-test T Pasteur. L'Eurobiologiste 1990;24:219-228.

17 Relyveld EH: Titrage in vivo des anticorps antidiphtériques et antitétaniques à plusieurs niveaux. J Biol Stand 1977;5:45-55.

18 Gupta RK, Maheshwari SC, Singh H: The titration of tetanus antitoxin II. A comparative evaluation of the indirect haemagglutination and toxin neutralization test. J Biol Stand 1984;12: 137-143.

19 Gupta RK, Maheshwari SC, Singh H: The titration of tetanus antitoxin. III. A comparative evaluation of indirect haemagglutination, and toxin neutralization titers of human sera. J Biol Stand 1984;12:145-149.

20 Gupta RK, Sharma SB, Ahuja S, Saxena SN: Indirect (passive) haemagglutination test for assay of antigen and antibody. Acta Microbiol Hung 1991;33:81-90.

21 Hagenaars AM, van Delft RW, Nagel J: Comparison of ELISA and toxin neutralization for the determination of tetanus antibodies. J Immunoassay 1984; 5: 1-11.

22 Simonsen O, Bentson MW, Heron I: ELISA for routine determination of antitoxic immunity to tetanus. J Biol Stand 1986;15:223-230.

23 Simonsen O, Schou C, Heron I: Modification of the ELISA for the estimation of tetanus antitoxin in human sera. J Biol Stand 1987;15:134-157.

24 OMS: Anticorps. Serie Rapports Techn 1994;840:5-6.

25 OMS: Programme des Vaccins: 1994; Mai: 1.

26 Champeau J, Arondel D, Bellanger P, et coll.: Tetanus immune coverage in France, in Nisticò G, Bizzini B, Bytchenko B, Triau R (eds): Eight International Conference on Tetanus. Roma, Milan, Phytagora Press, 1989, pp 574-583.

27 Lery L: Vaccitestez-vous! Le Quotidien du Médecin 1988;4147:28.

28 Vaccinum Tetanicum. Pharmacopée Européenne. 57 Sainte Ruffine, France, Maisonneuve S.A., 1977, Supplément au volume III, pp 184-189.

29 Huet M: La standardization des vaccins tétaniques, in Fondation Mérieux (ed): Proc 6th Int Conf on Tetanus Lyon 1981, pp 425-433.

30 Relyveld E, Bengounia A, Huet M, Kreeftenberg JG: Antibody response of pregnant women to two different adsorbed tetanus toxoids. Vaccine 1991;9:369-372.

31 Hardegree MC, Formwald RE, Farber J, London WT, Fave P, Kessler MJ, Rastogisc: Titration of tetanus toxoids in international units: relationship to antitoxin responses of rhesus monkeys, in Fondation Mérieux (ed): Proc 6th Int Conf on Tetanus Lyon 1981, pp 409-423.

32 Lyng J, Nyerger G: The second international Standard for tetanus toxin adsorbed. J Biol Stand 1984;12:121-130.

33 Hennessen W: The mode of action of mineral adjuvants. Progr Immunobiol Stand 1965;2: 71-79.

34 Relyveld EH: Immunological, prophylactic and standardization aspects in tetanus, in Nisticò G, Mastroeni P, Pitzurra M (eds): Seventh International Conference on Tetanus. Roma, Gangemi, 1985, pp 215-227.

35 Cabau N, Levy FM, Relyveld EH, Labusquière R, Poirier A, Ravisse P, Chambon L: Vaccination antidiphtérique-antitétanique par anatoxines adsorbées sur phosphate de calcium en deux injections à un an d'intervalle. Ann Inst Pasteur 1970;119:663-670.

36 Relyveld EH, Labusquière R, Gateff C, Le Bourthe F, Ravisse P, Lemarinier G, Chambon L: Antitetanus vaccination and neonatal protection in developing countries. Prog Immunobiol Stand 1972;5:517-527.

37 Sureau P, Fabre J, Bedaya N'Garo S, Come Butor S, Poulougou M, Relyveld EH: Vaccination simultanée de nourrissons en milieu tropical contre le tétanos et la poliomyélite. Bull Org Mond Santé 1977;55:739-746.

38 Relyveld EH: Résultats de calendriers de vaccinations simplifiées dans les pays en voie de développement. Dev Biol Stand 1978;41:295-299.

39 Gupta RK, Siber GR: Comparison of adjuvant activities of aluminium phosphate, calcium phosphate and stearyl tyrosine for tetanus toxoid. Biologicals 1994;22:53-63.

40 Aggerbeck H, Heron I: Potency test of diphtheria, tetanus and combined DT vaccines, suggestions for simplified potency assays. Dev Biol Stand 1995;86:295-299.

41 Huet M, Relyveld E, Camps S: Methode simple de contrôle de l'activité des anatoxines tétaniques absorbées. Biologicals 1990;18:61-67.

42 Huet M, Relyveld E, Camps S: Simplified evaluation of several tetanus vaccines. Biologicals 1992;20:35-43.

43 MMWR: Diphtheria outbreak. Russian Federation 1990-1993. 1993;42:840-847.

44 Rich V: Diphtheria in Russia. Lancet 1994;343:169.

45 Ivanov VA: Diphtheria in Russia. Lancet 1994;343:675.

46 Sasse A, Malfait P, Padron T, Erikashvili M, Freixa E, Moren A: Outbreak of diphtheria in Republic of Georgia. Lancet 1994; 343: 1358-1359.

47 Weekly Epid Rec: Communicable diseases – Epidemiological Situation in 1990, Algeria. 1992; 67:75-76.

48 Weekly Epid Rec: Expanded Programme on Immunization, Diphtheria epidemic. 1994;69:253-258.

49 Lery L, Ducomet P, Trabaud MA: Immunité humorale antidiphthérique dans une population âgée: mise en évidence par un test d'hémagglutination. Méd Mal Infect 1994;24:650-654.

50 Relyveld EH, Hénocq E, Raynaud M: Etude de la réaction de Schick à l'aide d'une toxine pure. Bull Acad nat Méd 1962;146:101-109.

51 Msellati P: Infection par le VIH chez l'enfant et vaccinations de routine. Thèse de médecine n° 3104, Université de Bordeaux II, 1990.

52 Berrah A, Colonna P, Berrah A, Relyveld EH: Vaccination antitétanique au cours des leucémies chroniques. Bulletin du Cancer 1989;76:468.

53 Berrah A, Relyveld EH, Colonna P, Berrah A: Immune response to antitetanic immunization and occurence of infections in chronic leukemias, in Merino J (ed): 12 th Congress of the European Association of Internal Medicine. Madrid Ediciones Libro del Ano, 1995, pp 342-345.

54 Relyveld EH, Ben-Efraim S: Preparation of vaccines by the action of glutaraldehyde on toxins, bacteria, viruses, allergens and cells, in Langone JJ, Van Vunakis (eds): Methods in enzymology, Immunochemical Techniques part F. New York, London, Academic Press, 1983, vol 93, pp 24-60.

55 Hendriksen CFM, Woltjes J, Akkermans AM, Van der Gun JW, Marsman FR, Verschure MH, Veldman K: Interlaboratory validation of in vitro serological assay systems to assess the potency of tetanus toxoid in vaccines for veterinary use. Biologicals 1994;22:257-268.

Correspondence and reprints requests to:

Prof. E. Relyveld, 6, rue du Sergent Maginot, 75016 Paris, France

Brown F, Cussler K, Hendriksen C (eds): Replacement, Reduction and Refinement of Animal Experiments in the Development and Control of Biological Products.
Dev Biol Stand. Basel, Karger, 1996, vol 86, p 243

WORKSHOP

Serological Methods and Cell Cultures

Chairmen: P. Vannier (Ploufragan, France)
R. Gupta (Boston, USA)

Brown F, Cussler K, Hendriksen C (eds): Replacement, Reduction and Refinement of Animal Experiments in the Development and Control of Biological Products.
Dev Biol Stand. Basel, Karger, 1996, vol 86, pp 245-260

Comparison of in Vivo and in Vitro Methods for Determining Unitage of Diphtheria Antitoxin in Adsorbed Diphtheria-Tetanus (DT) and Diphtheria-Tetanus-Pertussis (DTP) Vaccines

Y.P. Shinde, S.S. Jadhav

Serum Institute of India Ltd., Pune M.S., India

Abstract: In view of the current efforts to find a reliable in vitro method which can suitably act as an alternative for determining the potency of the diphtheria component in a combined vaccine, we have analysed experimental batches by the method proposed by WHO [1] i.e. challenge method in guinea pigs. The same batches were also analysed by the alternative antibody induction method as suggested in the Indian Pharmacopoeia (I.P.) [2] which is similar to the old method suggested in the British Pharmacopoeia (B.P.) 1973. As per I.P. the initial part of raising the antibodies remains unaltered but the actual titration of diphtheria antitoxin from the immunised guinea pigs was performed by using the following in vitro methods:

a) indirect haemagglutination test using human «O» red blood cells to coat diphtheria toxoid using chromic chloride as the coupling agent [3];
b) toxin neutralisation test using Vero cells [4];
c) a double diffusion technique in agar gel for titration of diphtheria antitoxin [5].

Our findings show clearly that the results of two in vivo methods i.e. Challenge Test, Alternative I.P. Method and the above-mentioned three in vitro methods are comparable and would certainly reduce the number of animals required by making a combination of in vivo and in vitro techniques to give us an assessment of the potency of the vaccine to be tested.

Introduction

The test for estimating the potency of the diphtheria component in adsorbed vaccines recommended by WHO requires a large number of animals. In India the potency test of the diphtheria component of a vaccine (DT, DTP) is performed by the antibody induction method in guinea pigs as described in I.P. 1985, which involves the immunisation of 10 guinea pigs with 1/50th dilution of a single human dose of vaccine on two occasions, separated by an interval of four weeks. Three weeks later the sera of immunised animals are analysed by the toxin neutralisation test in guinea pigs. In this study we have explored the use of some antigen-antibody reaction techniques to quantify the immune response of diphtheria component by agglutination, toxin neutralisation using Vero cell assay and double diffusion in agar. We have found very good correlation amongst them and also between in vivo tests.

Materials and Methods

Vaccines: Adsorbed Diphtheria Tetanus (DT) & Diphtheria Tetanus Pertussis (DTP) Vaccines, manufactured by the Serum Institute of India. The vaccines had the following composition per ml.:

	DT	DTP
Diphtheria toxoid	50 Lf	50 Lf
Tetanus toxoid	10 Lf	10 Lf
Bordetella pertussis	—	32×10^9 org.
AlPO4	5 mg	3.5 mg
Single human dose	0.5 ml	0.5 ml

Standard toxoid

Freeze-dried WHO reference standard of Adsorbed Diphtheria Toxoid [134 IU/ml].

Standard antitoxin

The National Reference Standard for Diphtheria Antitoxin 10 IU/ml calibrated against the International Standard for Diphtheria Antitoxin was used for the standardisation of the antibody titrations.

Diphtheria toxin

Lot No. 533, 300 Lf/ml was supplied by the Diphtheria Division of the Serum Institute of India Ltd., and used throughout this study.

Diphtheria toxoid

Purified Diphtheria Toxoid, Lot No. 424, 2800 Lf/ml was obtained from the Serum Institute of India Ltd. and used throughout the study.

Animals

Guinea pigs weighing 250-350g were used for immunisation. Adult guinea pigs weighing 600-700g were used for intradermal neutralisation tests.

Erythrocytes

Red blood cells were obtained from healthy human donors with blood group «O» (preferably Rh-ve).

Reagents

Bovine serum albumin (Sigma)-Fraction V, 96.99% albumin 4% solution was prepared in normal saline.

Stock chromic chloride was prepared by taking chromic chloride hexahydrate (Sigma) 157 mg in 10 ml normal saline. A 1:10 dilution was used.

Borate Buffer – 2 g of NaOH, 9 g of H_3BO_3 at pH 8.6 was used. Noble agar (1%) was made in borate buffer.

Diluent – Physiological saline was used for diluting the antitoxin, sera, vaccine and toxin.

Lethal Challenge Test in Guinea Pigs as per WHO Specifications

Each group of 16 guinea pigs was immunised subcutaneously with three different two-fold dilutions of the WHO reference preparation and the vaccine under test. Twenty-eight days later all the animals were challenged by subcutaneous injection with 100 LD50 of diphtheria toxin and the number of surviving animals was recorded up to five days. The potency was calculated by Probit analysis using the Marsman Programme.

Alternate Indian Pharmacopoeial Method

Ten guinea pigs were injected with a 1/50 dilution of a single human dose of vaccine on two occasions, separated by an interval of four weeks. Three weeks after the second dose, sera of immunised guinea pigs were collected and used for analysis. A toxin neutralisation test was performed using Lr+/100 dose of toxin in the depilated backs of guinea pigs. Erythematous reaction was compared with standard reference antitoxin. The geometric mean of the sera was taken for calculating the potency. To pass the batch Geometeric Mean should not be less than 2 IU/ml.

Table 1 shows comparative results of the WHO challenge method and the alternative IP Method.

Passive Haemagglutination Method

Sensitization of human red blood cells: Diphtheria toxoid-coated cells were prepared freshly by taking 0.2 ml of a 50% suspension of thoroughly washed «O» Rh-ve RBC's, 1 ml of diphtheria toxoid at 700 Lf/ml and 0.4 ml of 1:10 diluted chromic chloride and mixed for five minutes. After coating, cells were centrifuged and mixed in 3 ml normal saline containing 4% bovine serum albumin.

Table 1: Review of existing potency methods (Indian Pharmacopoeial/World Health Organisation Method).

Batch No.	W.H.O. method Challenge method	I.P. method Antibody induction method
	Diphtheria component I.U./Dose	Diphtheria component I.U./ml of sera
DT-70	87	GM is 4.0
DPT-268	73	GM is >4.0
DPT-274	91	GM is >4.0
DPT-280	73	GM is >4.0
DPT-291	86	GM is >4.0
DPT-269	107	GM is >4.0
DPT-270	80	GM is >4.0
DPT-271	60	GM is >4.0
DPT-272	95	GM is >4.0
DPT-273	66	GM is >4.0
DPT-291	133	GM is >4.0
DPT-292	112	GM is >4.0
DPT-293	91	GM is >4.0
DPT-294	94	GM is >4.0
DPT-295	120	GM is >4.0

Passive Haemagglutination Test

Serial two-fold dilutions of guinea pig sera (25 µl) were prepared in «V» bottomed microtitre plates and to this 25 µl of sensitised RBC were added. The mixture was incubated at 37°C for one hour. Reference diphtheria antitoxin (10 IU/ml) was introduced in the test; Table 2 shows the protocol of the test. A positive reaction was recognised on the basis of a firm button or mat, while negative results were seen as a flowing button (Figs. 1 and 2).

The comparison of the Passive Haemagglutination (PHA) and Intradermal Neutralisation (IDN) tests in guinea pigs are shown in Table 3.

Vero Cell Assay to Determine Diphtheria Antitoxin Content in Guinea Pig Sera

Vero cells were seeded in Roux bottles (200 cm^2) containing Eagle's Base medum (MEM) supplemented with Hanks salt L-glutamine at pH 7.0-7.2, filter sterilized and supplemented with 10% foetal bovine serum. After six to seven days a confluent monolayer is obtained which is trypsinised and the cell count is adjusted to 4×10^5 cells/ml with a haemocytometer.

The Vero cell assay has been used to determine the concentration of diphtheria antitoxin in guinea pig sera [4-6]. Serum samples from immunised guinea pigs were inactivated for 30 minutes at 56°C and stored at –20°C until assayed. Sera were initially diluted 1:5 times and then two-fold dilutions of each of the ten serum samples were made; an antitoxin standard was included as shown in Table 4. The toxin neutralisation test was carried out at Ltc+/100 dose of toxin. The excess amount of toxin which remains after neutralisation will show metabolic inhibition as observed

Table 2: Determining diphtheria anti-toxin by passive haemagglutination test using group «O» human red cells & chromic chloride.

Product: Batch No:
Date of 1st immunisation: Date of 2nd immunisation:
Date of bleeding: Date of titration:
Diluent: N. Saline

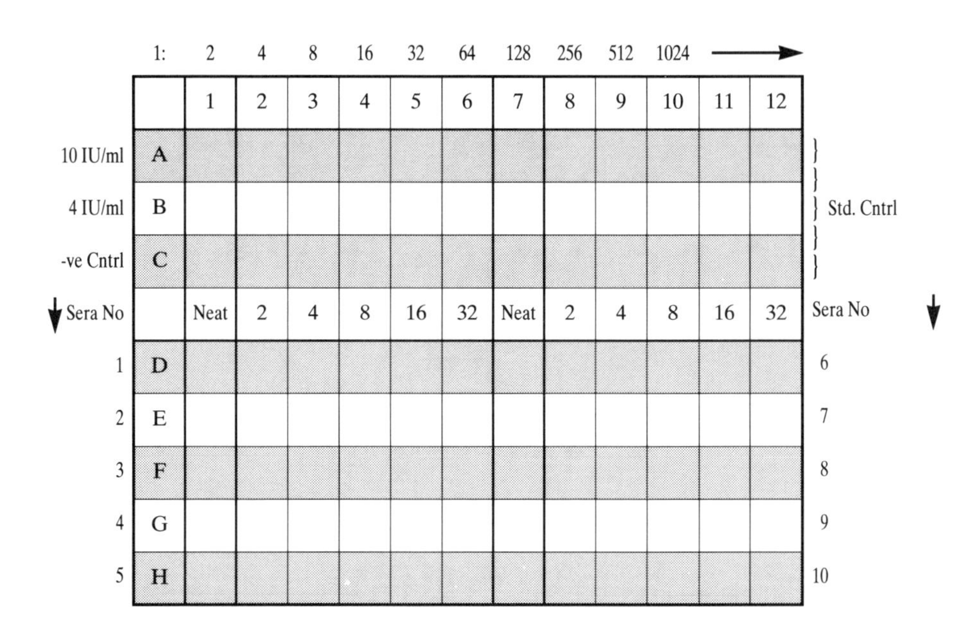

	1:	2	4	8	16	32	64	128	256	512	1024 →			
		1	2	3	4	5	6	7	8	9	10	11	12	
10 IU/ml	A													Std. Cntrl
4 IU/ml	B													Std. Cntrl
-ve Cntrl	C													Std. Cntrl
↓ Sera No		Neat	2	4	8	16	32	Neat	2	4	8	16	32	Sera No ↓
1	D													6
2	E													7
3	F													8
4	G													9
5	H													10

Standard ADS – 10 IU/ml & 4 IU/ml
Titre for 10 IU std. –
Lowest detectable Titre –

Results: → Agglutinated mat = + reaction, Button formation -ve reaction

Sera	Titre IU/ml	Sera	Titre IU/ml
1		6	
2		7	
3		8	
4		9	
5		10	

Remarks:

in the Figures 3, 4, and 5. Table 5 compares the results between the Vero cell assay and intradermal toxin neutralisation test.

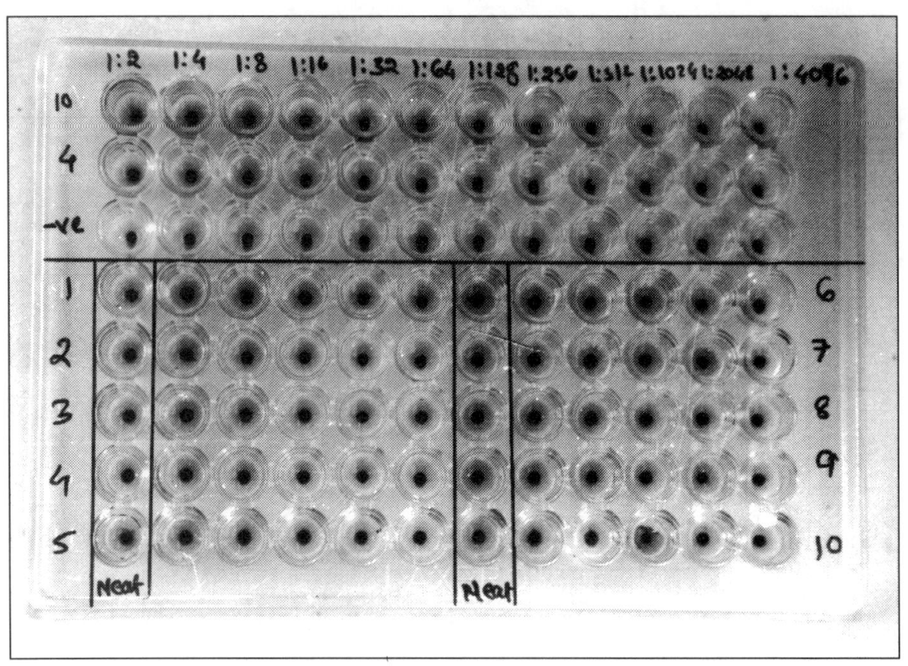

Fig. 1: Passive haemagglutination test.

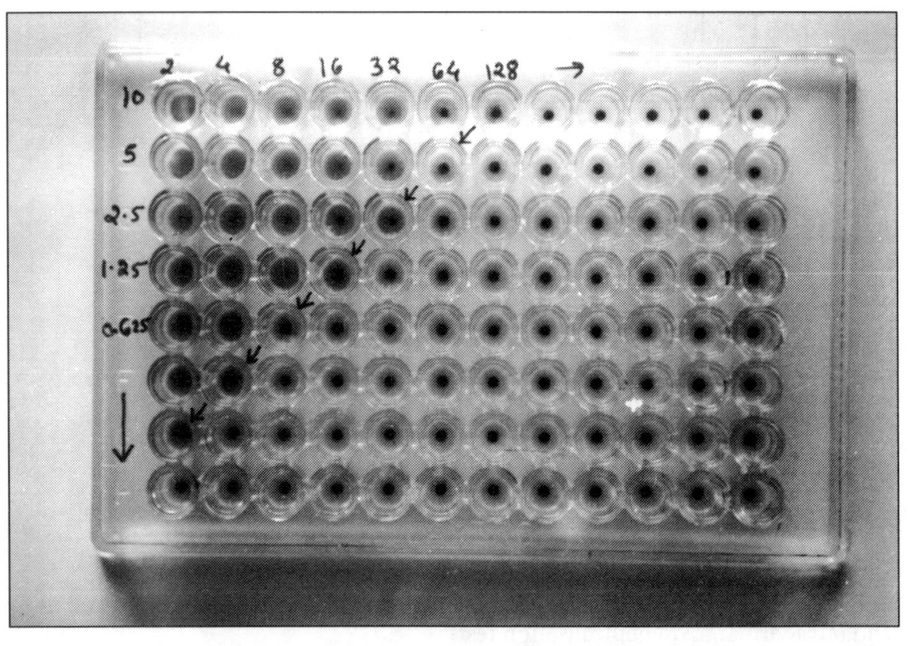

Fig. 2: Diagonal relationship between IU/ml & haemagglutination titre.

Table 3: Determining the end point of diphtheria antitoxin titres by toxin neutralisation and passive haemogglutination test.

Batch No.	Sera No.	IU/ml by PHA	IU/ml by TN	Batch No.	Sera No.	IU/ml by PHA	IU/ml by TN
DTP 245	1.	10	7.5	DTP 246	1.	10	8.0
	2.	10	8.0		2.	10	8.0
	3.	10	7.0		3.	5	6.0
	4.	5	3.5		4.	10	8.5
	5.	10	7.0		5.	5	5.5
	6.	10	14.0		6.	5	6.0
	7.	5	4.0		7.	10	8.0
	8.	5	6.0		8.	10	12.0
	9.	10	8.0		9.	10	12.0
	10.	10	7.5		10.	10	8.0
Co-efficient of correlation = 0.6658				Co-efficient of correlation = 0.7352			

Batch No.	Sera No.	IU/ml by PHA	IU/ml by TN	Batch No.	Sera No.	IU/ml by PHA	IU/ml by TN
DT 58	1.	10	8.0	DT 59	1.	5	3.5
	2.	5	8.0		2.	5	5.0
	3.	10	8.0		3.	10	15.0
	4.	20	16.0		4.	10	8.0
	5.	10	8.5		5.	5	4.0
	6.	10	10.0		6.	5	4.5
	7.	10	7.5		7.	5	5.0
	8.	5	6.0		8.	10	8.5
	9.	10	12.0		9.	10	6.5
	10.	2.5	1.5		10.	10	8.0
Co-efficient of correlation = 0.8937				Co-efficient of correlation = 0.6695			

Batch No.	Sera No.	IU/ml by PHA	IU/ml by TN	Batch No.	Sera No.	IU/ml by PHA	IU/ml by TN
DT 61	1.	2.5	2.0	DTP 251	1.	10	12.0
	2.	5	4.0		2.	10	12.0
	3.	2.5	6.0		3.	20	16.0
	4.	5	6.0		4.	10	8.0
	5.	10	8.0		5.	10	14.0
	6.	2.5	2.0		6.	5	6.0
	7.	10	8.0		7.	10	10.0
	8.	10	12.0		8.	5	8.0
	9.	2.5	2.0		9.	10	14.0
	10.	5	8.0		10.	10	12.0
Co-efficient of correlation = 0.8324				Co-efficient of correlation = 0.777			

Serum samples from six batches immunised against diphtheria component were titrated to determine the end point of diphtheria antitoxin content. Table shows units of diphtheria antitoxin per ml, in guinea pig sera detected by TN as well as the passive haemagglutination test.

Table 4: Determining diphtheria antitoxin using Vero cells.

Product:
Date of 1st immunisation:
Date of bleeding:
Cell count: per ml

Batch No.:
Date of 2nd immunisation:
Date of titration:

SERA (heat inactivated) DILUTION 1:5 IN SALINE. Two fold dilution of the sera from A to H.

← Serum sample No.: 1 to 10 →											Std. controls		
	1	2	3	4	5	6	7	8	9	10	11	12	Standard ADS Kasauli
A													→ 0.04 IU/50 µl
B													→ 0.02 IU/50 µl
C													→ 0.01 IU/50 µl
D													→ 0.005 IU/50 µl
E													} Toxin control
F													}
G													} Cell control
H													}

Standard ADS (K) Dilution (10 IU/ml) – N.S. 2.3 ml + 0.2 ml ADS = 0.8 IU/ml
i.e. 0.04 IU/50 µl

Toxin dilution at Ltc/100 level (Toxin lot No. 553) – N.S. 9.9 ml + 0.1 ml toxin
N.S. 18 ml + 2 ml of 1:100 dil'n
i.e. 1:1000 dil'n

Results: → + Toxic effect – Healthy cells

Sera	Titre IU/ml	Sera	Titre IU/ml
1		6	
2		7	
3		8	
4		9	
5		10	

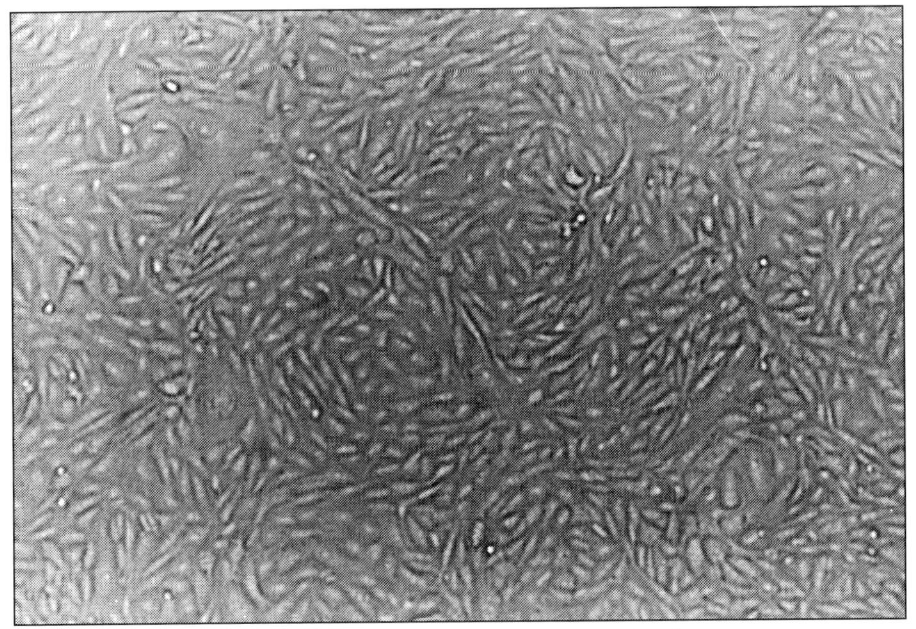

Fig. 3: Vero cells – Healthy mat (negative control).

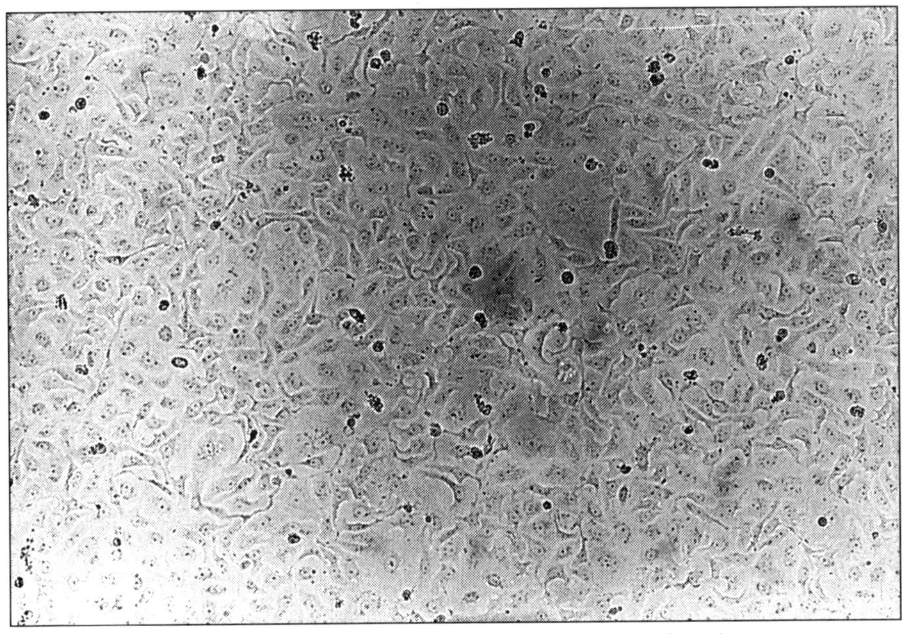

Fig. 4: Vero cells – Initiation of toxic efffect of diphtheria toxin.

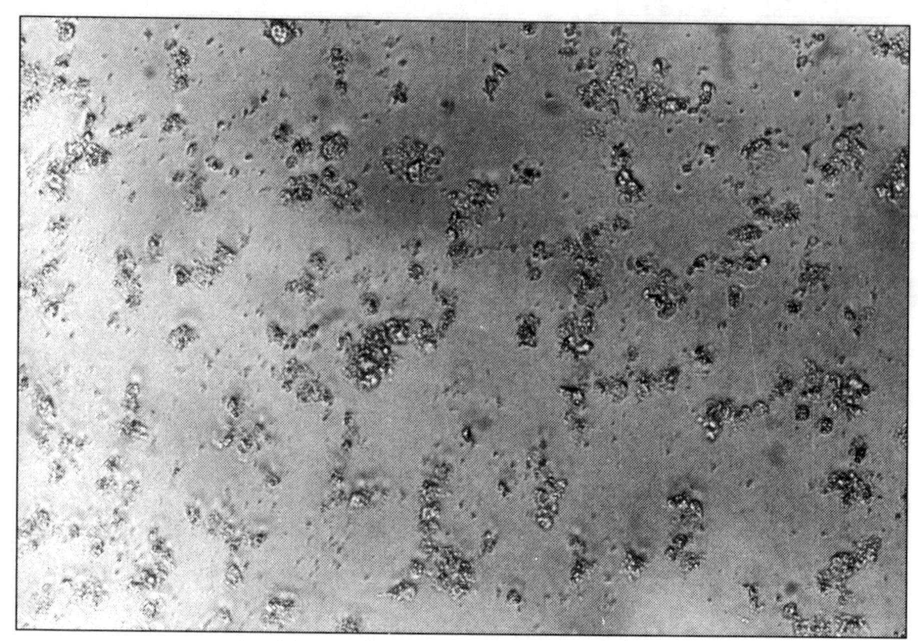

Fig. 5: Vero cells – necrotic cells (positive control).

Double Diffusion Technique for Determination of Diphtheria Antitoxin

The method of titration [7], with few modifications, was adapted to test the antitoxin content in the immunised guinea pig sera.

Noble agar (1%) was prepared in borate buffer 17 ml agar was poured into 100 mm dia. petri dish. A seven well pattern as shown in Figure 1 was used. These wells were of 5.3 mm in diameter and had been placed 2.4 mm apart. The central well is loaded with toxin and the side well with sera. Initially we needed to determine the dose of toxin which precipitated 2 IU/ml (15 Lf/ml of toxin) and 4 IU/ml (30 Lf/ml of toxin) of standard diphtheria antitoxin.

In the test, 35 µl of undiluted serum from immunised guinea pigs was placed in side wells and 35 ml of corresponding toxin in the central well. Plates were incubated at 20-22°C for 24 hours and observed for precipitation lines. Test results for two different batches are shown in Figures 6, 7 and 8.

Table 6 gives the detailed comparison of the results obtained by the three different in vitro methods with the results of the toxin neutralisation test as per Indian Pharmacopoeia.

Discussion

In this study we have tried to determine whether it will be possible to reduce the number of animals for determining the potency of the diphtheria component in

Table 5: Determining the end point of diphtheria antitoxin titres by toxin neutralisation and Vero cell assay method.

Batch No.	Sera No.	IU/ml by Vero	IU/ml by TN	Batch No.	Sera No.	IU/ml by Vero	IU/ml by TN
DTP 245	1.	8	7.5	DTP 246	1.	8	8.0
	2.	8	8.0		2.	8	8.0
	3.	8	7.0		3.	8	6.0
	4.	4	3.5		4.	8	8.5
	5.	8	7.0		5.	8	5.5
	6.	16	14.0		6.	4	6.0
	7.	4	4.0		7.	8	8.0
	8.	8	6.0		8.	16	12.0
	9.	8	8.0		9.	16	12.0
	10.	8	7.5		10.	8	8.0
Co-efficient of correlation = 0.9787				Co-efficient of correlation = 0.9287			

Batch No.	Sera No.	IU/ml by Vero	IU/ml by TN	Batch No.	Sera No.	IU/ml by Vero	IU/ml by TN
DT 58	1.	8	8.0	DT 59	1.	4	3.5
	2.	8	8.0		2.	4	5.0
	3.	8	8.0		3.	16	15.0
	4.	16	16.0		4.	8	8.0
	5.	8	8.5		5.	4	4.0
	6.	8	10.0		6.	4	4.5
	7.	8	7.5		7.	4	5.0
	8.	8	6.0		8.	8	8.5
	9.	16	12.0		9.	8	6.5
	10.	2	1.5		10.	8	8.0
Co-efficient of correlation = 0.9236				Co-efficient of correlation = 0.9747			

Batch No.	Sera No.	IU/ml by Vero	IU/ml by TN	Batch No.	Sera No.	IU/ml by Vero	IU/ml by TN
DT 61	1.	2	2.0	DTP 251	1.	8	12.0
	2.	4	4.0		2.	16	12.0
	3.	8	6.0		3.	16	16.0
	4.	8	6.0		4.	8	8.0
	5.	8	8.0		5.	16	14.0
	6.	2	2.0		6.	8	6.0
	7.	8	8.0		7.	8	10.0
	8.	16	12.0		8.	8	8.0
	9.	2	2.0		9.	16	14.0
	10.	8	8.0		10.	16	12.0
Co-efficient of correlation = 0.9459				Co-efficient of correlation = 0.8026			

Serum samples from six batches immunised against diphtheria component were titrated to determine the end point diphtheria antitoxin content. Table shows units of diphtheria antitoxin per ml, in guinea pig sera detected by TN as well as the Vero cell assay method.

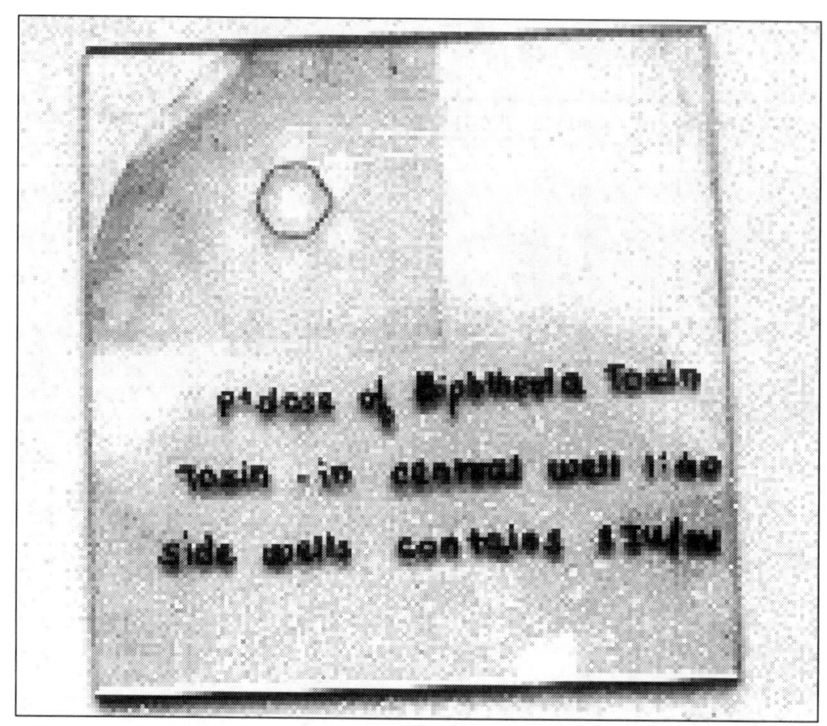

Fig. 6: Double diffusion in agar for 1IU/ml of diphtheria antitoxin with diphtheria toxin.

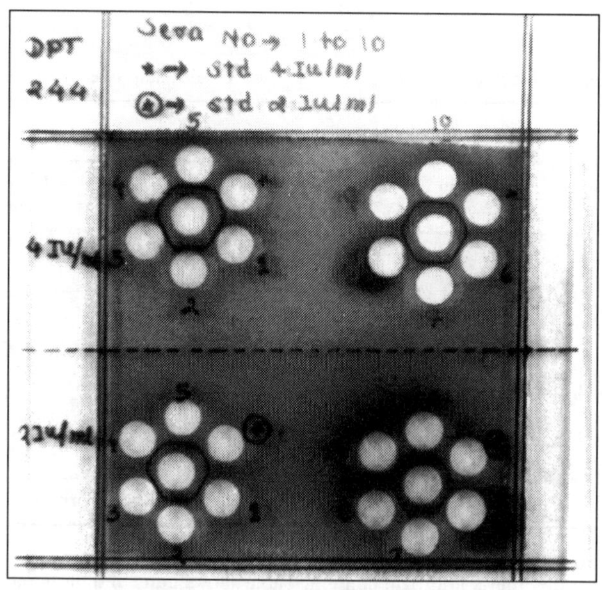

Fig. 7: Test for DPT batch showing titre more than 2 & 4 IU/ml.

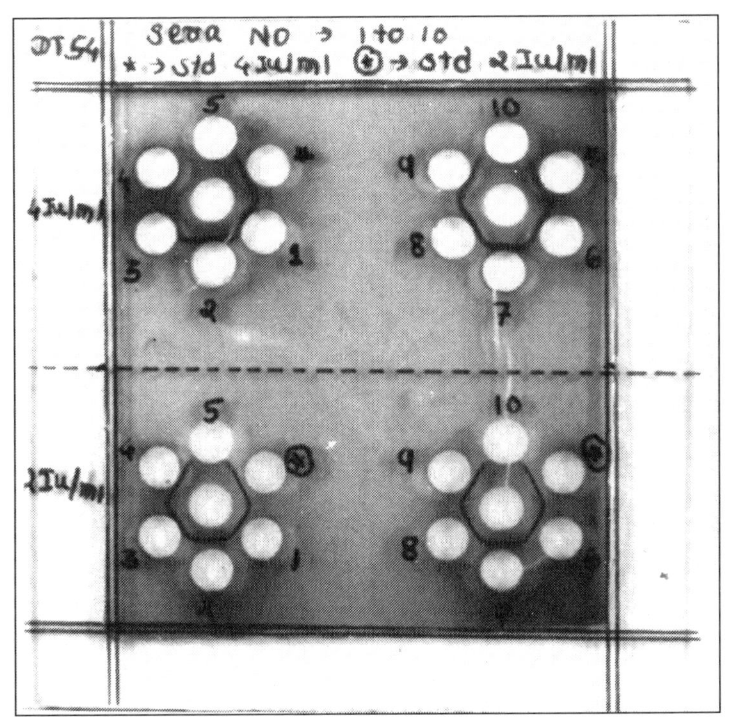

Fig. 8: Test for DT batch showing precipitation line for 2IU & 4IU/ml for different sera.

vaccines. At present, we require 116 guinea pigs for the WHO Challenge Method to analyse one sample of the test batch, whereas, in the case of the antibody-induction method as described in the Indian Pharmacopoeia, we require 10 guinea pigs for the induction of antibody response and an additional two guinea pigs for the determination of the potency of individual samples. Both these methods can be very well correlated and the results can be relied upon.

The in vitro methods which were already reported and also standardised in our laboratory are sufficient for determining the potency of the antitoxin content of the serum of the immunised guinea pigs. In other words the requirement of the number of animals if alternative tests can be correlated with the WHO Method remains the same. We propose that we should use a combination of in vivo and in vitro method to achieve the goal of reducing the number of animals. This is possible if we combine the antibody induction method as per I.P. for immunising the animals and its serological evaluation by any one of the three in vitro methods described above. This reduces the number of animals required as well as the distress and discomfort suffered by them. In the I.P. method we require 10 guinea pigs for one test and if we use a standard vaccine preparation we require an additional 10 guinea pigs. Serological analysis can be performed using one of the alternate methods.

We are fully aware that the study will require the collaboration of different laboratories to come to the same conclusion and also to decide which of the three in vitro tests gives more reliable results.

Table 6: Comparaison of diphtheria antitoxin titres by different methods

G. pig sera No.	Diphtheria antitoxin vaccine				G. pig sera No.	Diphtheria antitoxin vaccine				G. pig sera No.	Diphtheria antitoxin vaccine				G. pig sera No.	Diphtheria antitoxin vaccine				G. pig sera No.	Diphtheria antitoxin vaccine			
	PHA	TN	Vero	PPT		PHA	TN	Vero	PPT		PHA	TN	Vero	PPT		PHA	TN	Vero	PPT		PHA	TN	Vero	PPT
Vaccine batch No. A					Vaccine batch No. B					Vaccine batch No. C					Vaccine batch No. D					Vaccine batch No. E				
1	> 4	> 4	16	> 4	1	> 4	> 4	16	> 4	1	> 4	> 4	16	> 4	1	> 4	> 4	8	> 4	1	> 4	> 4	8	> 4
2	> 4	> 4	16	> 4	2	> 4	> 4	32	> 4	2	> 4	> 4	16	> 4	2	> 4	> 4	8	> 4	2	> 4	> 4	8	> 4
3	> 4	> 4	8	> 4	3	> 4	> 4	16	> 4	3	> 4	> 4	16	> 4	3	> 4	> 4	8	> 4	3	> 4	> 4	16	> 4
4	> 4	> 4	16	> 4	4	> 4	> 4	16	> 4	4	> 4	> 4	16	> 4	4	> 4	> 4	8	> 4	4	> 4	> 4	16	> 4
5	> 4	> 4	16	> 4	5	> 4	> 4	16	> 4	5	> 4	> 4	16	> 4	5	> 4	> 4	8	> 4	5	> 4	> 4	8	> 4
6	> 4	> 4	16	> 4	6	> 4	> 4	16	> 4	6	> 4	> 4	16	> 4	6	> 4	> 4	8	> 4	6	> 4	> 4	8	> 4
7	> 4	> 4	16	> 4	7	> 4	> 4	16	> 4	7	> 4	> 4	16	> 4	7	> 4	> 4	8	> 4	7	> 4	> 4	8	> 4
8	> 4	> 4	16	> 4	8	> 4	> 4	16	> 4	8	> 4	> 4	16	> 4	8	> 4	> 4	16	> 4	8	> 4	> 4	8	> 4
9	> 4	> 4	16	> 4	9	> 4	> 4	16	> 4	9	> 4	> 4	16	> 4	9	> 4	> 4	8	> 4	9	> 4	> 4	8	> 4
10	> 4	> 4	16	> 4	10	> 4	> 4	8	> 4	10	> 4	> 4	16	> 4	10	2	2	2	2	10	> 4	> 4	8	> 4
Vaccine batch No. F					Vaccine batch No. G					Vaccine batch No. H					Vaccine batch No. I					Vaccine batch No. J				
1	> 4	> 4	8	> 4	1	> 4	> 4	16	> 4	1	> 4	> 4	8	> 4	1	2	2	2	2	1	> 4	> 4	32	> 4
2	> 4	> 4	8	> 4	2	> 4	> 4	8	> 4	2	> 4	> 4	32	> 4	2	4	4	4	> 4	2	> 4	> 4	32	> 4
3	> 4	> 4	8	> 4	3	> 4	> 4	8	> 4	3	> 4	> 4	16	> 4	3	> 4	> 4	8	> 4	3	> 4	> 4	16	> 4
4	> 4	> 4	8	> 4	4	> 4	> 4	8	> 4	4	> 4	> 4	16	> 4	4	> 4	> 4	8	> 4	4	> 4	> 4	16	> 4

G. pig sera No.	Diphtheria antitoxin vaccine				G. pig sera No.	Diphtheria antitoxin vaccine				G. pig sera No.	Diphtheria antitoxin vaccine				G. pig sera No.	Diphtheria antitoxin vaccine				G. pig sera No.	Diphtheria antitoxin vaccine			
	PHA	TN	Vero	PPT		PHA	TN	Vero	PPT		PHA	TN	Vero	PPT		PHA	TN	Vero	PPT		PHA	TN	Vero	PPT
Vaccine batch No. F					Vaccine batch No. G					Vaccine batch No. H					Vaccine batch No. I					Vaccine batch No. J				
5	> 4	> 4	8	> 4	5	> 4	> 4	8	> 4	5	> 4	> 4	8	> 4	5	2	2	2	2	5	> 4	> 4	32	> 4
6	> 4	> 4	8	> 4	6	> 4	> 4	8	> 4	6	> 4	> 4	8	> 4	6	> 4	> 4	8	> 4	6	> 4	> 4	32	> 4
7	> 4	> 4	8	> 4	7	> 4	> 4	16	> 4	7	> 4	> 4	16	> 4	7	> 4	> 4	8	> 4	7	> 4	> 4	32	> 4
8	> 4	> 4	8	> 4	8	> 4	> 4	8	> 4	8	> 4	> 4	8	> 4	8	> 4	> 4	8	> 4	8	> 4	> 4	16	> 4
9	> 4	> 4	8	> 4	9	> 4	> 4	8	> 4	9	> 4	> 4	8	> 4	9	2	2	2	2	9	> 4	> 4	32	> 4
10	> 4	> 4	8	> 4	10	> 4	> 4	8	> 4	10	> 4	> 4	8	> 4	10	> 4	> 4	8	> 4	10	> 4	> 4	32	> 4
Vaccine batch No. K					Vaccine batch No. L					Vaccine batch No. M					Vaccine batch No. N					Vaccine batch No. O				
1	> 4	> 4	16	> 4	1	> 4	> 4	16	> 4	1	> 4	> 4	8	> 4	1	> 4	> 4	32	> 4	1	> 4	> 4	16	> 4
2	> 4	> 4	16	> 4	2	> 4	> 4	8	> 4	2	> 4	> 4	4	> 4	2	> 4	> 4	32	> 4	2	> 4	> 4	16	> 4
3	> 4	> 4	16	> 4	3	> 4	> 4	8	> 4	3	> 4	> 4	16	> 4	3	> 4	> 4	16	> 4	3	> 4	> 4	32	> 4
4	> 4	> 4	16	> 4	4	> 4	> 4	8	> 4	4	> 4	> 4	16	> 4	4	> 4	> 4	16	> 4	4	> 4	> 4	16	> 4
5	> 4	> 4	8	> 4	5	> 4	> 4	8	> 4	5	> 4	> 4	16	> 4	5	> 4	> 4	16	> 4	5	> 4	> 4	16	> 4
6	> 4	> 4	16	> 4	6	> 4	> 4	16	> 4	6	> 4	> 4	16	> 4	6	> 4	> 4	16	> 4	6	> 4	> 4	16	> 4
7	> 4	> 4	16	> 4	7	> 4	> 4	8	> 4	7	> 4	> 4	16	> 4	7	> 4	> 4	16	> 4	7	> 4	> 4	16	> 4
8	> 4	> 4	16	> 4	8	> 4	> 4	8	> 4	8	> 4	> 4	16	> 4	8	> 4	> 4	16	> 4	8	> 4	> 4	16	> 4
9	> 4	> 4	16	> 4	9	> 4	> 4	8	> 4	9	> 4	> 4	16	> 4	9	> 4	> 4	16	> 4	9	> 4	> 4	16	> 4
10	> 4	> 4	16	> 4	10	> 4	> 4	8	> 4	10	> 4	> 4	16	> 4	10	> 4	> 4	16	> 4	10	> 4	> 4	16	> 4

References

1 World Health Organisation. Technical Report Series No. 800, 1990.

2 Indian Pharmacopoeia: Monograph on DTP Vaccine (Adsorbed), DT Vaccine (Adsorbed) and Tetanus Vaccine (Adsorbed), Vol. I, 3rd Edition, 1985.

3 Shinde YP, Jadhav SS: Indirect haemagglutination test using Human «O» red blood cells to coat diphtheria toxoid using chromic chloride as a coupling agent. Proceeding of First National Conference of Indian Academy of Vaccinology and Immunobiology, July 1991.

4 Kreeftenberg JG, van der Gun JW, Hendriksen CFH: Potency Determination of the Diphtheria Component of DT & DTP Vaccines in mice by serum neutralisation of Diphtheria Toxin in Vero cell cultures. Bulletin of World Health Organisation: BS/89 1613, 1989.

5 Sharma SB, Sharma R, Maheshwari, Gupta SC, Bhandari RK, Ahuja SK, Saxena SN: An in vitro method for titration of diphtheria antitoxin on vero cells. J Commun Dis 1985;17:177-180.

6 Kreeftenberg JG, van der Gun J, Marsman FR, Sekhuis VM, Bhandari SK, Maheshwari SC: An investigation of a mouse model to estimate the potency of the diphtheria component in vaccines. J Biol Stand 1985;229-234.

7 Beys Hoest B: J Pharmacol Belg 1968;23:60. Cited from Svenson SB, Larsen K. J Immunol Methods 1977;17:249-256.

Dr. S.S. Jadhav, Serum Institute of India Ltd., 212/2 Hadapsar, Pune M.S. 411028, India

Brown F, Cussler K, Hendriksen C (eds): Replacement, Reduction and Refinement of Animal Experiments in the Development and Control of Biological Products.
Dev Biol Stand. Basel, Karger, 1996, vol 86, p 261

SESSION VIII

Bacterial Vaccines

Chairmen: S. Houghton (Milton Keynes, U.K.)
E. Relyveld (Marnes-la-Coquette, France)

Brown F, Cussler K, Hendriksen C (eds): Replacement, Reduction and Refinement of Animal Experiments in the Development and Control of Biological Products.
Dev Biol Stand. Basel, Karger, 1996, vol 86, pp 263-270

Developments in Pertussis Vaccines Leading to Reduction and Replacement of in vivo Testing

K. Redhead

Division of Bacteriology, National Institute for Biological Standards and Control, Potters Bar, Herts., UK

Key words: Pertussis vaccine, control testing, LAL

Abstract: The control of bacterial vaccines, in common with other biological products, requires that they conform to specified standards of purity, safety and efficacy. The specifications and the methods of assessment can vary between the different types of vaccines. Highly defined vaccines can be largely evaluated by physico-chemical methods. However, the control of whole-cell bacterial vaccines can be a very different proposition owing to the complex nature of the materials. Safety and efficacy testing of such products can involve a large number of in vivo assays. Our growing understanding of the pathogenesis of, and protective immune responses to, bacterial diseases makes it possible to devise more effective and better-defined vaccines. In addition, it allows the development of sensitive and precise in vitro assays. As a result, the requirement for in vivo control tests can be reduced and in some cases eliminated. This phenomenon is well illustrated by the recent advances in the development and control of pertussis vaccines. Certain in vitro tests, such as the LAL and CHO-cell assays, are proving valuable in checking for specific toxicities which can be associated with pertussis vaccines. Although some problems still exist in the development of a suitable potency assay for the acellular vaccines, as we gain more information on the contributions of the different aspects of the immune system towards protection from pertussis, the prospects for further improvements in both the vaccine and its control evaluation are promising.

Introduction

Pertussis vaccines come under the general classification of bacterial vaccines. This is a grouping which includes live attenuated bacteria, killed whole-cell preparations, toxoids and extracted acellular components. Bacterial vaccines have a relatively long history of routine use in human medicine, starting in the 1920s with immunization campaigns against tuberculosis using BCG [1]. Recently there has been an upsurge in research to improve established bacterial vaccines and to develop new products. Table 1 gives an indication of the extent and variety of current bacterial vaccines.

Table 1: Bacterial vaccines in widespread use.

Type of vaccine	Infections
Killed whole-cell	Pertussis. typhoid, cholera, typhus, plague
Live attenuated	Tuberculosis, typhoid
Toxoid	Diphtheria, tetanus, botulinum
Polysaccharide	Haemophilus influenzae type b, pneumococcus
Acellular	Pertussis

The control of bacterial vaccines, in common with other biological products, requires that they conform to specified standards of purity, safety, efficacy and stability. The hazards and complications which can arise owing to incomplete or inappropriate control evaluation can vary in both nature and severity. While the areas of most concern have always been purity and safety, the problems which can result from failures in efficacy or stability, although not as immediately obvious, are potentially just as serious.

The specifications and methods of control assessment can vary greatly between different types of vaccine. Highly defined vaccines such as the *Haemophilus influenzae* type b vaccines, which comprise capsular polyribose phosphate conjugated to protein carriers, can now be largely evaluated by physico-chemical means. Such testing ensures that the product is pure and consistent with preparations that have been shown to be stable and efficacious in clinical studies. Few in vivo assays are necessary except to check for the absence of abnormal toxicity and, where a toxoided protein carrier is employed, reversion to specific toxicity. The position is very different in the case of whole-cell bacterial vaccines. Here the complex and relatively undefined nature of the materials may require the use of several different in vivo assays to check all the potentially relevant aspects of safety and efficacy.

The usual reason for employing a whole-cell vaccine is uncertainty as to the identity of the specific protective antigens and/or how best to present these antigens to the immune system. Improvements in our understanding of the pathogenesis of, and protective immune responses to, a bacterial disease makes it possible to focus and refine the control testing of the vaccine. This increasing knowledge also promotes the development of improved and better defined vaccines, which in turn may allow the application of more sensitive and precise in vitro assays for product evaluation. As a result, the requirement for in vivo tests can be reduced and in some cases eliminated. A good example of this sequence of events is the history of the development of pertussis vaccine, with its allied improvements and reduction in animal usage, in its control evaluation.

Pertussis Vaccine Development and Control Evaluation

Attempts to produce a pertussis vaccine quickly followed the first isolation, in 1906 [2], of the causative organism, now known as *Bordetella pertussis.* The majority of these early vaccines contained intact bacteria, killed and inactivated by physical or chemical methods. This form became known generically as whole-cell pertussis vaccine (WCPV). For 25 years clinical trials of pertussis vaccines in children gave poor or variable estimates of efficacy. Then in 1931 the stepwise phase degradation that *B.pertussis* undergoes in vitro was described [3]. During repeated passage on culture media, initially virulent phase I isolates were shown to degrade to avirulent phase IV strains. Only the original phase I form was found to be suitable for the production of effective vaccines. Subsequently the origins and histories of the *B.pertussis* strains used for vaccine production were monitored. As a result the quality and consistency of WCPVs greatly improved.

The Medical Research Council (MRC) field studies reported in 1959 [4] not only demonstrated that pertussis vaccines could confer substantial protection but also established the main basis for pertussis vaccine control evaluation. Laboratory tests, conducted in parallel with the trials, examined the predictive values of several putative potency assays. As a result of these studies the intracerebral challenge mouse protection test (ICMPT) was adopted as the potency assay and a British Standard for Pertussis Vaccine was prepared from one of the efficacious vaccine batches. In the years that followed, the phenomenon of antigenic modulation was described [5], an international reference vaccine was established and the advantages of including serotypes 1, 2, 3 in the vaccine were proposed [6]. Then, in 1974, adverse media publicity suggesting possible severe side-effects associated with pertussis immunisation in the UK precipitated a dramatic fall in acceptance rates. As part of an attempt to discover whether there was any scientific basis for these concerns and to allay public fears, the control testing was expanded to assess a wider range of the biological activities known to be associated with WCPVs. Thus, by 1977, the full control evaluation of WCPVs at NIBSC comprised a total of 12 tests, as shown in Table 2, of which the nine underlined assays were conducted in vivo.

As part of the continuing research investigation into the pathogenesis of *B.pertussis,* the component now known as pertussis toxin (PT) was purified to homogeneity [7]. This one protein was found to possess most of the biological activities, including islet activation, histamine sensitisation, leukocytosis, adjuvanticity and mitogenicity, ascribed to the whole bacterium. This discovery advanced the development of acellular pertussis vaccines, by identifying PT as a potential major immunogen and helped to reduce the amount of necessary in vivo control testing. With the development of the in vitro Chinese hamster ovary cell (CHO-cell) assay for active PT [8], the control evaluation process was further refined and, by 1994, the testing of WCPVs at NIBSC comprised eight assays of which only three involved the direct use of animals (Table 2). If clinical trials prove successful and defined acellular vaccines become widely accepted their control evaluation is likely to become more physico-chemically orientated and use even fewer in vivo assays. However, this process will involve ensuring that the selected in vitro assays are applicable in the contexts in which they are used. As the following examples illustrate this may not always be a simple task.

Pyrogenicity Assessed by Limulus Amoebocyte Lysate

The potential pyrogenicity of many biologicals has traditionally been estimated by measuring the temperature rises they induce when administered to rabbits. This has always presented something of a problem with Gram-negative bacterial whole-cell vaccines such as WCPVs. When examined in the rabbit pyrogenicity assay it is

Table 2: Control evaluation of pertussis vaccines.

Property	Assays		
	1977	1994	Future acellular vaccines
Potency	ICMPT	ICMPT	Humoral and cellular immunogenecity?
Serotype	Agglutination	Agglutination	
Antigen content	Immunodiffusion	Immunodiffusion	Immunosorbent assay/ HPLC/FPLC
Sterility	Culture	Culture	Culture
Endotoxin	Rabbit pyrogenicity	LAL	LAL
General toxicity	Abnormal toxicity		
	Mouse weight gain	Mouse weight gain	Mouse weight gain?
Specific toxicities	Histamine sensitivity	Histamine sensitivity	
	Leukocytosis	CHO cell	CHO cell
	Hyperinsulinaemia		
	Hypoglycaemia		
	Rat paw œdema		

possible to distinguish between such preparations only when they possess very marked differences in their inherently high endotoxin levels. It is therefore no surprise that, soon after its appearance, the Limulus amoebocyte lysate (LAL) assay was adopted by several laboratories as the preferred method for monitoring endotoxin content of Gram-negative bacterial vaccines.

It has been suggested that the results of the LAL assay may not necessarily correlate with pyrogenicity in vaccine recipients. However, in 1989, Baraff et al. [9] published the findings of an investigation in which they examined 25 lots of commercially available adsorbed diphtheria-tetanus-pertussis (DTP) vaccine by LAL, pertussis vaccine potency and mouse weight gain, and compared the results with the rates of local and systemic reactions when the vaccines were routinely administered to infants. There was a significant positive association between endotoxin content, as measured by LAL and the percentage of vaccine recipients who developed fever, suggesting that the LAL has a predictive value for at least one possible systemic adverse reaction to DTP.

During the course of vaccine evaluation at NIBSC several batches of one manufacturer's vaccines, containing a whole-cell pertussis component, were subjected to a battery of control assays, including endotoxin assessment by a micro-LAL assay, over a period of seven years from 1986 to 1992. Analysis of the results of this testing (Table 3) showed a very significant difference between the detectable endotoxin contents of plain and adsorbed vaccines, with the plain vaccines containing on average more than ten times as much endotoxin than the adsorbed preparations. Both types of vaccine used bacteria prepared in the same way and in some cases from the same harvest. This suggested that the presence of the aluminium hydroxide adsorbent may have interfered with the LAL reaction. However, experiments where the LAL reaction mixture was spiked with aluminium hydroxide showed that this was not the case.

A study was performed, between 1978 and 1980, to assess relative reactogenicities of DT and DTP vaccines in infant recipients [10]. The results of the study had been further analysed to distinguish between plain and adsorbed vaccines (Table 4). The vaccines used were from the same manufacturer as those tested at NIBSC and there had been no significant changes in the method of production between 1978 and 1992. The data show a significantly higher incidence of certain systemic adverse effects, particularly the frequency of fevers, to be associated with plain rather than adsorbed pertussis containing vaccines. This confirms that the endotoxin content of pertussis vaccines, as assessed by LAL, correlates with the frequency of fevers in recipients and supports earlier findings that adsorbed vaccine is not only more potent than plain vaccine but also associated with fewer reactions [11].

The mechanism by which aluminium hydroxide adsorption reduces levels of detectable endotoxin in the LAL and lowers the frequency of endotoxin-associated reactions in vaccine recipients is currently under investigation. However, it does seem probable that the adsorption of soluble endotoxin to aluminium hydroxide renders it inactive in the LAL assay and may reduce its availability for the stimulation of typical reactions in the human body. These findings suggest that the incorporation of aluminium carriers in other Gram-negative bacterial whole-cell vaccines could have beneficial effects in reducing their intrinsic reactogenicities. They also demonstrate how, under the right circumstances, an in vitro assay can not only adequately replace an in vivo assay but may prove to be superior and more advantageous. However, this is not always the case when attempts are made to refine or replace in vivo control testing.

Replacement of the Pertussis Vaccine Potency Assay

Increased understanding of the relevance of *B.pertussis* virulence factors has promoted the development of a new generation of defined acellular vaccines. These vaccines are based on differing combinations of selected purified components including toxoided forms of PT. Control evaluation has shown the preparations to be almost completely devoid of endotoxin, PT activity and all other

Table 3: Mean values from control assays of plain and adsorbed whole-cell pertussis vaccines, 1986-1992.

Type of vaccine	No. tested	Potency IU/dose	Active PT ng/dose	Endotoxin IU/dose
Plain	12	4.5	113	103800
Adsorbed	37	7.2	187	9335
P value	–	<.001	>.1	<.0001

Table 4: Symptoms within 48 hours of first dose of vaccine.

Vaccine	Total followed up	Crying more than usual		Screaming attack		Fever	
		No.	(%)	No.	(%)	No.	(%)
DTP plain	338	197	(58.2)	28	(8.2)	100	(29.6)
DTP ads.	1674	481	(28.7)	45	(2.7)	157	(9.4)
DT ads.	1121	266	(23.7)	44	(3.9)	68	(6.1)

toxicities associated with *B.pertussis*. These vaccines have been found to induce strong antibody responses, usually greater than those elicited by WCPVs, in mice and in humans. However, there is one problem. Virtually none of these materials has been shown to pass the potency assay, the ICMPT, reproducibly.

The ICMPT, also known as the active mouse protection test or the Kendrick test, was first developed in 1947 [12]. The assay involves comparing the ability of the test vaccine to protect mice against a virulent *B.pertussis* intracerebral challenge with that of a standard vaccine possessing an assigned unitage. Since the MRC pertussis field trials of the 1950s, when it was shown that this assay correlated with vaccine efficacy in children, it has been routinely used to monitor vaccine potency with apparent success. Few attempts have been made to analyse the mechanisms involved in this assay. This lack of insight into the underlying nature of the test made it difficult to determine its applicability to acellular vaccines. It was, then, somewhat arbitrarily, decided that acellular vaccines were efficacious and therefore the ICMPT could not be an appropriate potency assay.

Several years have been spent searching for an assay to replace the ICMPT for the testing of acellular vaccines. Candidates have included a modified ICMPT with an extended interval between immunisation and challenge, protection against respiratory infection, and the measurement of serological responses by immunosorbent assay or neutralisation of PT activities in vivo or in vitro. The most popular approach has been to immunise mice and measure the subsequent antibody responses to the appropriate antigens by ELISAs. While this method reduces animal usage and appears to be a more refined form of assessment certain doubts must be expressed about its applicability. Serological responses have not been shown to correlate with protection, the measurement of antibody levels in terms of their antigen-binding capacity may not reflect their functional activity in protection, any protective contribution of secretory antibodies is overlooked, and no assessment is made of the possible role of cellular immune responses.

Cell-mediated immunity (CMI) may have a crucial influence in protection against pertussis infection. Studies have described the induction of *B.pertussis*-specific proliferative $CD4^+$ T cell responses in human subjects and mice after immunisation or infection [13-15]. In an experimental murine respiratory infection model it has been shown that, to a large extent, protection correlates with the presence of $CD4^+$ Th1 cell responses to *B.pertussis* antigens [16]. Such cellular immune responses can be generated by infection or immunisation with WCPV or acellular vaccine containing genetically detoxified PT (rPT) but not with chemically detoxified acellular vaccines. Interestingly, when examined in the ICMPT, genetically detoxified acellular vaccines perform far better than their chemically detoxified counterparts and appear to be almost as potent as WCPV (Table 5). These findings suggest the possibility that CMI may play a role in both efficacy in humans and potency in the ICMPT, in which case the ICMPT could be a more accurate test of acellular vaccine potency than has been supposed and certainly a more applicable potency assay than the measurement of serum antibody responses.

Further investigations into the basis of human immunity to pertussis infection and the mechanisms operating in the potential assays, particularly the ICMPT, are necessary before a truly informed choice of potency test can be made. The whole-cell and acellular vaccines in the current Swedish phase III efficacy trial are being monitored for the humoral and cellular immune responses they elicit, and have also been examined in various putative potency assays. It is to be hoped that this trial will not only generate data on the vaccine efficacies but also provide immune correlates of protection and identify an appropriate potency assay or at least furnish the scientific basis for its development.

Conclusions

Increased information on the causative organism, pathogenesis and immunity involved in an infectious disease allows the development of new and improved vaccines. In the case of bacterial vaccines, improvements generally tend towards the production of more defined preparations with reduced reactogenicities and enhanced efficacies. The combination of advances in knowledge and defined vaccine composition provides the best basis for the refinement, reduction and replacement of in vivo testing in the control evaluation of bacterial vaccines. However, all in vitro assays must be carefully examined for their applicability to the role in which they are to be used.

Acknowledgements

Thanks are due to E. Miller, P. Farrington, A.M. Attwell, A. Barnard and K. Mills for the data and assistance they provided.

Table 5: Performance of cellular pertussis vaccines in protection models.

Vaccine composition	% Protection against aerosol challenge	Protection against i.c. challenge (IU/SHD)
Whole-cell	100	4.0
PT + FHA	24	0.4
PT + FHA + Aggs 2,3	33	1.25
PT + FHA + Aggs 2,3 + Pertactin	38	0.2
rPT + FHA	72	2.64

References

1 Griffith AH: Achievements in Europe, in Perkins FT (ed): Symp Series Immunobiol Standard. Basel, Karger, 1973, vol 22, pp13-24.

2 Bordet J, Gengou O: Le microbe de la coqueluche. Ann Inst Pasteur 1906;20:731-741.

3 Leslie PH, Gardner AD: The phases of *Haemophilus pertussis*. J Hyg Camb 1931;31:423-434.

4 Medical Research Council: Vaccination against whooping cough: The final report to the Whooping Cough Immunization Committee of the Medical Research Council and to the Medical Officers of Health for Battersea and Wandsworth, Bradford, Liverpool and Newcastle. Brit Med J 1959;I:994-1000.

5 Lacey BW: Antigenic modulation of *Bordetella pertussis*. J Hyg Camb 1960;58:57-93.

6 Preston NW: Effectiveness of pertussis vaccines. Brit Med J 1965;II:11-13.

7 Yajima M, Hosoda K, Kanbayashi Y, Nakamura T, Nogimori K, Nakase Y, Ui M: Islet-activating protein in *Bordetella pertussis* that potentiates insulin secretory responses of rat. Purification and characterization. J Biochem 1978;83:295-303.

8 Gillenius P, Jaatmaa E, Askelof P, Granstrom M, Tiru M: The standardisation of an assay for pertussis toxin and antitoxin in microplate culture of Chinese hamster ovary cells. J Biol Standard 1985;13:61-66.

9 Baraff LJ, Manclark CR, Cherry JD, Christenson P, Marcy SM: Analyses of adverse reactions to diphtheria and tetanus toxoids and pertussis vaccine by vaccine lot, endotoxin content, pertussis vaccine potency and percentage of mouse weight gain. Pediatr Infect Dis J 1989;8:502-507.

10 Pollock TM, Miller E, Mortimer JY, Smith G: Symptoms after primary immunisation with DTP and with DT vaccine. Lancet 1984;II:146-149.

11 Butler NR, Voyce MA, Burland WL, Hilton ML: Advantages of aluminium hydroxide adsorbed combined diphtheria, tetanus and pertussis vaccines for the immunization of infants. Brit Med J 1969;I:663-666.

12 Kendrick PL, Eldering G, Dixon MK, Misner J: Mouse protection tests in the study of pertussis vaccine. Amer J Publ Hlth 1947;37:803-810.

13 De Magistris MT, Romano M, Nuti S, Rappuoli R, Tagliabue A: Dissecting human T cell responses against *Bordetella* species. J Exp Med 1988;168:1351-1362.

14 Gearing AJH, Bird CR, Redhead K, Thomas M: Human cellular immune responses to *Bordetella pertussis* infection. FEMS Microbiol Immunol 1989;1:205-211.

15 Mills KHG, Barnard A, Watkins J, Redhead K: Cell-mediated immunity to *Bordetella pertussis*: Role of Th1 cells in bacterial clearance in a murine respiratory infection model. Infect Immun 1993;61:399-410.

16 Redhead K, Watkins J, Barnard A, Mills KHG: Effective immunization against *Bordetella pertussis* respiratory infection in mice is dependent on induction of cell-mediated immunity. Infect Immun 1993;61:3190-3198.

Dr. K. Redhead, Hoechst UK Ldt., Animal Health Business Unit, Walton Manor, Walton, Milton Keynes, Bucks MK7 7AJ, UK

Brown F, Cussler K, Hendriksen C (eds): Replacement, Reduction and Refinement of Animal Experiments in the Development and Control of Biological Products.
Dev Biol Stand. Basel, Karger, 1996, vol 86, pp 271-281

Pertussis Serological Potency Test as an Alternative to the Intracerebral Mouse Protection Test

A. van der Ark, I. van Straaten-van de Kappelle, C. Hendriksen, H. van de Donk

Laboratory for Control of Biological Products, National Institute of Public Health and Environmental Protection, Bilthoven, The Netherlands

Abstract: The current potency test for pertussis vaccines, the intracerebral mouse protection test (MPT), is still the only mandatory laboratory model available. This test, however, is a valid, but inhumane and imprecise test and therefore a good candidate for replacement. Recently we have developed the Pertussis Serological Potency Test (PSPT) as an alternative for the MPT. The PSPT is based on in vitro assessment of the humoral immune response against the whole range of surface-antigens of *B. pertussis* in mice after immunisation with Whole Cell Vaccine (WCV). We have demonstrated a relationship between the mean pertussis antibody concentration at the day of challenge and the proportion of surviving mice at each vaccine dose in the MPT (R = 0.91). The PSPT is a model in which mice (20-24 g) are immunised i.p. with graded doses of vaccine and bled after four weeks. Sera are titrated in a whole cell ELISA and potency based on the vaccine dose-dependent antibody response is estimated by means of a parallel line analysis.
In an in-house validation study 13 WCVs were tested in the PSPT and MPT. Homogeneity of both tests was proven by means of the chi-square test; potencies were significantly similar (p = 0.95). Compared to the MPT, the PSPT is more reproducible as is indicated by its smaller 95% confidence intervals. Moreover, by using the PSPT the animal distress can be reduced to an acceptable level and the PSPT also results in a reduction of more than 25% in use of mice. Additional experiments showed that estimation of WCV-potency in the PSPT based on specific antibody responses against protective antigens (PT, FHA, 69- and 92-kDa OMPs) was not possible or did not correlate with protection in MPT. Sera obtained from the PSPT showed a correlation between pertussis antibody levels and complement-mediated killing by pertussis antibodies in in vitro assays. In conclusion, the PSPT is a promising substitute for the MPT though further validation and additional studies on functional validity should finally warrant replacement of the MPT by this serological model.

Introduction

The intracerebral (i.c.) mouse protection test (MPT) developed by Kendrick et al [1] is still the only mandatory potency assay for pertussis vaccines which showed a correlation between this test and protection in children [2]. The MPT is therefore a valid test, but it is costly, inhumane and imprecise. For that reason we have developed the Pertussis Serological Potency Test (PSPT) with the ultimate goal of refining and reducing the use of animals as an alternative for the MPT. Potency of vaccines might be assayed more conveniently by measuring the ability of the vaccines to induce antibodies with demonstrable relationship to protection. Wiertz et al [3] demonstrated that protection of mice in the MPT is T-cell dependent, as is the humoral response to several «protective» pertussis antigens such as pertussis toxin (PT), filamentous haemagglutinin (FHA), outer membrane proteins (OMPs) and pili. The variety of protective antigens in whole cell vaccines (WCVs) makes it difficult to discriminate between the augmentative effect of specific antibodies [4]. The research therefore focused on the in vitro assessment of humoral immune response against the whole range of surface antigens of *B.pertussis* in mice after immunisation with WCV.

Materials and Methods

Mice

N:NIH/RIVM outbred mice weighing 10-14 grams or 20-24 grams were used.

Vaccines

The lyophilised *B.pertussis* plain whole cell in-house reference Kh 85/1 (6 IU/ml) was used after reconstitution in 5 ml PBS. Pertussis WCVs were obtained from different manufacturers. Diphtheria-pertussis-tetanus-polio (DPT-polio) vaccines were routinely produced in our Institute and contain 16 OU/ml. DPT vaccines supplied by seven other manufacturers contain 32-40 OU/ml.

Mouse Protection Test (MPT)

The MPT was performed according to the E.P. requirements.

Serology

Mouse IgG antibodies against the whole range of pertussis surface antigens were measured in whole cell ELISAs (WCE) as described recently [5]. Two different pertussis whole cell coatings, reference Kh 85/1 (Kh 85/1-WCE) and the International challenge strain 18323 (18323-WCE), were used. The latter was used to exclude the effect of whole cell strain-specific determinants. Antibodies to PT, FHA, 69-, and 92-kDa OMP were determined by an indirect ELISA technique. The PT, FHA and 69-kDa OMP were kindly provided by R. Rappuoli (Sclavo, Italy) and the 92-kDa OMP was purified in our Institute (manuscript in preparation).

Results

Relation of antibody concentration to survival of mice in MPT

The relation between the antibody concentration (EU/ml) and the survival of mice after i.c. challenge was investigated in a slightly modified MPT. Mice (10-14 g) were immunised intraperitoneally (i.p.) with graded doses of vaccine. At day 14, just before i.c. challenge, blood samples were taken. The number of dead mice was recorded daily from day 18 until day 28. The induction of pertussis antibody was determined by Kh 85/1-WCE and correlated with the actual survival of the corresponding mice. In a regression analysis a correlation co-efficient of 0.91 was found.

Development of the Pertussis Serological Potency Test (PSPT)

In preliminary experiments the induction of pertussis antibodies per vaccine dose was measured in Kh 85/1-WCE up to week six after i.p. immunisation. Mice (10-14 g) were immunised with graded vaccine doses and blood samples were taken weekly (Fig. 1). To induce antibody concentrations within the linear part of the sigmoid curve, the graded vaccine doses were fixed at 50, 25, 12.5 and 6.25 µl for reference Kh 85/1 and at 80, 40, 20 and 10 µl for DPT-polio vaccines. Potency was calculated by means of a parallel line analysis with log transformation of the antibody concentrations (Table 1). We chose to bleed the mice at day 28. Log transformation was chosen to obtain a normal distribution of the antibody concentrations per group of mice. Statistical calculations of preliminary results fixed the number of mice at 12 animals per vaccine dose, allowing a standard deviation of the antibody concentration equal to the average of the antibody concentration induced per vaccine dose.

The optimal route of immunisation in the PSPT was evaluated within two groups of mice (10-14 g). One group was immunised i.p. and the other group sub-

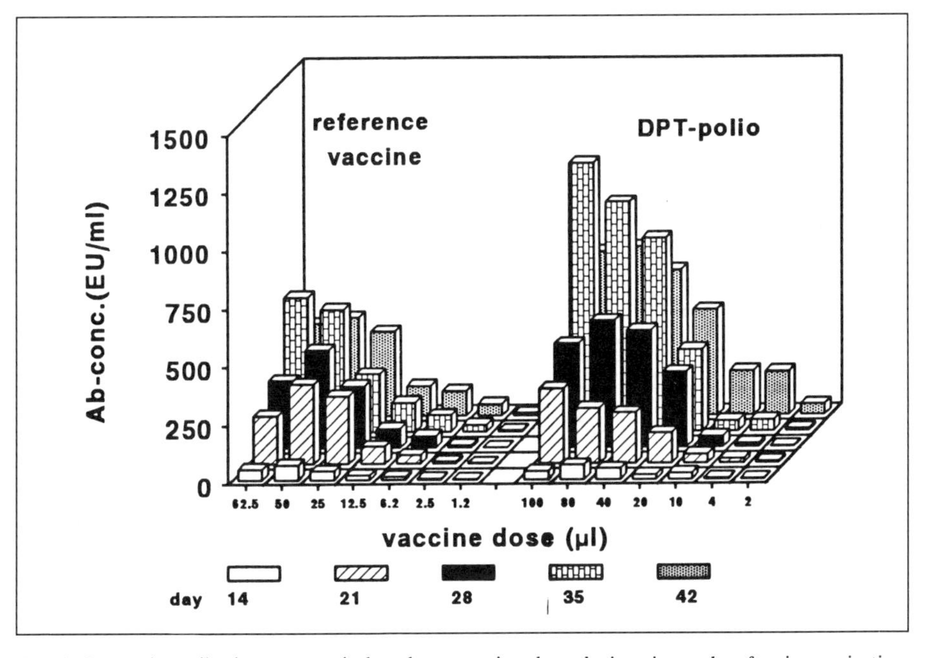

Fig. 1: Pertussis antibody response induced per vaccine dose during six weeks after immunisation (i.p.) with WCV.

Table 1: Potency of DPT-polio vaccines estimated in the PSPT and MPT. Potencies were calculated with the mean antibody concentrations per vaccine dose of serum samples obtained during six weeks after immunisation with WCV.

PSPT*	**Test 1**		**Test 2**	
	DPT-polio A	**DPT-polio B**	**DPT-polio B**	**DPT-polio A**
day 14	5.5 (3.0-11.4)	9.2 (5.0-21.9)	4.7 (3.8-5.9)	4.3 (2.6-7.6)
day 21	3.7 (2.6-5.3)		5.0 (3.1-7.6)	
day 28	6.1 (3.8-10.8)	7.7 (4.7-14.4)	8.8 (6.1-12.7)	5.2 (3.3-8.2)
day 35	6.1 (4.5-8.7)		7.2 (5.0-10.5)	
day 42	8.9 (5.2-19.3)		8.8 (5.7-14.2)	
MPT**	**test 1**		**test 2**	
	DPT-polio A	**DPT-polio B**	**DPT-polio B**	**DPT-polio A**
	4.6 (1.6-12.2)	8.3 (4.2-16.3)	8.6 (3.1-18.0)	4.5 (2.0-10.2)

* potency (IU) with 95% confidence interval estimated by means of parallel analysis
** potency (IU) with 95% confidence interval estimated by means of probit analysis

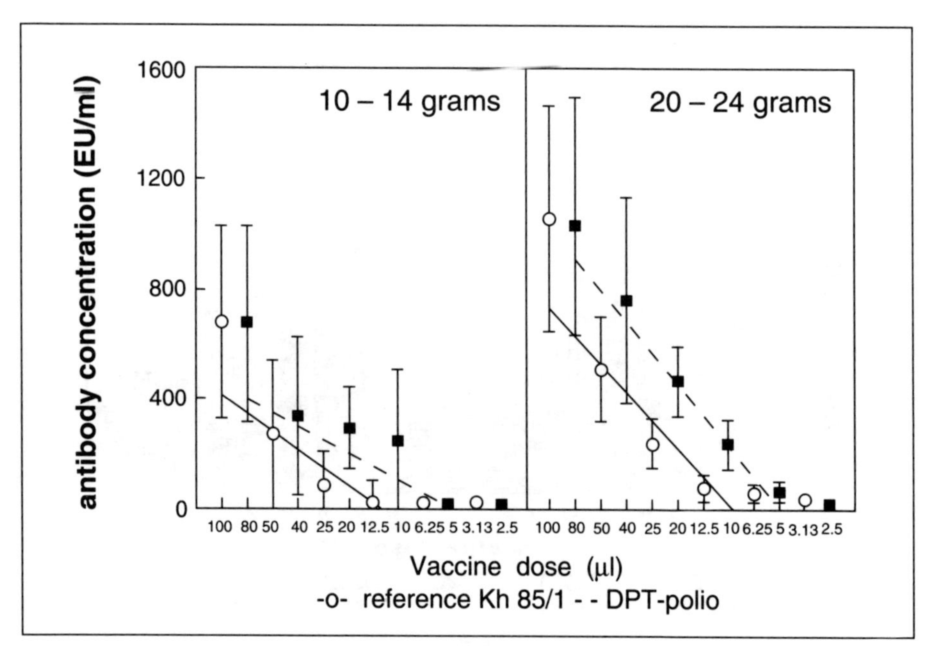

Fig. 2: Effect of body weight in antibody response. Mice of 10-14 g or 20-24 g were tested simultaneously in the PSPT.

cutaneously (s.c.). Antibody concentrations were significantly lower in mice immunised s.c. compared with the i.p. injected group and variation in antibody induction was increased (data not shown). The effect of body weight of mice on the range of induced antibody concentrations per vaccine dose in the PSPT was examined with mice of 10-14 g and 20-24 g in weight (Fig. 2). Mice of 20-24 g induced higher antibody concentrations, and variation of antibody concentrations per vaccine dose was decreased compared to mice of 10-14 g.

The reproducibility of the PSPT and MPT was calculated by means of an analysis of variances. In a series of six experiments the potency of DPT-polio vaccine was estimated in the MPT and PSPT, with mice of 10-14 g and 20-24 g. The PSPT with mice of either weight-range was considered to be more reproducible than the MPT (Table 2). Due to the decrease of variance, the 95% confidence interval of the potencies in the PSPT was clearly smaller than in the MPT.

In-house validation of the PSPT

Thirteen pertussis WCVs were tested in the MPT and PSPT. In the PSPT, mice were immunised with different doses vaccine: for RIVM-vaccines, doses of 80, 40, 20 and 10 µl were used, for the other vaccines, doses were 50, 25, 12.5 and 6.2 µl. Sera were tested in 18323-WCE, Kh 85/1-WCE, PT-, FHA-, 92- and 69-kDa OMP-ELISAs. WCVs were prepared in different ways by several manufacturers and were of widely differing potencies. The Kh 85/1-WCE and 18323-WCE were used to exclude a bias in the results obtained with the RIVM DPT-polio vaccines. Potency of the tested vaccines obtained by the MPT and PSPT were significantly similar in X^2-test (Table 3).

Table 2: Reproducibility of MPT and PSPT. DPT-polio vaccine was tested in the MPT and PSPT in a series of six experiments.

exp.	MPT*	PSPT*	
	mice of 10-14 g	mice of 10-14 g	mice of 20-24 g
1	7.7 (2.8-14.3)	7.7 (4.7-14.5)	12.0 (10.1-14.1)
2	13.9 (4.0-26.9)	9.8 (7.2-13.8)	9.9 (5.7-17.1)
3	9.7 (2.8-32.0)	6.0 (3.5-11.9)	7.8 (4.3-9.8)
4	2.9 (0.8-9.1)°	9.0 (6.2-14.5)	9.8 (6.5-16.8)
5	10.3 (2.2-75.5)°°	10.3 (7.4-15.5)	10.8 (8.3-14.2)
6	7.0 (1.0-60.3)°°	14.3 (10.1-21.0)	7.4 (4.6-10.5)
m.v.**	0.098	0.008	0.009

* potency (IU) with 95% confidence interval
** mean variance
° technically invalid test
°° statistically invalid test, 95% confidence interval out of range of 44-244%

The WCVs induced a distinct antibody response against 92-kDa OMP, but differed widely in responses against 69-kDa OMP and FHA and induced hardly antibodies against PT (Table 4). The antibody responses are expressed as the mean antibody concentration per vaccine dose, after log transformation. Hyper- and non-responders were mainly found in the anti-PT and anti-FHA responses. It is

Table 3: In-house validation of the PSPT. a) Potency of 13 WCVs were estimated in the MPT and PSPT, which was based on antibody concentrations measured in Kh 85/1-WCE, 18323-WCE or 92-kDa OMP-ELISA. b) Homogeneity of the test systems was estimated by means of a modified chi-square test.

a) Potency (IU) with 95% confidence interval per human dose				
Vaccine	**MPT**	**PSPT**		
		Kh 85/1-WCE	**18323-WCE**	**92-kDa OMP-E**
A	25.8 (10.9-66.0)	18.7 (14.1-26.3)	31.8 (24.7-43.0)	8.2 (6.7-10.1)
B	44.2 (16.3-147.5)	36.1 (23.8-64.1)	24.7 (19.1-33.5)	19.7 (15.1-27.0)
C	15.4 (5.8-41.2)	11.0 (7.6-15.4)	16.4 (13.5-20.4)	9.3 (7.3-12.0)
D	11.8 (2.8-58.7)	7.4 (6.5-16.8)	11.4 (4.9-12.3)	3.9 (2.9-5.1)
E	4.1 (1.6-10.2)	3.5 (2.6-4.7)	2.1 (1.0-3.9)	0.1 (0.0-0.3)
F	9.7 (3.6-25.3)	10.9 (8.5-14.3)	10.3 (8.2-13.4)	10.2 (7.9-13.6)
G	17.1 (7.3-40.5)	9.3 (7.5 11.9)	13.5 (10.7-17.6)	3.4 (2.2-3.1)
H	5.5 (0.5-27.4)	6.4 (5.1-8.0)	8.2 (6.2-10.6)	2.3 (1.7-3.1)
I	19.3 (5.3-88.4)	19.7 (13.6-31.9)	25.2 (15.5-50.9)	35.5 (19.6-92.3)
J	10.0 (8.8-11.8)*	9.8 (6.5-16.8)*	7.4 (4.9-12.3)*	9.8 (5.8-20.8)*
K	4.5 (3.0-6.0)*	4.6 (3.8-5.6)*	5.1 (4.2-6.1)*	3.8 (3.1- 4.7)*
L	6.7 (1.9-25.6)	8.5 (6.2- 12.3)	10.6 (7.8-15.3)	7.2 (4.7-12.2)
M	9.2 (4.0-21.9)	9.2 (7.4-11.9)	10.2 (7.9-13.9)	9.1 (6.6-13.6)
b) Homogeneity MPT and PSPT				
X^2-test		**Kh 85/1-WCE**	**18323-WCE**	**92-kDa OMP-E**
X^2		2.66	5.32	28.66
ratio		1.085	1.042	1.523
		(0.864-1.362)	(0.831-1.306)	(1.192-1.946)
p-value		0.99	0.95	>0.001

* geometric mean

noticeable that our in-house reference induced hardly any antibodies against 69-kDa OMP, FHA and PT. Consequently it is not possible to estimate the potency of the tested vaccines based on these antibody responses in the PSPT. The estimated potencies based on the antibody response against 92-kDa OMP were statistically invalid - 95% confidence intervals were out of the 44-244% range and did not correlate with the MPT.

Biological capacity of pertussis antibodies induced by WCV in PSPT

Sera from the in-house validation study were pooled in equal volumes per vaccine dose and tested in several in vitro assays for their biological activity. None of the WCVs induced measurable amounts of PT-neutralising antibodies in the CHO-neutralisation assay [6]. The ability of pertussis antibodies to activate the classical pathway of the complement system in an ELISA (18323-CAE) was assessed by measuring the antibody-mediated C_3-depositions on the whole cell coating (manuscript in preparation). The preliminary results indicated that the proportion of pertussis antibody-mediated complement activation correlates well with the antibody concentrations found with the 18323-WCE. Subsequently, the capacity of these sera to enhance the complement-mediated killing of virulent *B.pertussis* strain 18323 was tested in a bactericidal antibody in vitro assay [7] and correlated also with the total amount of pertussis antibodies (Fig. 3).

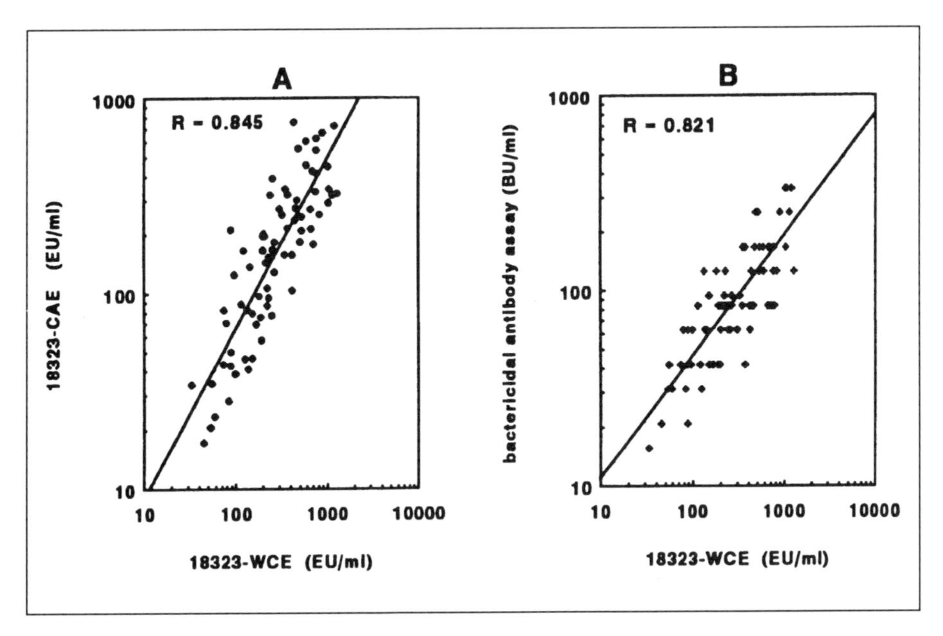

Fig. 3: Correlation between antibody concentration in 18323-WCE and the capacity of sera to activate the complement system. Sera were tested in 18323-WCE, 18323-CAE (A) and bactericidal antibody assay (B).

Table 4: Pertussis antigen specific antibody responses induced by WCVs in the PSPT. Sera were tested in several ELISAs and expressed as the mean antibody concentration per vaccine dose, after log transformation (EU/ml).

Test	Vaccine	Dose (µl)	Kh 85/1-WCE	18323-WCE	92-kDa OMP ELISA	69-kDa OMP ELISA	FHA-ELISA	PT-ELISA
1	reference	62.5	396.0	494.0	104.6	7.9	0.3	0.2
		31.2	265.0	246.0	34.2	1.3	0.3	0.0
		15.6	200.0	118.0	14.0	0.2	0.1	0.0
		7.8	74.0	49.0	5.8	0.1	0.2	0.0
	A	50.0	1219.0	1298.0	74.4	67.7	126.9	4.4
		25.0	568.0	906.0	48.0	21.7	93.3	1.0
		12.5	306.0	639.0	29.3	7.7	37.4	1.9
		6.2	168.0	231.0	4.5	0.3	7.1	0.2
	B	50.0	1412.0	984.0	157.1	28.6	1.1	2.4
		25.0	673.0	615.0	91.7	15.9	1.0	2.0
		12.5	471.0	494.0	68.8	1.7	0.4	1.1
		6.2	329.0	191.0	23.0	0.5	0.2	0.3
	C	50.0	734.0	1429.0	96.2	107.5	2.4	0.1
		25.0	425.0	761.0	68.7	32.8	2.0	0.0
		12.5	167.0	283.0	32.0	0.8	1.1	0.0
		6.2	112.0	98.0	4.7	0.0	0.3	0.0
	D	50.0	363.0	877.0	38.2	32.3	1.1	5.7
		25.0	278.0	29.0	23.9	24.4	0.8	3.3
		12.5	187.0	152.0	5.2	0.5	0.2	1.2
		6.2	78.0	106.0	1.1	0.2	0.1	0.2
2	reference	50.0	356.0	184.0	31.4	0.4	0.4	0.5
		25.0	262.0	84.0	37.6	0.3	0.1	0.1
		12.5	106.0	65.0	16.7	0.1	0.0	0.0
		6.2	57.0	31.0	6.3	0.1	0.0	0.0
	E	100.0	508.0	185.0	5.1	3.7	10.9	3.1
		50.0	303.0	85.0	0.2	0.7	1.5	2.0
		25.0	142.0	61.0	0.1	0.7	0.3	0.2

Test	Vaccine	Dose (μl)	Kh 85/1-WCE	18323-WCE	92-kDa OMP ELISA	69-kDa OMP ELISA	FHA-ELISA	PT-ELISA
	E	12.5	51.0	22.0	0.0	0.3	0.1	0.0
3	reference	50.0	1065.0	873.0	352.7	7.8	2.7	1.7
		25.0	475.0	370.0	97.8	0.5	1.5	0.2
		12.5	127.0	110.0	73.8	0.8	0.1	0.0
		6.2	90.0	66.0	22.2	0.1	0.1	0.0
	F	50.0	1672.0	1368.0	307.8	27.4	7.5	0.3
		25.0	1065.0	700.0	440.0	17.9	3.5	0.2
		12.5	287.0	237.0	140.3	7.9	1.0	0.1
		6.2	192.0	142.0	36.0	2.0	0.2	0.1
	G	50.0	723.0	741.0	82.1	42.3	10.9	1.9
		25.0	612.0	615.0	69.8	16.4	11.9	2.4
		12.5	360.0	411.0	37.7	23.8	4.2	0.7
		6.2	195.0	244.0	39.1	0.5	2.5	0.2
	H	50.0	806.0	629.0	75.1	12.9	18.3	1.4
		25.0	394.0	447.0	54.9	7.2	13.5	0.6
		12.5	169.0	212.0	19.1	1.9	1.3	1.1
		6.2	141.0	144.0	10.7	0.2	0.9	0.9
4	reference	50.0	336.0	231.0	99.6	0.8	0.3	0.1
		25.0	243.0	144.0	54.0	0.0	0.3	0.0
		12.5	114.0	84.0	22.5	1.0	0.0	0.0
		6.2	69.0	47.0	20.1	0.8	0.0	0.0
	I	50.0	768.0	681.0	369.0	17.3	1.6	2.5
		25.0	445.0	364.0	141.5	12.1	2.1	1.2
		12.5	292.0	310.0	174.5	2.6	1.2	0.3
		6.2	198.0	130.0	64.5	2.1	1.3	0.5
	J	80.0	618.0	547.0	188.7	5.4	0.6	0.0
		40.0	519.0	509.0	221.3	4.7	0.1	0.0
		20.0	211.0	149.0	36.1	1.9	0.0	0.0
		10.0	38.0	30.0	4.0	0.9	0.0	0.0
	K	80.0	581.0	706.0	212.6	7.5	0.1	0.0

Test	Vaccine	Dose (µl)	Kh 85/1-WCE	18323-WCE	92-kDa OMP ELISA	69-kDa OMP ELISA	FHA-ELISA	PT-ELISA
	K	40.0	742.0	550.0	270.8	5.8	0.0	0.0
		20.0	421.0	149.0	49.5	1.0	1.8	0.0
		10.0	67.0	70.0	15.3	0.1	0.1	0.0
	L	80.0	813.0	512.0	191.8	11.4	1.4	0.1
		40.0	672.0	412.0	268.4	1.1	0.0	0.0
		20.0	309.0	209.0	71.0	10.6	0.9	0.0
		10.0	99.0	85.0	21.6	5.8	0.0	0.0
5	reference	50.0	227.0	333.0	66.6	0.6	1.0	2.4
		25.0	184.0	208.0	13.1	0.4	0.4	0.6
		12.5	52.0	47.0	4.7	0.1	0.0	0.5
		6.2	44.0	47.0	3.2	0.1	0.0	0.1
	M	80.0	750.0	371.0	215.5	0.9	1.0	0.3
		40.0	202.0	621.0	32.2	0.2	0.9	0.2
		20.0	79.0	113.0	15.0	0.0	0.1	0.3
		10.0	31.0	84.0	3.8	0.1	0.1	0.1

Discussion

As the humoral response against pertussis is considered to play an important role in the MPT, a serological test which correlates with this test might be an alternative. We showed that the mean pertussis antibody concentration at the day of challenge was related to the proportion of surviving mice at each vaccine dose. We assumed the protection of mice in the MPT is related to a synergistic effect of the WCV-induced humoral immune response against several protective antigens. The PSPT has therefore been developed and compared to the MPT. The PSPT is based on the vaccine dose-dependent induction of antibodies against the whole range of virulent *B. pertussis* surface-antigens. Mice of 20-24 grams in weight are immunised i.p. with a twofold dilution range of the reference Kh 85/1 and vaccines under study.We have chosen to use mice of 20-24 g in weight because these young adult mice have a more mature immune system and induce higher antibody levels with a significantly lower variation per group compared to mice of 10-14 grams (Fig. 2). After a four-week interval mice are bled and sera are titrated in Kh 85/1-WCE or 18323-WCE. The means of the log-transformed antibody concentrations induced per vaccine dose are used to estimate the potency by means of a parallel line analysis.

The PSPT corresponds well with the MPT in the in-house validation study with differentWCVs. Potencies obtained by both tests were significantly similar in X^2-test, which shows homogeneity between the two tests. In contrast to Dellipiane [8] we were not able to estimate the potency of WCVs based on the antibody response against PT in our serological model. Neither could we do that for FHA, 69-kDa OMP antibody responses. Although we were able to estimate potency based on serology against 92-kDa OMP these results were statistically invalid and did not correlate with the MPT. This confirmed our assumption that protection in the MPT is related to a synergistic effect of the humoral responses against a range of protective antigens. Preliminary results indicated that the antibodies measured in the 18323-WCE are involved in the complement-mediated immune response against virulent *B. pertussis* and could underpin the correlation between protection in the MPT and WCV induced humoral immunity.

Compared to the MPT, the PSPT has a number of advantages; the i.c. challenge becomes redundant and consequently the animal distress is reduced to an acceptable level. The PSPT reduces the number of mice by more than 25 percent. Twelve instead of 16 mice per vaccine dose are used and the virulence control of 70 mice becomes superfluous. Moreover, the lower mean variance of the PSPT indicates better reproducibility, which results in smaller 95% confidence interval.

The PSPT is a promising substitute for the MPT. Further validation and additional studies on functional validity should finally warrant replacement of the MPT by this serological test.

References

1 Kendrick PL, Eldering G, Dixon MK: Misner J. Mouse protection tests in the study of pertussis vaccines: a comparative series using intracerebral route of challenge. Am J Public Health 1947;37:803-810.

2 Medical Research Council: Vaccination against whooping-cough: relation between protection in children and results of laboratory tests. Brit Med J 1956;2:454-462.

3 Wiertz EJHJ, Walvoort HC, van Loveren H, van Straaten-van de Kappelle I, van der Gun JW, Kreeftenberg JG: Acellular and whole cell pertussis vaccines protect against the lethal effects of intracerebral challenge by different T-cell dependent humoral routes. Biologicals 1990;18:173-180.

4 Novotny P, Chubb AP, Cownley K, Charles IG: Biologic andprotective properties of the 69-kDa outer membrane protein of Bordetella pertussis: A novel formulation for an acellular pertussis vaccine. J Inf Diseases 1991;164:114-122.

5 Van der Ark A, v Straaten-vd Kappelle I, Akkermans A, Hendriksen C, vd Donk H: Development of Pertussis Serological Potency Test: Serological assessment of antibody response induced by whole cell vaccine as an alternative to mouse protection in an intracerebral challenge model. Biologicals 1994;22: 233-244.

6 Gillenius P, Jäätmaa E, Aslelöf P, Tiru M, Granström M: The standardization of an assay for pertussis toxin and antitoxin in microplates culture of Chinese hamster ovary cells. Biol Stand 1985;13:61-66.

7 Ackers JP, Dolby JM: The antigen of Bordetella pertussis that induces bactericidal antibody and its relationship to protection of mice. J Gen Microbiol 1972;70:371-382.

8 Dellepiane NI, Manghi MA, Erikson PV, di Poala G, Cangelosi A: Pertussis whole cell vaccine: Relation between intracerebral protection in mice and antibody response to Pertussis Toxin, Filamentous Hemagglutinin and Adenylate Cyclase. Zbl Bakt 1992;277:65-73.

A. van der Ark, Laboratory for Control of Biological Products, National Institute of Public Health and Environmental Protection, P.O. Box 1, 3720 BA Bilthoven, The Netherlands

Brown F, Cussler K, Hendriksen C (eds): Replacement, Reduction and Refinement of Animal Experiments in the Development and Control of Biological Products.
Dev Biol Stand. Basel, Karger, 1996, vol 86, pp 283-296

Development of a Guinea-Pig Model for Potency/Immunogenicity Evaluation of Diphtheria, Tetanus Acellular Pertussis (DTaP) and *Haemophilus influenzae* Type b Polysaccharide Conjugate Vaccines

R.K. Gupta, R. Anderson, D. Cecchini, B. Rost, P. Griffin, Jr., K. Benscoter, J. Xu, L. Montanez-Ortiz, G.R. Siber

Massachusetts Public Health Biologic Laboratories, State Laboratory Institute, Boston, MA, USA

Key words: Guinea-pig, potency test, tetanus toxoid, diphtheria toxoid, acellular pertussis vaccine, *Haemophilus influenzae* type b conjugate vaccine, combined vaccines.

Abstract: We have evaluated a guinea pig model for assessing the immunogenicity of *Haemophilus influenzae* type b (Hib) polysaccharide-protein conjugate vaccines, acellular pertussis vaccine and combination vaccines consisting of tetanus toxoid (TT), diphtheria toxoid (DT), acellular pertussis vaccine and Hib-TT (Hib-T) conjugate vaccine. The model was based on the United States (US) potency test for TT and DT which requires injection of guinea pigs with a single dose of undiluted vaccine. Guinea pigs showed dose-dependent antibody responses to pertussis toxoid (PTxd) and filamentous haemagglutinin (FHA), two important components of acellular pertussis vaccine. Antibody responses of guinea pigs to commercially available Hib conjugate vaccines qualitatively resembled those of human infants. Unconjugated polyribosylribitolphosphate (PRP) was not immunogenic; PRP-D conjugate produced a low antibody response, HbOC, PRP-T (Merieux) and Hib-T (MPHBL) produced a low response to the first dose and a strong anamnestic response to the booster dose. PRP-OMP uniquely produced a strong response after the first dose which was boosted by the second dose. In preliminary experiments, injection of guinea pigs with the combined vaccine formulations consisting of TT, DT, whole cell or acellular pertussis vaccine (PTxd and FHA) and Hib-T conjugate showed that these vaccines were immunogenic when combined, with some effects on the antibody responses of certain components. This model for testing potency/immunogenicity of combined vaccines substantially reduces the number of animals needed to test each lot of vaccine. To reduce the use of animals in testing vaccines further, we propose the use of a Vero cell assay for titrating diphtheria antitoxin and ELISA for measuring IgG antibody to tetanus toxin. The guinea pig model may also be useful for evaluating combination vaccines.

Introduction

Vaccines against tetanus and diphtheria are among the most successful vaccines developed so far and have virtually eliminated these diseases in the developed countries. Despite the success of these vaccines, there is still no universally accepted method for testing their potency/immunogenicity. The World Health Organization (WHO) and many of the European countries use a very cumbersome method of titrating these vaccines against a reference preparation [1-3]. Despite the use of a large number of animals (>100 guinea pigs to test one lot of diphtheria toxoid [DT] and more than 100 mice or guinea pigs to test one lot of tetanus toxoid [TT]) [2], this test has not been shown to correlate with immunogenicity in humans [4-6]. The test has several other drawbacks which have long been debated [7-14] and there have been efforts to develop alternative tests which do not require the use of such large numbers of animals [8, 9, 13, 14, 15-20]. On the other hand, a simple potency test using a small number of animals is performed for the vaccines licensed in the United States (US) [21-23]. It has been criticized for the lack of reference preparation, analysis of pooled sera, lack of end-point titration and lack of statistical analysis of results. The US potency test requires only four guinea pigs for immunization and another four to six guinea pigs and 12-15 mice for titration of pooled sera of immunized animals for diphtheria and tetanus antitoxins, respectively [21-23]. The titrations could be performed by in vitro assays [24-26], thus saving further on animals. Vaccines tested by both the WHO and US methods have been effective in controlling tetanus and diphtheria. The US potency test has an advantage over the WHO or the European method in that it uses undiluted vaccines for immunization in contrast to vaccines diluted in saline. Dilutions of adjuvanted/adsorbed vaccines may not reflect the true immunogenicity of the preparations [10, 11, 27-30].

There are no animal models which can reliably predict the immunogenicity of acellular pertussis vaccine and *Haemophilus influenzae* type b (Hib) polysaccharide-protein conjugate vaccines in human infants. Immunogenicity and potency of these vaccines, individually or in combinations, have been evaluated in mice [31, 32]. Because wide variations in potency and immunogenicity were observed in different strains of mice [33-35], and to minimize the use of animals, we evaluated the immunogenicity of acellular pertussis vaccine and Hib conjugate vaccines individually or in combination with TT and DT in guinea pigs using doses of vaccines consistent with the US minimum requirements for TT and DT (half the total human dose or 1.5 times the single human dose [SHD]).

Materials and Methods

Vaccines

The following vaccines were purchased: polyribosylribitolphosphate (PRP) polysaccharide vaccine (Hib-Immune), manufactured by the Lederle Laboratories, Pearl River, NY; PRP-D (ProHiBit), manufactured by Connaught Laboratories Inc., Swiftwater, PA; HbOC (HibTITER), manufactured by the Lederle-Praxis Biologics, Rochester, NY; PRP-OMP (PedvaxHIB), manufactured by Merck Sharp & Dohme, West Point, PA; PRP-T (ActHib) manufactured by the Pasteur-Mérieux Vaccines & Serums, Lyon, France; and Hib-T, manufactured at our laboratories using the method described by Chu et al [36]. The characteristics of these vaccines are given in Table 1.

Table 1: Details of Hib polysaccharide and Hib-conjugate vaccines.

Vaccine	**Manufacturer**	**Carrier protein**	**Polysaccharide**	
			Nature	**Human dose (μg)**
Hib-Immune	Lederle	None	Native	25
ProHiBit	Connaught	Diphtheria toxoid	Size-reduced	25
HibTITER	LederlePraxis	Diphtheria CRM_{197} Protein	Oligosaccharide	10
PedvaxHIB	Merck	Outer membrane protein of Neisseria meningitidis Group B	Medium	15
ActHib	Pasteur-Mérieux	Tetanus toxoid	Native	10
Hib-TT	MPHBL	Tetanus toxoid	Native	10

From Siber et al [47] with kind permission from Butterworth-Heinemann journals.

Adsorbed diphtheria, tetanus and pertussis (DTP) vaccine was our licensed product containing (per dose) five Lf of TT, 10 Lf of DT, four protective units of pertussis vaccine and 0.68 mg of aluminium phosphate. Pertussis toxoid (PTxd) was prepared by detoxifying purified pertussis toxin (PT) with tetranitromethane [37] and filamentous haemagglutinin (FHA) was purified from culture supernatants of *Bordetella pertussis* by hydroxyapatite chromatography and treating mildly with formaldehyde to stabilize it. PTxd and FHA at different concentrations were adsorbed on to 4 mg/ml of aluminium phosphate as described [38]. Three formulations of acellular pertussis vaccine were prepared with different concentrations of PTxd and FHA (composition per SHD of 0.5 ml: lot 5, PTxd 25 μg, FHA 3 μg; Lot 6, PTxd 8 μg and FHA 1 μg; Lot 7, PTxd 2.5 μg and FHA 0.3 μg) and adsorbed on to 4 mg/ml of aluminium phosphate. Several investigational lots of diphtheria-tetanus-acellular pertussis (DTaP) vaccine with or without Hib-T conjugate vaccine were formulated.

Various DTaP-HibT formulations, lots 1 to 4 and DTaP vaccine had the same amounts of TT (10 Lf/ml), PTxd (50 μg/ml), FHA (6 μg/ml) and aluminium phosphate (4 mg/ml). The concentration of DT in DTaP-HIbT-1 and DTaP was 20 Lf/ml; in DTaP-HibT-2, and DTaP-HibT-4 it was 30 Lf/ml; and in DTaP-HibT-3 it was 40 Lf/ml. The concentration of PRP as a Hib-T conjugate in DTaP-HibT-1, DTaP-HibT-2 and DTaP-HibT-3 was 20 μg/ml and in DTaP-HibT-4 it was 10 μg/ml.

Animals

White female guinea pigs of Hartley strain weighing 450-550 g and outbred mice (CD-1) weighing 17-20 g were purchased from Charles River, Wilmington, MA.

Immunization of guinea pigs

In a preliminary experiment, groups of four guinea pigs were injected subcutaneously with various doses of PTxd (5, 12.5, 25 and 37.5 μg per guinea pig) and FHA (5 and 15 μg per guinea pig) separately. The guinea pigs were bled at four and six weeks and boosted with the same dose at six weeks. The guinea pigs were bled again two weeks after the second injection. For the acellular pertussis vaccine formulations (Lots 5, 6 and 7), groups of eight guinea pigs were injected with 1.5 SHD subcutaneously and bled at four and six weeks. The guinea pigs were boosted with the same dose at six weeks and bled two weeks later.

For a comparison of various Hib vaccines, groups of eight guinea pigs were injected subcutaneously with 1.5 SHD twice at an interval of six or eight weeks. The guinea pigs were bled six weeks after the first injection and two weeks after the second injection.

To see the interaction of adsorbed DTP vaccine with Hib-T conjugate vaccine, groups of guinea pigs were injected with Hib-T vaccine alone and conjugate vaccine mixed (before injection) with adsorbed DTP vaccine. In another experiment, Hib-T conjugate vaccine was mixed with adsorbed DTP vaccine or aluminium phosphate adjuvant (4 mg/ml) and stored at 4°C for one month before injection in to guinea pigs.

To study the interaction between diphtheria-tetanus-acellular pertussis (DTaP) vaccine components and Hib-T, the formulations (DTaP-Hib-T, DTaP, Hib-T and aluminium phosphate adsorbed Hib-T) were injected into guinea pigs subcutaneously at 1.5 SHD and the guinea pigs were bled at four and six weeks. The guinea pigs were boosted with the same dose at six weeks and bled two weeks later.

The sera of Guinea pigs were separated and stored frozen at –20°C or below until tested.

Serologic evaluations

Pooled four-week and six-week sera from at least four guinea pigs were assayed for diphtheria antitoxin and tetanus antitoxin respectively, by toxin neutralization tests [21, 22, 39]. Sera from individual guinea pigs at four weeks were assayed for diphtheria antitoxin by Vero cell assay at the Lcd/1 dose level of diphtheria toxin [24]. Individual sera at four, six and/or eight weeks were tested for PT IgG and FHA IgG antibodies by ELISA and neutralizing antibodies to PT. IgG antibodies against PT and FHA were determined by ELISA by coating the plates with purified antigens as described for human sera [40]. Neutralizing antibodies to PT were determined by the TN test in Chinese Hamster Ovary (CHO) cells [41]. Serum samples at six and eight or ten weeks from guinea pigs injected with Hib vaccines were assayed for total antibody to PRP polysaccharide by a Farr-type radio-immunoassay [42, 43], using ^{3}H-labelled PRP kindly provided by Dr. Porter Anderson, University of Rochester Medical Center, Rochester, NY.

Results and Discussion

Immunogenicity to acellular pertussis vaccine components

The immunogenicity of acellular pertussis vaccine has usually been evaluated in mice mainly due to the cost and ease of handling them. To obtain a response of mice on the steep part of the dose response curve, vaccine has to be diluted as we could not obtain a dose response in mice when our DTaP vaccine was injected into mice undiluted in volumes ranging from 25 µl to 500 µl (unpublished data). The dilution of adsorbed vaccines for the immunogenicity tests is not desirable [7, 10, 11, 27-30] as the diluted aluminium adjuvants may not be able to form a depot at the site of injection [30]. Therefore we evaluated a dose response with undiluted vaccine in guinea pigs and found that the guinea pigs showed a dose response for both anti-PT IgG and neutralizing antibodies to PT, particularly at four weeks when the animals were injected with a dose of five to 37.5 µg of PTxd (Table 2). The highest dose selected in this experiment represented 1.5 times the SHD of PTxd in our acellular pertussis vaccine.

We have found that FHA is highly immunogenic in mice as well as in guinea pigs (unpublished data). When immunogenicity of FHA was evaluated at 5 and 15 µg per guinea pig, we did not observe a dose response (Table 3). A dose of 5 µg per guinea pig would be the maximum dose injected based upon 1.5 times the SHD of FHA (3 µg) in our acellular pertussis vaccine. When acellular vaccines were formulated at different doses, the guinea pigs showed a clear dose response for both PTxd and FHA components (Table 4). Based on this limited number of experi-

Table 2: Antibody response of guinea pigs injected with various doses of aluminium phosphate adsorbed pertussis toxoid. Groups of four guinea pigs were injected subcutaneously with various doses of pertussis toxoids twice at an interval of six weeks. The guinea pigs were bled at four, six and eight weeks.

Dose (µg)	**Antibody response**	**Geometric mean PT antibodies at**		
		four weeks	**six weeks**	**eight weeks**
5.0	PT IgG (µg/ml)	13.31 (6.21-28.53)[a]	18.15 (7.92-41.62)[e]	78.88 (50.63-122.90)
	PT Nt (Titre)	19.03 (10.96-33.03)[a']	64.00 (64.00-64.00)	430.54 (107.49-1724.5)
12.5	PT IgG (µg/ml)	20.63 (13.23-32.17)[b]	22.26 (13.59-36.46)	61.75 (0.48-7929.5)
	PT Nt (Titre)	26.91 (5.15-140.7)[b']	80.63 (29.84-217.9)	362.04 (4.42-29593)
25.0	PT IgG (µg/ml)	28.87 (16.42-50.74)[c]	23.00 (11.63-45.52)	154.04 (75.02-316.27)
	PT Nt (Titre)	38.05 (21.92-66.05)[c']	76.11 (43.85-132.1)	1024.0 (416.1-2519)
37.5	PT IgG (µg/ml)	36.51 (22.03-60.51)[d]	42.57 (19.10-94.87)[f]	166.96 (95.79-290.99)
	PT Nt (Titre)	53.82 (31.01-93.01)[d']	76.11 (43.85-132.1)	1217.7 (701.6-2114)

Figures in parentheses show 95% confidence intervals of geometric mean.
a versus c and d; b versus d; e versus f; a' versus c' and d', $P<0.05$ by *t*-test.

ments, it appears that guinea pigs at four weeks after immunization showed a better dose response than at six weeks (Table 2). Guinea pigs had an anamnestic response to both PTxd and FHA after a booster injection which also increased with increasing dose (Table 4). Based on earlier observations that it is not possible to detect small differences among vaccines after booster injection (7,38), we recommend that the antibody response to pertussis components may be evaluated after a single injection. Further studies may be required to evaluate the optimal timing of bleeding after a single injection.

Table 3: Antibody response of guinea pigs injected with various doses of aluminium phosphate adsorbed filamentous haemagglutinin (FHA). Groups of four guinea pigs were injected subcutaneously with various doses of FHA twice at an interval of six weeks. The guinea pigs were bled at four, six and eight weeks.

Dose (µg)	**Geometric mean Anti-FHA IgG (µg/ml) at**		
	four weeks	**six weeks**	**eight weeks**
5.0	305.9 (241.4-387.7)	354.0 (192.9-649.5)	1360.1 (728.3-2539.8)
15.0	308.0 (147.0-645.5)	258.0 (165.5-402.2)	978.2 (473.9-2019.2)

Figures in parentheses show 95% confidence intervals of geometric mean.

Table 4: Antibody response of guinea pigs injected with 0.75 ml (1.5 times single human dose) of various formulations of aluminium phosphate absorbed acellular pertussis vaccine. Groups of eight guinea pigs were injected subcutaneously with various formulations containing varying amounts of pertussis toxoid and FHA twice at an interval of six weeks. The guinea pigs were bled at four, six and eight weeks.

Vaccine lot	Dose (μg) PTxd FHA	Antibody response	Geometric mean antibody levels at		
			four weeks	six weeks	eight weeks
5	37.5 4.5	PT IgG (μg/ml)	53.6 (36.8-78.0)[a]	50.3 (31.0-81.7)[a']	374.8 (203.1-691.7)
		PT Nt (Titre)	39.6 (27.3-57.4)[d]	105.2 (49.5-223.6)[d']	Not Done
		FHA IgG (μg/ml)	164.7 (112.9-240.4)[g]	252.9 (231.6-276.3)[g']	759.0 (666.3-864.5)
6	12.0 1.5	PT IgG (μg/ml)	23.8 (16.7-33.9)[b]	21.3 (14.3-32.0)[b']	153.5 (93.9-250.8)
		PT Nt (Titre)	7.9 (2.0-31.5)[e]	29.6 (17.1-51.4)[e']	Not Done
		FHA IgG (μg/ml)	80.4 (37.9-170.6)[h]	125.0 (66.4-235.1)[h']	1055.2 (661.7-1682.8)
7	3.8 0.5	PT IgG (μg/ml)	11.2 (7.0-18.2)[c]	10.3 (4.9-21.6)[c']	19.8 (10.6-37.0)
		PT Nt (Titre)	8.8 (3.7-20.9)[f]	15.2 (5.8-39.8)[f']	285.4 (122.3-665.8)
		FHA IgG (μg/ml)	32.6 (20.6-51.7)[i]	69.6 (47.0-102.6)[i']	400.3 (222.2-721.1)

Figures in parentheses show 95% confidence intervals of geometric mean.
a versus b and c; b versus c; a' versus b' and c', P≤0.01 by *t*-test.
d versus f; d' versus e' and f', P<0.01; d versus e, P<0.05 by *t*-test.
g versus i; g' and i', P<0.001; h versus i; g' versus h', P<0.05 by *t*-test.

Immunogenicity to *Haemophilus influenzae* type b Conjugate Vaccines

Like acellular pertussis vaccine, mice have been used for evaluating the immunogenicity of polysaccharide-protein conjugate vaccines including Hib conjugate vaccines [32, 44, 45]. Mice typically do not respond to the polysaccharide component of the conjugate after a single dose and sometimes even after two doses [44, 46]. There are usually a few non-responder mice to the polysaccharide component even after three doses [44, 46]. Various strains of mice show variations in their antibody responses to polysaccharides [44] and protein antigens [33-35].

We have described a guinea pig model for the immunogenicity of Hib conjugate vaccines [47] which showed qualitative responses of guinea pigs to various Hib conjugates very similar to those of human infants (Table 5). Guinea pigs did not respond to pure polysaccharide vaccine, PRP, after both injections. PRP has

Table 5: Antibody responses of guinea pigs to various kinds of PRP polysaccharide and PRP-conjugate vaccines. The guinea pigs were injected with 1.5 times the single human dose of the vaccine twice at an interval of eight weeks. Guinea pigs were bled six weeks after the first dose and two weeks after the booster dose.

Vaccine	Lot n°	Dose of PRP (µg)	PRP antibodies (µg/ml)				Responders*/ Total	
			Geometric mean after		Pooled sera			
			1st dose	2nd dose	1st dose	2nd dose	1st dose	2nd dose
PRP	262-941	37.5	0.04 (0.04-0.04)	0.04 (0.04-0.04)	0.04	0.06	0/8	0/8
PRP-D	8Co1059	37.5	0.09 (0.03-0.28)	0.10 (0.04-0.27)	0.13	0.09	2/8	3/8
HbOC	M660FB	15.0	0.29 (0.10-0.85)	3.23 (0.22-47.04)	0.58	56.00	6/8	7/8
	M190EB	15.0	0.15 (0.05-0.44)	3.96 (0.33-46.71)	0.56	2.50	4/8	7/8
PRP-OMP	44297	22.5	6.86 (2.02-23.31)	82.78 (20.61-332.5)	27.0	86.0	8/8	8/8
	1221T	22.5	20.63 (1.09-388.7)	242.65 43.6-1349.5)	ND	ND	5/5	5/5
HiB-T (Mérieux)	1982	15.0	0.28 (0.09-0.88)	78.86 (6.82-911.3)	2.80	220.0	5/8	7/7
	2102	15.0	0.55 (0.11-2.83)	3.98 (0.48-33.15)	ND	ND	6/8	6/7
PRP-T (MPHBL)	T1a	15.0	0.84 (0.11-6.39)	13.13 (0.42-408.6)	ND	ND	5/7	5/6
	T2	15.0	0.46 (0.10-2.20)	7.25 (0.52-100.6)†	ND	ND	6/8	7/8

From Siber et al [47] with kind permission from Butterworth-Heinemann journals.
ND = Not done
Figures in parentheses show 95% confidence intervals of geometric mean.
* A guinea pig eliciting 0.15 µg/ml of PRP antibodies was taken as responder.
† For this lot, the guinea pigs were boosted at six weeks and bled two weeks later.

not been found effective and immunogenic in infants and children younger than 18 months [48-52]. The response of guinea pigs to various conjugate vaccines resembles that of infants and young children [48, 53, 54]. PRP-D (ProHiBit), the first conjugate vaccine licensed for use in the US, was found efficacious in preventing diseases caused by Hib, but was less immunogenic when compared to other conjugate vaccines developed later [48, 53]. In our guinea pig model, PRP-D was the least immunogenic among all the conjugates tested in this study (Table 5). PRP-OMP has been found to be most immunogenic in infants after a single injection [48, 54-56]. In the guinea pig model also, this vaccine showed high antibody levels after the first injection which were similar or even higher than those elicited by two injections of HbOC, PRP-T and Hib-T vaccines. These findings are similar to those found in infants [54]. HbOC and PRP-T conjugate vaccines show similar immunogenicity in infants [48, 54] and also behaved similarly in guinea pigs. Our own conjugate vaccine Hib-T, prepared using the same conjugation reactions used for making PRP-T, showed similar immunogenicity. The variations among individual guinea pigs were as great as those observed in infants injected with the same preparation, HbOC, at different times from the same or different lots [54, 57]. Up to three guinea pigs out of eight did not respond after the first immunization and after the second immunization occasionally one animal did not respond. Despite the variations, the guinea pig model was found useful for discriminating Hib conjugate preparations with low levels of conjugation [47]. Therefore the guinea pig model can be used to assess the immunogenicity of polysaccharide-protein conjugates.

Immunogenicity/Potency of Combined Vaccines

As the new vaccines are developed, it would be desirable to reduce the number of injections by mixing vaccines in the same syringe at the time of injection or preferably by making combination vaccines. New combined formulations require laboratory and clinical demonstration of safety and efficacy before licencing and laboratory evaluation to control production. We therefore used our guinea pig model to evaluate Hib-T formulated with DTP or DTaP vaccines. Initially, we evaluated mixing our DTP vaccine with Hib-T vaccine either immediately at the time of injection or one month earlier. Preliminary results indicate that PRP antibody responses were similar for Hib-T alone and Hib-T mixed with DTP vaccine immediately before injection (Table 6). Adsorption of Hib-T conjugate on aluminium phosphate or mixing with aluminium phosphate-adsorbed DTP vaccine and storing the formulations at 4°C for one month before injection resulted in enhanced primary and secondary responses to PRP. In infants, a combination of other manufacturers' Hib conjugate vaccine with DTP vaccine did not show any significant effects on the immunogenicity of PRP, tetanus and diphtheria components [58, 59] whereas other studies showed lower antibody responses to PRP [60] and pertussis antigens [61] in recipients of combination vaccines than recipients who received separate injections. The guinea pig model also showed that the combined vaccines passed the US potency tests for DT and TT. However, the combined vaccines induced slightly lower antibody responses to PTxd and TT with significantly lower response to DT (Table 6).

After immunization with combined formulations of DTaP and Hib-T, guinea pigs had lower secondary antibody response to PRP, than after Hib-T alone or Hib-T adsorbed on aluminium phosphate (Table 7). Lower PRP antibody responses have also been observed in infants injected with PRP-T and DTP combined vaccines than those injected with these vaccines separately [60]. This may be due to carrier specific epitope suppression or antigenic competition. In DTaP-HibT-4 formulation, when the dose of Hib-T was reduced by half, the animals showed a better secondary response than the other three formulations having full strength Hib-T conjugate. Certain combination formulations also showed reduction in antibody responses to PTxd and FHA. Infants injected with combination vaccines consisting of PRP-T and DTP showed lower responses to PT and pertussis agglutinogens than those injected with PRP-T and DTP at separate sites or with DTP alone [61]. The clinical significance of the lower responses may not be important but it appears that antibody responses of guinea pigs to combination vaccines may

Table 6: Antibody response of guinea pigs to Hib-T alone and mixed with adsorbed DTP vaccine. Groups of eight guinea pigs were injected with 0.75 ml (1.5 times the single human dose) twice at an interval of six weeks and bled at four, six and eight weeks.

Vaccine/ formulation	GM (95% CI) anti-PRP (μg/ml) after		GM (95% CI) anti-PT IgG (μg/ml) after		GM (95% CI) anti-DT	Tetanus antitoxin
	1st dose	2nd dose	1st dose	2nd dose	IgG (μg/ml) 4 weeks	(IU/μl) at 6 weeks
Mixed at time of use						
Hib-T	0.38 (0.05-19.8)	8.15 (0.51-129)	N/A	N/A	N/A	<0.5
Hib-T + DTP	0.67 (0.12-3.61)	8.95 (0.25-267)	0.02 (0.00-0.46)	nd	32.5 (20.5-51.6)	nd
DTP	N/A		0.86 (0.03-269)	nd	31.9 (27.0-51.7)	nd
Mixed formulation stored at 4°C for one month						
Hib-T	0.84 (0.11-6.39)	13.1 (0.42-409)	N/A	N/A	N/A	<2
Hib-T adsorbed	11.4 (0.26-506)	66.7 (1.95-2280)	N/A	N/A	N/A	2
Hib-T + DTP 278	29.7 (3.78-233)	40.8 (3.87-430)*	0.41 (0.04-4.46)	0.94 (0.10-8.95)	44.7 (26.9-74.2)†	4
Hib-T + DTP 276	2.43 (0.19-30.8)	21.4 (5.44-83.8)	0.70 (0.08-6.29)	2.42 (0.23-25.2)	42.2 (24.7-72.0)†	4
DTP 278	N/A		0.43 (0.06-3.22)	1.13 (0.07-17.5)	69.8 (42.8-112)	6
DTP 276	N/A		1.48 (0.19-11.8)	8.37 (1.20-58.5)	72.1 (51.5-101)	6

From Siber et al [47] with kind permission from Butterworth-Heinemann journals.
N/A = Not applicable; nd = Not done.
* Significantly different from Hib-T alone by a *t*-test ($P<0.05$).
† Significantly different from corresponding lot of DTP alone by *t*-test ($P<0.01$).

show antigenic competition similar to that observed in infants. Therefore, it appears that the guinea pig immunogenicity test, which is simple and requires only small numbers of animals, may be useful in the development of combination vaccines.

Conclusions

Potency testing of vaccines containing adsorbed DT and TT according to the US Minimum Requirements is simple and requires only a small number of animals.

Table 7: Antibody response of guinea pigs to various formulations of DTaP-HibT vaccine, Hib-T alone and DTaP alone. Groups of eight guinea pigs were injected with 0.75 ml (1.5 times the single human dose) twice at an interval of six weeks and bled at four, six and eight weeks.

Vaccine formulation*	GM anti-PT at six weeks		GM anti-FHA IgG	Tetanus antitoxin	Diphtheria anti-toxin at 4 weeks		GM anti-PRP (Farr), μg/ml	
	IgG (μg/ml)	Nt (Titre)	(μg/ml), six weeks	(IU/ml), six weeks	in vivo, IU/ml	GM Vero IU/ml	Primary, six weeks	Boost, 8 weeks
DTaP-HibT-1	21.52	74.45	75.37	≥6	3	1.90	2.01	1.72
DTaP-HibT-2	16.67	43.60	111.02	5	3	1.64	0.81	2.91
DTaP-HibT-3	38.18	60.02	107.18	≥6	≥4	2.33	3.36	3.21
DTaP-HibT-4	13.23	35.18	204.01	5	≥4	2.44	2.14	7.92
DTaP	35.23	63.82	357.16	≥6	≥4	2.10	N/A	N/A
HibT-, adsorbed	N/A	N/A	N/A	≥6	N/A	N/A	4.01	38.15
HibT	N/A	N/A	N/A	<2	N/A	N/A	1.52	26.08

N/A = Not applicable;
* Various formulations differed with regard to amount of components (see Materials & Methods).

Vaccines tested by this method have been effective in controlling tetanus and diphtheria in North America. As this test requires the use of undiluted vaccines for immunization of animals, it may better reflect the immunogenicity of the final formulation of vaccine injected into humans. Based on the US potency test, we developed immunogenicity tests for acellular pertussis vaccine and Hib conjugate vaccines so that a combined vaccine containing all these components may be evaluated in a single test. For this test, groups of six to eight guinea pigs (450-550 g) were injected subcutaneously with 1.5 times the SHD and bled 1) at four weeks for potency of diphtheria component; 2) at six weeks for potency of tetanus component; 3) at four and six weeks for assessing immunogenicity to PTxd and FHA and 4) at six weeks to assess the primary response to Hib conjugate vaccine. Since evaluation of the secondary or anamnestic response to polysaccharide component is important for Hib conjugate vaccines, the guinea pigs were boosted at six weeks with the same dose as the first injection and bled two weeks later. This model provided information on the immunogenicity of vaccines, particularly Hib conjugates, and on the immunological interaction between various components of combined vaccines. The combined test substantially reduced the number of animals needed. To reduce the use of animals further, we propose the Vero cell assay for titration of diphtheria antitoxin and ELISA for IgG antibodies to tetanus toxin. We recommend further evaluation of the guinea pig model for assessing the potency of DTaP and DTaP-Hib combination vaccines.

REFERENCES

1 World Health Organization Expert committee on Biological Standardization, Fortieth Report: Requirements for Diphtheria, Tetanus and Combined Vaccines. in: Technical Report Series, No. 800, World Health Organization, Geneva, 1990, pp 87-179.

2 World Health Organization Manual of details of tests required on final vaccines used in the WHO Expanded Programme on Immunization. Unpublished WHO document BLG/UNDP/82.1 Rev 1.

3 Dobbelaer R: A comparison between the World Health Organization and European Pharmacopoeia methods for the potency testing of adsorbed tetanus toxoids, in Manclark CR (ed): Proceedings of an Informal Consultation on the World Health Organization Requirements for Diphtheria, Tetanus, Pertussis and Combined Vaccines. Department of Health and Human Services, United States Public Health Service, Bethesda, MD, DHHS Publication No. [FDA] 91-1174, 1991, pp 65-69.

4 Relyveld E, Bengounia A, Huet M, Kreeftenberg JG: Antibody response of pregnant women to two different adsorbed tetanus toxoids. Vaccine 1991;9:369-372.

5 Bizzini B: Tetanus, in Germanier R (ed): Bacterial Vaccines, Academic Press, Orlando, 1984, pp 37-68.

6 Nyerges G, Adam MM, Gacs, P, Ring R: A comparison of combined diphtheria, tetanus and pertussis vaccines produced by different manufacturers in laboratory animals and in infants. J Biol Stand 1986;14:241-247.

7 Discussion on Laboratory testing of the diphtheria toxoid and tetanus toxoid components of tetanus toxoid vaccines, diphtheria and tetanus toxoid vaccines, and diphtheria and tetanus toxoids and pertussis vaccines, in Manclark CR (ed): Proceedings of an Informal Consultation on the World Health Organization Requirements for Diphtheria, Tetanus, Pertussis and Combined Vaccines. Department of Health and Human Services, United States Public Health Service, Bethesda, MD, DHHS Publication No. [FDA] 91-1174, 1991, pp 75-79.

8 Lyng J, Heron I, Ljungqvist L: Quantitative estimation of diphtheria and tetanus toxoids. 3. Comparative assays in mice and in guinea pigs of two tetanus toxoid preparations. Biologicals 1990;18:3-9.

9 Lyng J, Heron I: Quantitative estimation of diphtheria and tetanus toxoids. 5. Comparative assays in mice and in guinea pigs of two diphtheria toxoid preparations. Biologicals 1991;19:327-334.

10 Lyng J: The validity and meaning of potency assays of toxoid vaccines. Unpublished WHO document, BS/92.1691.

11 Lyng J, Haslov K: The validity and meaning of potency assays of toxoid vaccines II. Unpublished WHO document, BS/93.1748.

12 Lyng J: International standard preparations for diphtheria & tetanus toxoids. A historical review of the period 1947-1955 and a proposal for discontinuation of the two plain toxoid standard preparations. Unpublished WHO document, BS/93.1753.

13 Huet M, Relyveld E, Camps S: Methode simple de controle de l'active des anatoxines tetaniques adsorbees. Biologicals 1990;18:61-67.

14 Huet M, Relyveld E, Camps S: Simplified activity evaluation of several tetanus vaccines. Biologicals 1992;20:35-43.

15 Nyerges G, Virag E, Lutter J: The potency testing of diphtheria and tetanus toxoids as determined by the induction of antibody in mice. J Biol Stand 1983;11:99-103.

16 Kreeftenberg JG, van der Gun J, Marsman FR, Sekhuis VM, Bhandari SK, Maheshwari SC: An investigation of a mouse model to estimate the potency of the diphtheria component in vaccines. J Biol Stand 1985;13:229-234.

17 Knight PA, Roberts PAG: An evaluation of some proposals for a reduction in the number of animals used for the potency testing of diphtheria and tetanus vaccines. J Biol Stand 1987;15:165-175.

18 Hendriksen CFM, van der Gun J, Marsman FR, Kreeftenberg JG: The effects of reductions in the number of animals used for the potency assay of the diphtheria and tetanus components of adsorbed vaccines by the methods of the European Pharmacopoeia. J Biol stand 1987;15:353-362.

19 Knight PA: Guidelines for performing one-dilution tests for ensuring that potencies of diphtheria and tetanus containing vaccines are above the minimum required by WHO. Unpublished WHO document, BS/89.1618.

20 World Health Organization: Laboratory methods for the testing for potency of diphtheria (D), tetanus (T), pertussis (P) and combined vaccines. Unpublished WHO document, BLG/92.1.

21 United States Minimum Requirements: Diphtheria Toxoid, 1947. U.S. Department of Health, Education and Welfare, National Institutes of Health, Bethesda, MD, USA, 4th Revision.

22 United States Minimum Requirements: Tetanus toxoid, 1952. U.S. Department of Health, Education and Welfare, National Institutes of Health, Bethesda, MD, USA.

23 Fitzgerald EA: Overview of the methods for potency testing of diphtheria and tetanus toxoids in the United States, in Manclark CR (ed): Proceedings of an Informal Consultation on the World Health Organization Requirements for Diphtheria, Tetanus, Pertussis and Combined Vaccines. Department of Health and Human Services, United States Public Health Service, Bethesda, MD, DHHS Publication No. [FDA] 91-1174, 1991, pp 61-64.

24 Gupta RK, Higham S, Gupta CK, Rost B, Siber GR: Suitability of the Vero cell method for titration of diphtheria antitoxin in the United States potency test for diphtheria toxoid. Biologicals 1994;22:65-72.

25 Gupta RK, Siber GR: Comparative analysis of tetanus antitoxin titers of sera from immunized mice and guinea pigs determined by toxin neutralization test and enzyme-linked immunosorbent assay. Biologicals 1994;22:215-219.

26 Gupta RK, Siber GR: Use of in vitro Vero cell assay and ELISA in the United States potency test of vaccines containing adsorbed diphtheria and tetanus toxoids. Dev Biol Stand 1995;86:207-215.

27 Hennessen W: The mode of action of mineral adjuvants. Progr Immunobiol Stand 1965;2:71-79.

28 Hennessen W: Mode of action and consequences for standardization of adjuvanted vaccines. Symp Series Immunobiol Stand 1967;6:319-326.

29 Relyveld EH: Immunological, prophylactic and standardization aspects in tetanus, in Nistico G, Maestroni P, Pitzurra M (eds): Proceedings of the Seventh International Conference on Tetanus. Roma, Gangemi Publ. Co., 1985, pp 215-227.

30 Gupta RK, Rost BE, Relyveld E, Siber GR: Adjuvant properties of aluminium and calcium compounds, in Powell MF, Newman MJ (eds): Vaccine Design, New York, Plenum Press, 1995, pp 229-248.

31 Kreeftenberg JG: Collaborative study on the candidate reference materials JNIH-3, JNIH-4, and JNIH-5 for the assay of acellular pertussis vaccines. Unpublished WHO document, BS/88.1586.

32 Giri L, Vincent-Falquet J: Quality control, in Ellis RW, Granoff DM (eds): Development and Clinical Uses of Haemophilus b Conjugate Vaccines, New York, Marcel Dekker, Inc., 1994, pp 91-109.

33 Hardegree MC, Pittman M, Maloney CJ: Influence of mouse strain on the assayed potency (unitage) of tetanus toxoid. Appl Microbiol 1972;24:120-126.

34 Lyng J, Nyerges G: The second international standard for tetanus toxoid (adsorbed). J Biol Stand 1984;12:121-130.

35 Matuhasi T: Influence of mouse strain on the results of potency testing for diphtheria and tetanus toxoid components in D, T, DT and DTP vaccines, in Manclark CR (ed): Proceedings of an Informal Consultation on the World Health Organization Requirements for Diphtheria, Tetanus, Pertussis and Combined Vaccines. Department of Health and Human Services, United States Public Health Service, Bethesda, MD, DHHS Publication No. [FDA] 91-1174, 1991, pp 55-58.

36 Chu C, Schneerson R, Robbins JB, Rastogi SC: Further studies on the immunogenicity of Haemophilus influenzae type b and pneumococcal type 6A polysaccharide-protein conjugates. Infect Immun 1983;40:245-256.

37 Winberry L, Walker R, Cohen N, Todd C, Sentissi A, Siber G: Evaluation of a new method for inactivating pertussis toxin with tetranitromethane. in International Workshop on Bordetella pertussis, Rocky Mountain Laboratories, Hamilton, Montana, 18-20 August, 1988.

38 Gupta RK, Siber GR: Comparison of adjuvant activities of aluminium phosphate, calcium phosphate and stearyl tyrosine for tetanus toxoid. Biologicals 1994;22:53-63.

39 Gupta RK, Maheshwari SC, Singh H: The titration of tetanus antitoxin IV. Studies on the sensitivity and reproducibility of the toxin neutralization test. J Biol Stand 1985;13:143-149.

40 Siber GR, Thakrar N, Yancey BA, Herzog L, Todd C, Cohen N, Sekura RD, Lowe CU: Safety and immunogenicity of hydrogen peroxide-inactivated pertussis toxoid in 18-month-old children. Vaccine 1991;9:735-740.

41 Gupta RK, Siber GR: Need for a reference preparation of pertussis antitoxin for Chinese hamster ovary cell assay. Biologicals 1995;23:71-73.

42 Farr RS: A quantitative immunochemical measure of the primary interaction between I*BSA and antibody. J Infect Dis 1958;103:239-262.

43 Anderson P, Insel RA, Porcelli S, Ward JI: Immunochemical variables affecting radioantigen-binding assays of antibody to Haemophilus influenzae type b capsular polysaccharide in children's sera. J Infect Dis 1987;156:582-590.

44 Schneerson R, Barrera O, Sutton A, Robbins JB: Preparation, characterization, and immunogenicity of *Haemophilus influenzae* type b polysaccharide-protein conjugates. J Exp Med 1980;152:361-376.

45 World Health Organization Expert Committee on Biological Standardization, Forty First Report: Requirements for *Haemophilus* type b conjugate vaccines, in Technical Report Series No. 814, World Health Organization, Geneva, 1991, pp 15-37.

46 Gupta RK, Szu, SC, Finkelstein, RA, Robbins JB: Synthesis, characterization, and some immunological properties of conjugates composed of the detoxified lipopolysaccharide of *Vibrio cholerae* O1 serotype Inaba bound to cholera toxin. Infect Immun 1992;60:3201-3208.

47 Siber GR, Anderson R, Habafy M, Gupta RK: Development of a guinea-pig model to assess immunogenicity of *Haemophilus influenzae* type b capsular polysaccharide conjugate vaccines. Vaccine 1995;13:525-531.

48 Shapiro ED, Ward JI: The epidemiology and prevention of disease caused by *Hemophilus influenzae* type b. Epidemiol Rev 1991;13:113-142.

49 *Haemophilus influenzae* type b vaccines, in Stratton KR, Howe CJ, Johnston RB, Jr (eds): Adverse Events Associated with Childhood Vaccines. Evidence bearing on causality. Washington DC, National Academy Press, 1994, pp 236-273.

50 Shapiro ED, Murphy TV, Wald ER, Brady CA: The protective efficacy of *Haemophilus* b polysaccharide vaccine. J Am Med Assoc 1988;260:1419-1422.

51 Robbins JB, Schneerson R: Polysaccharide-protein conjugates: a new generation of vaccines. J Infect Dis 1990;161:821-832.

52 Robbins JB, Schneerson R, Pittman M: *Haemophilus influenzae* type b infections, in Germanier R (ed): Bacterial Vaccines. Orlando, Academic Press, 1984, pp 289-316.

53 Robbins JB, Schneerson R: Evaluating the *Haemophilus influenzae* type b conjugate vaccine PRP-D. New England J Med 1990;323:1415-1416.

54 Granoff DM, Anderson EL, Osterholm MT, Holmes SJ, McHugh JE, Belshe RB, Medley F, Murphy TV: Differences in the immunogenicity of three *Haemophilus influenzae* type b conjugate vaccines in infants. J Ped 1992;121:187-194.

55 Lenoir AA, Granoff PD, Granoff DM: Immunogenicity of *Haemophilus influenzae* type b polysaccharide-*Neisseria meningitidis* outer membrane conjugate vaccine in 2- to 6-month-old infants. Pediatrics 1987;80:283-287.

56 Campbell H, Byass P, Ahonkhai VI, Vella PP, Greenwood BM: Serologic responses to an *Haemophilus influenzae* type b polysaccharide-*Neisseria meningitidis* outer membrane protein conjugate vaccine in very young Gambian infants. Pediatrics 1990; 86:102-107.

57 Weinberg GA, Granoff DM: Polysaccharide-protein conjugate vaccines for the prevention of *Haemophilus influenzae* type b disease. J Pediatr 1988;113:621-631.

58 Scheifele D, Bjornson G, Barreto L, Meekison W, Guasparini R: Controlled trial of *Haemophilus influenzae* type b diphtheria toxoid conjugate combined with diphtheria, tetanus and per-

tussis vaccines, in 18-month old children, including comparison of arm versus thigh injection. Vaccine 1992;10:455-460.

59 Mulholland EK, Ahonkhai VI, Greenwood AM, Jonas LC, Lukacs LJ, Mink CM, Staub JM, Todd J, Vella PO, Greenwood BM: Safety and immunogenicity of *Haemophilus influenzae* type B-*Neisseria meningitidis* Group B outer membrane protein complex conjugate vaccine mixed in the syringe with diphtheria-tetanus-pertussis vaccine in young Gambian infants. Pediatr Infect Dis 1993;12:632-637.

60 Ferreccio C, Clemens J, Avendano A, Horwitz I, Flores C, Avila L, Cayazzo M, Fritzell B, Cadoz M, Levine M: The clinical and immunologic response of Chilean infants to *Haemophilus influenzae* type b polysaccharide-tetanus protein conjugate vaccine coadministered in the same syringe with diphtheria-tetanus toxoids-pertussis vaccine at two, four and six months of age. Pediatr Infect Dis 1991;10:764-771.

61 Clemens JD, Ferreccio C, Levine MM, Horwitz I, Rao MR, Edwards KM, Fritzell B: Impact of *Haemophilus influenzae* type b polysaccharide-tetanus protein conjugate vaccine on responses to concurrently administered diphtheria-tetanus-pertussis vaccine. J Am Med Assoc 1992; 267:673-678.

Dr. R.K. Gupta, Massachusetts Public Health Biologic Laboratories, State Laboratory Institute, 305 South St., Boston, MA 02130, USA

Brown F, Cussler K, Hendriksen C (eds): Replacement, Reduction and Refinement of Animal Experiments in the Development and Control of Biological Products.
Dev Biol Stand. Basel, Karger, 1996, vol 86, pp 297-301

Potency Tests of Diphtheria, Tetanus and Combined Vaccines Suggestion for a Simplified Potency Assay

H. Aggerbeck, I. Heron

Statens Seruminstitut, Bacterial Vaccine Department, Copenhagen S, Denmark

Key words: Diphtheria, tetanus, vaccine, aluminium, calcium, potency.

Abstract: Two diphtheria-tetanus vaccines (DT), adsorbed to either aluminium hydroxide or calcium phosphate but identical with respect to toxoid origin and amounts, were compared in full potency assays in mice according to the European Pharmacopoeia (EP) and in a reduced potency assay in guinea-pigs using a double dose immunization schedule. The efficacy of the vaccines was compared in a clinical trial with revaccination of 313 military recruits. The reduced potency assay gave a better reflection of the efficacy of the two vaccines in humans than the required assays of the EP. For release of combined, final vaccine formulations the reduced potency assay suggested will reduce the number of animals in quality control.

INTRODUCTION

Potency assays of toxoid based vaccines may be done in animals in different ways depending on the regulations to which one is assigned. According to the requirements of WHO and the EP, groups of mice or guinea-pigs are immunized with different doses of vaccine diluted in saline. According to the U. S. Department of Health, Education and Welfare at least four guinea-pigs are immunized with undiluted vaccine in an amount corresponding to not more than one-half the volume recommended as the total human immunizing dose. However, none of these potency assays corresponds directly to the schedule used in humans where a full primary vaccination consists of three to four doses of undiluted vaccine given several weeks apart.

The aim of the study was to evaluate the present potency assays required by EP by testing two DT vaccines adsorbed to either calcium phosphate or aluminium hydroxide and eventually to compare the result of these assays with the revaccination response in humans.

Materials and Methods

Vaccines

Two DT vaccines were mixed in saline to contain 12 Lf ml^{-1} of diphtheria and tetanus toxoids, 100 mg l^{-1} thimerosal and adjuvant corresponding to a final concentration of 1 mg ml^{-1} Al or 1 mg ml^{-1} Ca. Diphtheria toxoid batch M, 1520 Lf ml^{-1} corresponding to an antigenic purity on 2171 Lf mg^{-1} protein nitrogen (PN) and tetanus toxoid batch 38, 620 Lf ml^{-1} corresponding to 1632 Lf mg^{-1} PN obtained from Statens Seruminstitut, Bacterial Vaccine Department were used. The toxoids were free of pyrogens and were made by treating purified toxins with formaldehyde.

Depending on whether a dose of 1.0 or 0.5 ml is used the aluminium containing vaccine is registered in Denmark for primary and booster vaccination of adults.

Animals

Outbred, female mice strain Ssc:CF1, 20-22 days old (16-18 g) or outbred, female guinea-pigs strain Ssc:AL, eight weeks old (450-550 g) from Statens Seruminstitut were used in potency assays.

Potency assays

Full potency assay

Determinations of the biological potency of the two DT vaccines were made in mice according to the WHO and the EP requirements, using immunizations with two-fold dilution series in saline of the test vaccines and a reference vaccine. In the diphtheria assay mice were bled after four weeks and antibody content in sera was determined in a Vero cell toxin neutralization assay [1]. The relative potencies were determined by parallel line calculation on the basis of these antibody responses. In the tetanus assay mice were challenged with tetanus toxin after four weeks and the relative potencies were determined by probit analysis on the basis of the proportion of animals without paralysis.

Reduced potency assay

Groups of six guinea-pigs were immunized twice four weeks apart with 250 µl undiluted DT vaccine. Blood was collected immediately before and two weeks after the second immunization and tested for antibody content [1, 2].

Efficacy in man

Military recruits (18 - 24 years of age) were randomized and re-vaccinated with 0.5 ml of either DT vaccine. Blood samples were taken before and four weeks after vaccination and tested for diphtheria and tetanus antibody content [1-3].

Results

Potency

Two DT vaccines varying only in the nature of the adjuvant were tested for potency following international regulations as described by WHO and EP. The calcium phosphate adsorbed vaccine was significantly less potent according to these criteria and only just passed the EP requirements (Table 1). In compliance with these the lower confidence limit of a vaccine for primary immunization shall be 30

IU for diphtheria and 40 IU for tetanus per human dose. The corresponding requirements if the vaccine is used as a booster are 2 and 20 IU per human dose for diphtheria and tetanus respectively.

A simplified potency assay in guinea-pigs immunized twice with combinations of the two DT vaccines showed no significant difference between the two vaccines (Fig. 1). The combination A/C was included to imitate the clinical study in which all the probands had received aluminium containing vaccines for their primary immunization.

Efficacy

The efficacy of the two DT vaccines were evaluated on the basis of the median diphtheria and tetanus antibody content in military recruits four weeks after revaccination. Contrary to the full potency assay of the EP, the calcium phosphate adsorbed vaccine induced a significantly higher antibody response against both antigens (Table 2).

Discussion

Standard procedures for potency assays for D and T with dilution of doses in saline was followed and showed that the adjuvanticity of calcium phosphate was about half that of aluminium hydroxide and the vaccine only just passed the international requirements of potencies of a vaccine for primary immunization. In humans, however, the better antibody response was found in the group revaccinated with calcium phosphate adsorbed vaccine. Nyerges et al [4] and Relyveld et al [5] also found no correlation between potency assays in animals and antibody responses in humans.

In different designs of potency assays Lyng and Heron [6] found that the potency result depended on the animal species and whether vaccine doses were diluted in saline or in adjuvant. They questioned the basic idea of the limiting dilution principle when more than one active substance is involved in the vaccine. In the present context a reduced test design was used which indicated no difference between the two vaccines.

Based on data collected over many years, the currently required potency assays in animals are clearly capable of preventing vaccines of inferior quality being used

Table 1: Relative potency in mice of DT vaccines according to WHO and EP.

	Diphtheria*	**Tetanus†**
Vaccine	IU ml^{-1}	IU ml^{-1}
Aluminium	136 (88-211)	117 (83-166)
Calcium	55 (32-96)	58 (40-85)

* Vero cell assay.
† Lethal challenge.

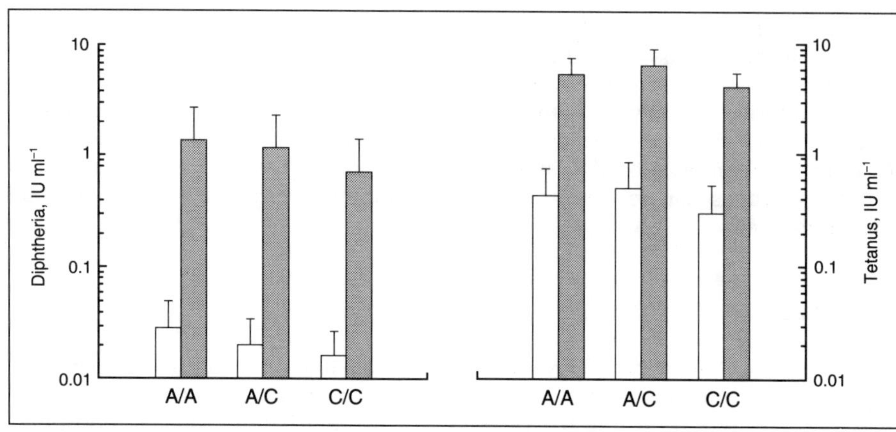

Fig. 1: Geometric mean antibody response and 95% confidence limit of six guinea-pigs four weeks after the first □ and two weeks after the second ▨ dose of 250 μl undiluted DT vaccine containing 12 Lf ml^{-1} of each toxoid. Aluminium hydroxide, 1 mg Al ml^{-1} (A) and calcium phosphate, 1 mg Ca ml^{-1} (C).

for immunization of humans, but they seem to be of limited value in predicting their efficacy in man. At least for the last decade pilot vaccines have been made from all bulk purified toxoids in the control laboratory at Statens Seruminstitut and tested in potency assays to ensure that the toxoids were suitable for use in final vaccine formulations. Final vaccines including D, T and combined DT vaccines made in production scale from toxoids previously tested in this way have all passed the requirements of the EP. A logical consequence of this would be to stop doing full potency assays on all final vaccine formulations but only to perform such assays on pilot vaccines to release bulk toxoids for use in subsequent vaccine formulations. Reference and pilot vaccine should contain the same amount of antigen (Lf ml^{-1}) and the same amount and type of adjuvant. The immunogenic power expressed in IU Lf^{-1} as well as the antigenic purity expressed in Lf PN^{-1} should be monitored. Such a design would be suitable to show consistency of production and to identify toxoid batches of inferior quality. The final filling lots may then be tested only for immunogenicity in simplified models.

Table 2: Median antibody level four weeks after re-vaccination in man.

Vaccine	N	Diphtheria (IU ml-1)	P	Tetanus (IU ml-1)	P
Aluminium	161	2.334		7.898	
		(1.452-3.909)	<0.001	(6.304-9.643)	<0.001
Calcium	152	4.104		10.132	
		(2.630-5.911)		(8.952-12.785)	

The present trend in vaccine formulation is to combine more and more antigens with D and T toxoids in a single vaccine. It creates considerable difficulties in assessing for potency of D, T and P according to the established methods, since the standards defining the units are not for similarly composed vaccines. We suggest that *final vaccines* might be tested in a reduced potency assay as described here by immunizing a few guinea-pigs twice with a fraction of a human dose. Serum samples from these animals could be used to test individual antibody levels against all the antigens by in vitro assays such as ELISA and Vero cell assay. The potency of a combined vaccine would be acceptable if the antibody titres reaches a certain level. This level may be determined on the basis of a lethal challenge test in animals and from the knowledge obtained with certain batches of vaccines used in clinical trials. Such a design has already been described and accepted for veterinary vaccines.

Lyng and Heron [7] have used a different approach to suggest a similar procedure for a reduced potency assay which in fact is a combination of the requirements by WHO and the EP and the minimum requirements of the U. S. Department of Health, Education and Welfare. This proposal would save a large number of animals and still ensure that only vaccines of sufficient potencies were used for the immunization of humans.

References

1 Aggerbeck H, Heron I: Improvement of a Vero cell assay to determine diphtheria antitoxin content in sera. Biologicals 1991;19:71-76.

2 Simonsen O, Bentzon MW, Heron I: ELISA for the routine determination of antitoxic immunity to tetanus. J Biol Stand 1986;14:231-239.

3 Ipsen, J: Systematische und zufällige Feflerquellen bei Messung kleiner Antitoxinmengen. Z f Immun Forsch Exp 1942;102:347-368.

4 Nyerges G, Ádám MM, Gács P, Ring R: A comparison of combined diphtheria, tetanus and pertussis vaccines produced by different manufacturers, in laboratory animals and in infants. J Biol Stand 1986;14:241-247.

5 Relyveld E, Bengounia A, Huet M, Kreeftenberg JG: Antibody response of pregnant women to two different adsorbed tetanus toxoids. Vaccine 1991;9:369-372.

6 Lyng J, Heron I: Quantitative estimation of diphtheria and tetanus toxoids. 3. Comparative assays in mice and in guinea-pigs of two diphtheria toxoid preparations. Biologicals 1990;18:3-9.

7 Lyng J, Heron I: Quantitative estimation of diphtheria and tetanus toxoids. 5. Comparative assays in mice and in guinea-pigs of two diphtheria toxoid preparations. Biologicals 1991;19:327-334.

Dr. H. Aggerbeck, Statens Seruminstitut, Bacterial Vaccine Department, Artillerivej 5, DK-2300 Copenhagen S, Denmark

Brown F, Cussler K, Hendriksen C (eds): Replacement, Reduction and Refinement of Animal Experiments in the Development and Control of Biological Products.
Dev Biol Stand. Basel, Karger, 1996, vol 86, pp 303-308

Potency Testing of Diphtheria and Tetanus Toxoids. A Reliable, Pragmatic and Economic Approach

D.W. Stainer

Stainer Associates, Richmond Hill, Ontario, Canada

Abstract: The aim of potency testing is to ensure that vaccines under test will provide protection when used in human subjects. To achieve this, many potency tests for diphtheria and tetanus toxoids have been devised. The potency tests listed in the European Pharmacopoeia and the W.H.O. Requirements call for the use of large numbers of either guinea pigs or mice. These quantal assays are extremely expensive to perform and alternative, more economic methods are clearly needed.The results presented here illustrate that, provided certain rules are followed, single point assays can provide the same degree of assurance of potency as the three-point assays currently in use.

The ideal method for testing a lot of vaccine would be to inject it into the population for whom it is intended and compare the response to a previously injected vaccine of the same type and concentration. This is obviously impractical, totally uneconomical and impossible to do. Animal models, which as far as possible will reflect the vaccine's immunogenicity and safety have therefore been designed.

In the preparation of vaccines, many methods can be used to show that the final product is potent, but the final bulk (or in some countries the final filled container) must be tested in animals. These tests over the years have became needlessly detailed, and extremely expensive. This presentation is concerned only with the potency tests for Diphtheria and Tetanus Toxoids, and how these tests have been introduced, modified, increased in detail, and finally simplified.

In the United States, the potency tests for the Toxoid vaccines are extremely simple, economical and reliable [1, 2] (Table 1). Similarly, the earlier B.P. tests, although a little more complicated, only called for a limited number of animals [3] (Table 2).

The introduction of the three-dose quantal assay and the incorporation of standard toxoids for potency testing by the European Pharmacopoeia (E.P.) and the W.H.O. greatly increased their precision but unfortunately, due to the very large numbers of animals involved, this led to full potency assays only being performed

Table 1: US potency tests for vaccines containing diphtheria & tetanus toxoids.

Diphtheria (Oct 11/1956)		**Tetanus (June 15/1953)**
Animal	Guinea pig	Guinea pig
Number	>4[1]	>4[1]
Weight	500g	500g
Volume injected	0.75 ml	0.5 ml[2]
Standard used?	NO	NO
Duration	4 weeks	6 weeks
Requirement	>2 units antitoxin[3]	>2 units antitoxin

[1] Additional animals needed for antitoxin determinations.
[2] Dose is 0.75 ml for DPT.
[3] > 0.5 units antitoxin for DT.

by large laboratories, with access to almost unlimited funds and quite complicated mathematical bioassay systems.

In response to pressure from manufacturers, regulatory agencies, and animal rights groups, there has been a major attempt to try and reduce the number of animals required for testing. This has resulted in many conferences and meetings devoted to this subject being held over the past several years [4, 5, 6, 7]. Much progress has been made and the most recent W.H.O. Requirements for Diphtheria, Tetanus, Pertussis and Combined Vaccines [8], while still recommending that a Reference Vaccine be included in all potency tests, nevertheless admitted that single dilution assays could be used to test for the potency of Diphtheria and Tetanus

Table 2: BP potency tests for vaccines containing diphtheria & tetanus toxoids.

Diphtheria		**Tetanus**
Animal	Guinea pig	Guinea pig
Number	10[1]	9[1]
Weight	250 g	250 – 350 g
Volume injected	2 x 1/50 dose in 1 ml[2]	(a) 0.5 ml OR/(b) 2 x 0.05 ml[2]
Standard used?	NO	NO
Duration	3 weeks	(a) 6 weeks OR/(b) 2 weeks after 2nd inj.
Requirement	>2 units antitoxin	>2/3 animals have > 0.05 unit antitoxin OR >1/3 animals have > 0.5 units antitoxin

[1] Additional animals needed for determination of antitoxin levels.
[2] Dilutions in saline.

in vaccines, provided that it could be demonstrated that the vaccine potencies were considerably above the minimal requirements for a single human dose [>30 I.U. for Diphtheria and 40 I.U. for Tetanus (or 60 I.U. if tested as DPT in mice)].

However, the E.P. still insists that the lower fiducial limits of the 95% confidence intervals of the estimated potencies is greater than the minimal requirements, and this would preclude the use of any single point assay procedures [9]. It is hoped that this contribution will assist in providing evidence that the E.P. position could be modified.

Materials and Methods

DPT, DT and TT Vaccines were prepared by the established procedures at Connaught Laboratories, Toronto, Canada, during the period 1984-1986. The potency determinations of the Diphtheria and the Tetanus components were performed using the full three-dose quantal assay procedures [10], employing House Standard Toxoids which had been standardized against the W.H.O. International Reference Preparations and assigned a value by the Canadian National Control Authorities. Diphtheria potencies were estimated using guinea pigs (250-350 g, 12 animals per group), while the tetanus potency determinations were performed in mice (14-20 g, 18 animals per group). Dilutions were made in non-pyrogenic, sterile saline. The numbers of survivors following challenge with standard diphtheria or tetanus toxin respectively were recorded and the results analysed by the Probit method.

Results and Discussion

Seventy-four lots of TT, eight lots of DT and 106 lots of DPT were tested (Tables 3, 4 and 5). Very good potencies were obtained for all three products, and it was decided to analyse the data further.

Table 3: Analysis of 74 lots of TT adsorbed vaccine.

No. of lots IU per single human dose				
60-70	70-80	80-90	90-100	100-110
33	16	14	8	3

Table 4: Analysis of eight lots of DT adsorbed vaccine.

	No. of lots		
Component	**IU per single human dose**		
	60 – 100	100 – 140	140 – 180
Diphtheria	2	4	2
Tetanus	6	1	1

Table 5: Analysis of 106 lots of DPT adsorbed vaccines.

			No. of lots				
Component			**IU per single human dose**				
	100	100-150	150-200	200-250	250-300	300-350	>350
Diphtheria	13	25	22	20	9	12	5
Tetanus	14	47	30	9	4	1	1

Table 6 represents a re-examination of the raw data obtained from numerous DPT potency tests. It can be seen from this Table that in the case of the diphtheria toxoid component, suitable dilutions of the test sample (e.g. 1:80) can be selected, which gave an average number of survivors of 10.9 animals. Compared with a single dose of the standard vaccine (e.g. 1.8 I.U.), an average survival of 3.5 animals resulted. The logical assumption could therefore be made that a 1/80 dilution of such a vaccine contains a greater number of I.U. than 1.8. This translates to an estimate of greater than 144 Units per single human dose, which is far in excess of the minimum requirements of 30 I.U.

Table 6: Analysis of raw data of DPT potency tests.

Diphtheria[1]				**Tetanus[2]**			
Standard		**Vaccine**		**Standard**		**Vaccine**	
Dose IU	S/12[3] (avge)	Dosc (diln)	S/12 (avge)	Dose IU	S/18 (avge)	Dose (diln)	S/18 (avge)
7.4	10.1	1/80	10.9	4.0	15.9	1/50	16.8
	(±1.8)		(±1.4)		(±2.0)		(±1.3)
3.7	7.1	1/160	5.9	2.0	11.5	1/100	10.8
	(±2.7)		(±2.5)		(±3.2)		(±3.2)
1.8	3.5	1/320	0.8	1.0	5.1	1/200	3.0
	(±2.1)		(±1.0)		(±2.8)		(±1.8)
0.9	0.6			0.5	1.1		
	(±0.9)				(±1.3)		

1 104 lots, 29 tests
2 72 lots, 31 tests
3 S = Survivors

Similarly for the tetanus component, a 1/100 dilution of a vaccine (average survival 10.8 animals) was used and the response compared to that observed with 1.0 I.U. of the Standard vaccine (average survival of 5.1 animals). This means that there are a greater number of I.U. in the 1/100 dilution of such a vaccine than 1.0

I.U, indicating a potency of greater than 100 I.U. per dose, again in excess of the minimum requirements of 60 IU.

Although it is not possible to obtain 95% limits when using a single dose, it is nevertheless possible to show what are the significant differences between the number of survivors observed for the standard vaccine of a known potency and the vaccine under test. This can be done using Chi-square analysis (Fig. 1). The regions where significant differences are guaranteed are shown in the shaded areas of the graph. For instance, in a tetanus test, if 10 of 18 animals survived at a particular dilution of vaccine and only three or four survived at the chosen dilution of the standard vaccine, then the minimal unitage of the unknown vaccine could be calculated. Similar calculations can be done for the diphtheria component.

The critical choice is of course, what doses should be used? This will depend on the quality of the animals being used, their immune responsiveness, and also on the basic experience obtained following many 3+3 tests. The important parameter is consistency in testing and this should be demonstrated to individual National Control Authorities before single-point assays are adopted as final release tests. A criticism might be raised regarding the heterogeneity between standards and unknown vaccine, but it is generally accepted that there will always be heterogeneity regardless of whether three doses or a single-point assay is used and the degree of heterogeneity might even be reduced with the single-point assay. After all, there is also a great deal of heterogeneity between the human subjects who are to receive vaccines.

It is essential to remember the reasons for doing the potency tests in the first place – to demonstrate that a vaccine will prove satisfactorily immunogenic and protective

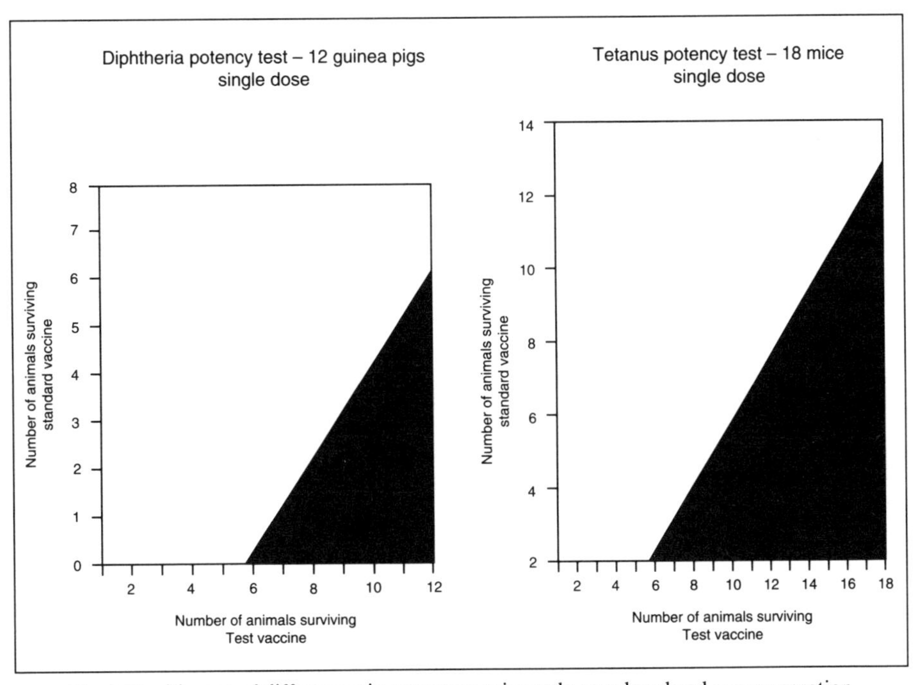

Fig. 1: Significance of differences in responses using only one dose level per preparation.

in the intended subjects.It has been established in human populations that antitoxin levels above 0.01 units per ml are indicative of protection against diphtheria or tetanus, but the minimal number of International Units of Tetanus or Diphtheria Toxoid needed to produce these antibody levels have not yet been satisfactorily answered.

A single-dose test has been used successfully in the U.S.A. for many years, without even using a standard vaccine and similarly, the B.P. test may also use only one dose to demonstrate potency. Studies reported by Hardegree [11] have indicated that toxoids meeting either the U.S. FDA or the BP requirements consistently produce antitoxin titres greater than 0.01 units. The epidemiological experience in both the U.S.A. and Great Britain over the past 45 years should certainly provide convincing evidence that single-dose tests can be both reliable and practical. The incorporation of a standard vaccine into the single-dose test serves to provide an even greater degree of confidence that, correctly used, only fully potent vaccines will be released for use in human populations.

References

1 Minimum Requirements: Diphtheria Toxoid.1954.U.S. Department of Health, Education and Welfare, Public Health Service, National Institutes of Health, Bethesda, Md. 5th Revision.

2 Minimum Requirements: Tetanus Toxoid. 1952.U.S. Department of Health, Education and Welfare, Public Health Service, National Institutes of Health, Bethesda, Md. 4th Revision.

3 British Pharmacopoeia, 1968.

4 W.H.O. ad hoc meeting, Geneva, December 12-14, 1983.

5 19th IABS Congress on Use and Standardization of Combined Vaccines, Amsterdam, The Netherlands. Dev Biol Stand 1986;65.

6 IABS Symposium on Reduction of Animal Usage in the Development and Control of Biological Products. London 1985. Dev Biol Stand 1986;64.

7 Proceedings of an Informal Consultation on the W.H.O. Requirements for Diphtheria, Tetanus Pertussis and Combined Vaccines. D.H.H.S., USPHS, Pub. No. FDA 91-1174, (edited by Manclark CR), 1988.

8 Requirements for Diphtheria, Tetanus, Pertussis and Combined Vaccines (revised 1989) 1990: WHO, TRS 800.

9 European Pharmacopoeia, 2nd Edition Part II, 1986.

10 Requirements for Diphtheria, Tetanus, Pertussis and Combined Vaccines (revised 1978) 1979: WHO, TRS 638

11 Hardegree MC: Human data justifying the use of the United States potency requirements for Diphtheria and Tetanus toxoids, in Manclark CR (ed): Proceedings of an Informal Consultation on the W.H.O. Requirements for Diphtheria, Tetanus Pertussis and Combined Vaccines. D.H.H.S., USPHS, Pub. No. FDA 91-1174, 1988, pp 132-135.

Dr. D.W. Stainer, Stainer Associates, 109, Regent Street, Richmond Hill, Ontario, Canada L4C 9P3

Brown F, Cussler K, Hendriksen C (eds): Replacement, Reduction and Refinement of Animal Experiments in the Development and Control of Biological Products.
Dev Biol Stand. Basel, Karger, 1996, vol 86, p 309

SESSION IX

Therapeutic Toxins and Monoclonal Antibodies

Chairmen: R. Dabbah (Rockville, USA)
H. van de Donk (Bilthoven, The Netherlands)

Brown F, Cussler K, Hendriksen C (eds): Replacement, Reduction and Refinement of Animal Experiments in the Development and Control of Biological Products.
Dev Biol Stand. Basel, Karger, 1996, vol 86, pp 311-318

Requirements for Valid Alternative Assays for Testing of Biological Therapeutic Agents

D. Sesardic

Division of Bacteriology, National Institute for Biological Standards and Control, Potters Bar, Hertfordshire, UK.

Key words: Therapeutic agents, bacterial toxins, alternative assays, immunoassays, bioassays

Abstract: Several novel toxin-derived biological therapeutic products are now available or under development. There is thus an increasing need to develop alternative assays for their efficacy testing, which currently still depends largely on in vivo methods. Scientific advances in the understanding of the mode of action of these substances has not only fuelled the pace of their development as therapeutic agents but has also provided the basis for improving upon conventional testing procedures.

INTRODUCTION

Biological substances used in medicine are usually chemically complex, often poorly defined and may contain many non-functional components, including degradation products. Thus in controlling such agents it is important to assess their biological activity [1]. Currently, those assays providing the specificity and sensitivity necessary for the testing of biological therapeutics are largely reliant on in vivo systems.

Bacterial exotoxins are the causative agents of a number of life threatening diseases. Amongst these, diphtheria and tetanus would be responsible for major loss of life if it were not for the success of vaccination against them. Whilst botulism is not so prevalent, the aetiological factor, botulinum toxin, is increasingly used as a therapeutic agent on the basis of its selective pharmacological property as a potent neuromuscular paralytic agent [2]. In addition, chimeric molecules made from bacterial toxin active components and monoclonal antibodies or suitable ligands, required for selectively targeting specific cell types, are increasingly being exploited as novel therapeutic agents in chemotherapy [3]. Current methods for estimating the potency of bacterial toxoid vaccines and antitoxins, and of botulinum toxin type A(BoNT/A)-haemagglutinin complex, require large number of animals.

Whilst efforts are being made to reduce the number and refine the biological end points used in potency testing of these therapeutics, the ultimate aim is to develop methods which would allow the entire replacement of current in vivo bioassays.

In the past decade, much effort has been devoted to developing alternative assays for the efficacy testing of a large array of biological therapeutics, including those derived from bacterial toxins. However, whilst these assays may provide alternatives to current testing methods, their validity is open to question. Many of the assays rely on non-functional characteristics such as immunoreactivity. In the light of current knowledge of the mode of action of bacterial toxins, and of their immunogenicity, it is quite clear that for an alternative assay to be valid it must be based on the relevant biological activity. The availability of in vitro systems in which the natural targets of the toxin are available may provide real prospects for the development of valid alternatives. For example, diphtheria toxin acts by killing susceptible cells, i.e. it is cytotoxic. Although it had been known for over twenty years that this is due to the selective inactivation of protein elongation factor 2, which results in the blockade of protein synthesis, it was only after a sensitive cell line, expressing the specific receptor for the toxin, had been identified that this information could be used for the development of a valid assay. Subsequently, alternative assays for both toxin and antitoxin have been developed, based on the same mode of action as that measured in the current in vivo methods.

Immunodetection of Bacterial Toxins and Antitoxins

Because of their simplicity, ease of use and speed, immuno-assays have been widely applied both for the identification of bacterial toxins and as alternative assays to toxin neutralisation in vivo, as a means of determining antibody responses to vaccines against bacterial toxins. Examples of the use of such assays include potency estimation of commercial preparations of equine diphtheria and tetanus antitoxins and anti-tetanus immunoglobulin [4], which are still occasionally used in the treatment of and prophylaxis against diseases caused by these bacterial toxins. In addition, modifications of enzyme-linked immunosorbent (ELISA) [5, 6, 7, 8], passive haemagglutination [9] and toxin binding [10] assays have been developed with the aim of replacing the toxin neutralisation (TN) assay in vivo for assessing the protective efficacy of antitoxins. Such assays have been applied to the assessment of protective immunity using human sera and to the control of vaccine potency using sera of immunised animals.

The inherent disadvantage of such immuno-assays is their inability to differentiate between biologically active domains and those which are non-functional. In the assay of toxins, most immuno-assays cannot distinguish between active toxin and that which is inactive, due either to point mutations or proteolytic degradation, but which is still antigenic. Following immunisation with vaccines against bacterial toxins (toxoid vaccines), antibodies against multiple epitopes are produced. Only some of these will be neutralising. Current immuno-assays cannot distinguish between neutralising and non-neutralising antibodies. For some vaccines, the neutralising epitope may also be immunodominant, but for others it may be only relatively weakly immunogenic.

In our studies, we have successfully developed an immuno-assay with sufficient sensitivity to detect the minute quantities of BoNT/A present in therapeutic preparations [11]. However, there was no correlation between immunodetectable BoNT/A and its biological potency, determined in vivo, in preparations from different sources. This was due presumably to variations in the ratio of immunoreactive, non-functional toxin to active toxin in different therapeutic preparations. Furthermore, exposure of therapeutic BoNT/A to higher than recommended storage temperature resulted in the complete loss of biological activity with no change in immunoreactivity (Fig. 1). Because of such limitations, immuno-assays could never serve as valid alternatives for the potency estimation of bacterial toxins present in therapeutic preparations, where it is absolutely essential to quantify the active component.

It may be possible to improve the validity of such assays by using antibodies, either monoclonal or anti-peptide antibodies [12], directed against specific functional domains of the toxin. However, this requires delineation of all the critical functional domains of the toxin and the production of antibodies that can discriminate between forms of the toxin that differ even subtly in their structure.

Many independent studies [5, 6, 10] have shown a very good correlation between the results of ELISA or TOBI and of TN assay in vivo, for the neutralising potential of human serum antibodies to tetanus toxin. This has been interpreted by many that such assays may be widely applicable to the assessment of

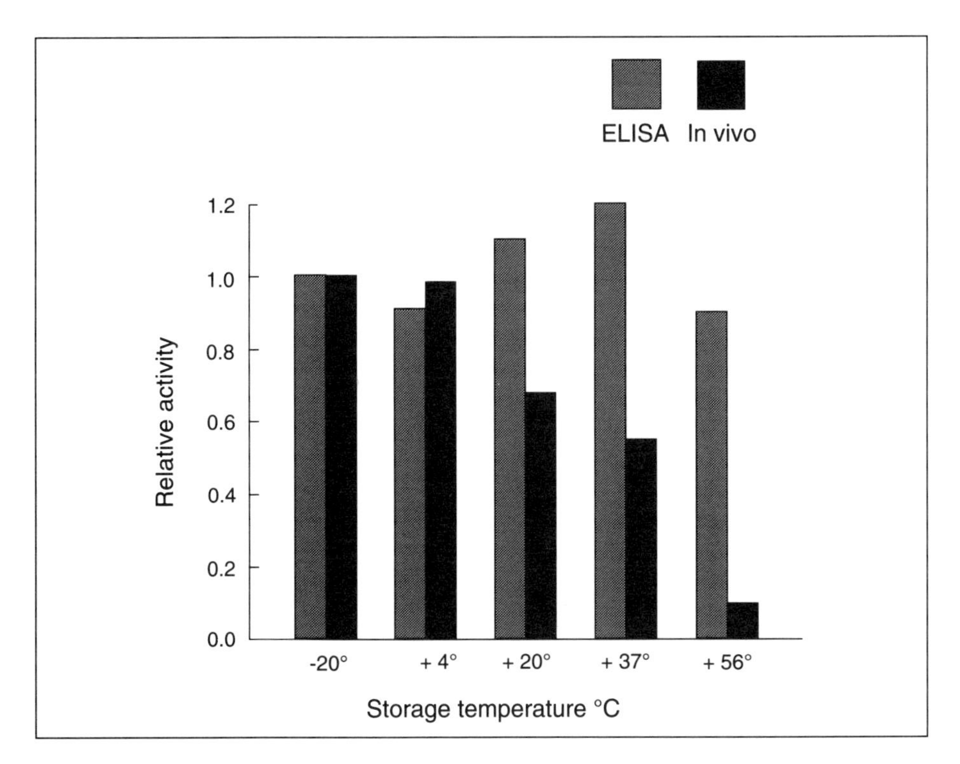

Fig. 1: Comparison of a relative activity (immunoreactivity) determined by ELISA (gray bars) with biological activity (black bars) of a therapeutic preparation of botulinum toxin type A, following heat denaturation.

vaccine efficacy. However, such assays are less reliable for the testing of sera with low immunity to tetanus [7, 8] and, perhaps even more importantly, are inadequate for the assay of diphtheria antitoxin [13, 14]. For example, ELISA estimation of antitoxin to diphtheria detects a high proportion of specific IgG that cannot neutralise the toxin, presumably as a consequence of the relative immunodominance of non-neutralising epitopes. Whereas a good correlation was obtained between ELISA and in vivo TN assay for vaccinated subjects with high antitoxin levels (≥ 0.1 IU/ml), the correlation for samples with low antitoxin levels was much poorer, with ELISA giving a higher mean value than the TN assay (Fig. 2).

Tetanus and diphtheria toxoid vaccines may elicit different proportions of protective and non-protective antibodies, which would explain why immuno-assays may be successful for one but not the other. For toxoid vaccines such as diphtheria and tetanus, measurement of vaccine efficacy must be based on the accurate determination of the toxin neutralising antibodies present in the serum of vaccinated subjects.

Assays for antitoxins may be improved by delineating the protective and immunodominant epitopes of the bacterial toxins. We have therefore mapped the epitopes for both protective and non-protective monoclonal and polyclonal antibodies to diphtheria toxin [15]. One neutralising epitope was identified. This was highly conformational and was located within the receptor-binding domain of the toxin. However, the most immunodominant regions of the toxin were located in other domains and were non-neutralising, supporting the view that immuno-assay of total antibody response will not necessarily correlate with protection.

Information derived from studies on the structure, function and immunogenicity of toxins may provide the basis for the development of alternative assays, more valid than those in current use, for testing the efficacy of therapeutic products derived from bacterial toxins. This applies to both toxins and vaccines as protective efficacy depends on the level of neutralising antibodies achieved. No available

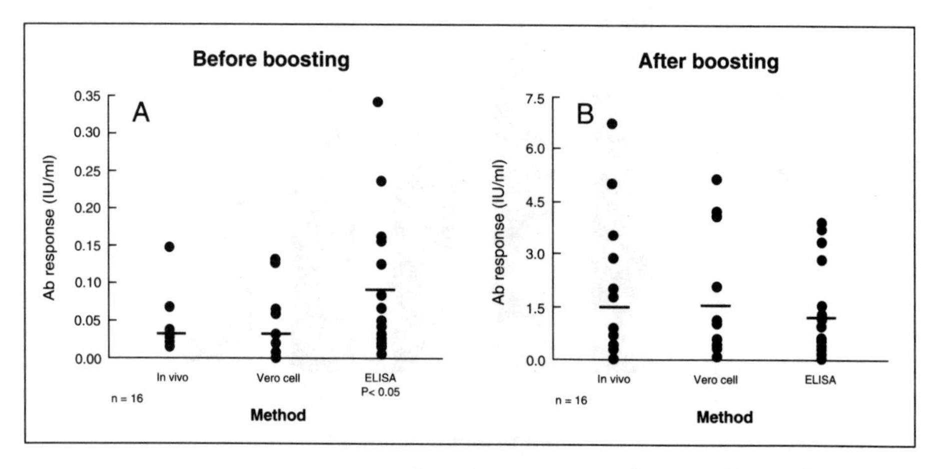

Fig. 2: Analysis of human sera samples for anti-tetanus antibodies (A) before and (B) after vaccination of subjects with diphtheria vaccine for adults. Antitoxin titres were determined by toxin neutralisation assay in vivo, by in vitro Vero cell toxin neutralisation assay and by enzyme-linked immunosorbent assay (ELISA).

immuno-assay has the ability to differentiate between biologically active and inactive antitoxins, although a reasonable correlation between the two methods can be obtained, as total antibody response may correlate with the protective antibody response. However, such positive correlations do not mean that the two methods are equivalent, as they are based on different end points. Clearly, the most effective approach to the development of valid alternatives should be based on the biological activity of the toxin.

In Vitro Bioassays for Bacterial Toxins and Antitoxins

There have been considerable advances in our knowledge of the mode of action of bacterial toxins over the last ten years. A number of toxins have a similar overall mode of action: they bind to a cell surface receptor, are internalised and then act on an intracellular target. Such toxins can be broadly classified into ADP-ribosylating toxins such as diphtheria, cholera and pertussis, and neurotoxins such as tetanus and botulinum [16]. Each of these functions of the toxins, receptor binding, internalisation and catalytic activity, is dependent on a different specific domain of the protein. Whilst toxin action depends upon the integrity of all three domains, neutralising anti-toxin activity appears to be directed against only the receptor binding domain. Based on this information, it should be possible to devise valid in vitro alternative assays.

Diphtheria toxin (DT) is cytotoxic, exerting its activity by selectively inactivating protein elongation factor 2 (EF-2), by ADP-ribosylating a specific diphthamide residue. This results in inhibition of protein synthesis and ultimately cell death [17]. Although the intracellular target for DT is present in every single mammalian cell capable of protein synthesis, sensitivity to the toxin is dependent on the presence of a specific cell surface receptor, recently identified as a precursor of heparin-binding growth factor [18]. Vero cells possess a large number of DT receptors [19] and for this reason such cells have proven suitable for the development of alternative assays for both toxin and antitoxin [20, 21]. In a recent study performed in our laboratory, several methods for detection of diphtheria toxin among isolates of pathogenic corynebacteria were compared. The Vero cell assay was identified as the most sensitive and reliable alternative method for accurate assessment of toxigenicity [unpublished observations].

The neutralising potential of serum antibodies from human, guinea pigs, rabbits, mice and rats immunised with diphtheria vaccine were compared using the Vero cell cytotoxicity assay and TN assay in vivo. There was an excellent correlation amongst all species. In addition human sera were analysed before and after vaccination with diphtheria vaccine (adult dose) and an excellent correlation was obtained for both low and high titre samples, in contrast to the ELISA which showed a good correlation for only the high titre samples (Fig. 2). Thus, the Vero cell has all of the characteristics necessary for valid in vitro alternative assay for the detection of biologically active DT and for assessing protective immunity.

Clostridium botulinum toxins are potent neuroparalytic agents which act presynaptically at the neuromuscular junction, their primary site of action, where they induce flaccid paralysis by blocking acetylcholine release. The structurally related tetanus toxin causes a spastic paralysis due to its retrograde transport to the spinal cord where it blocks the release of neurotransmitters GABA and glycine [22]. Until recently, lack of knowledge of the mode of action of these neurotoxins at the molecular level has prevented the effective development of functional in vitro assays for therapeutic agents derived from toxins. The recent identification of the specific intracellular targets for these bacterial neurotoxins provides the possibility of developing alternative methods for testing these products, which would reflect a blockade of neurotransmitter release and/or cleavage of the target proteins involved in neuro-exocytosis.

Like cytotoxic ribosylating toxins, neurotoxins bind to specific cell surface receptors following which they undergo cell entry, presumably via endocytosis. The receptor for neurotoxins has not been identified but there appear to be multiple interactions between the toxin receptor binding domain and both protein and lipid-binding sites on the plasma membrane. Very recently, it was discovered that clostridia neurotoxins block the release of neurotransmitters via a Zn^{2+}-dependent endopeptidase activity [22]. Each neurotoxin targets one of the components of the synaptic vesicle-synaptic membrane docking system. Since each neurotoxin cleaves a specific bond within the target protein, this is a highly selective mechanism of action. Tetanus and botulinum type B toxins cleave vesicle-associated membrane protein VAMP, also known as synaptobrevin [23]. Botulinum toxin type A cleaves a 25-kDa synaptosomal-associated protein known as SNAP25, at residues Gln_{197}-Arg_{198} and type E toxin selectively cleaves the same protein at residues Arg_{180}-Ile_{181} [24].

Tetanus and botulinum type B neurotoxins were the first toxins shown to block neurotransmission by acting at the level of synaptic vesicles. This information was soon used as the basis of an in vitro assay in which the rate of proteolytic cleavage of synthetic peptide fragments of synaptobrevin by type B botulinum toxin was determined [25]. This study confirmed that it was possible to use a synthetic substrate to demonstrate toxin endopeptidase activity in vitro. Preliminary data from my own laboratory on the use of a synthetic peptide of SNAP25 (made available from Dr Shone, CAMR) as a substrate in the assay of BoNT/A look promising. The rate of cleavage of the peptide correlated with the activity for several therapeutic preparations of the toxin, as determined by in vivo bioassay. The validity of this assay in assessing the biological potency of BoNT/A will depend upon whether peptide cleavage is rate-limiting in the action of the toxin. This can only be determined by further studies of toxin binding to cell surface receptors and its translocation across the membrane. Even if this assay proved valid for neurotoxin determination, it would not necessarily be suitable for assessing protective immunity as this may depend upon interaction of antitoxin with the receptor binding domain.

The initial step in both botulinum and tetanus neurotoxicity is binding to specific cell surface receptors on peripheral neuronal membranes. Hence, a neuronal-derived cell line expressing such receptors might provide the basis for the development of a functional in vitro assay. Unfortunately, no such cell line is currently known. The adrenal medullary chromaffin cell line (PC12), derived from a rat pheochromocytoma, is closely related to neuronal cells. Although this cell line lacks constitutive expression of the receptors for tetanus and botulinum neurotoxins it does possess the intracellular target for the toxins [26]. When these cells are differentiated by culture with nerve growth factor, they express a number of neurone-like properties, including stimulus-dependent release of acetylcholine. Hence, a neuronal-derived cell, line such as PC12 cells, may provide a suitable model for developing a cellular assay for clostridia neurotoxins, if necessary following transfection of the specific toxin receptor.

Conclusions

Valid alternative assays for estimating the efficacy of biological products must be based on end points comparable to those responsible for in vivo activity. While detailed knowledge of toxin structure, function and immunogenicity may enable more relevant immuno-assays to be developed, it is still unlikely that these will be entirely valid, for the reasons discussed above. Although correlations have been obtained with some non-functional immunological assays, this is often only when a sub-set of samples is used. In view of this and the fact that such assays are based on different principles from those involved in toxin effect, these cannot be regarded as generally valid.

Knowledge of the mode of action and biological targets of bacterial toxins has made the development of meaningful alternative assays a realistic possibility. While for some toxins suitable cell lines already exist, for others it may be necessary to genetically engineer a cell line that is constitutively deficient in some part of the toxin effector pathway.

While such assays may avoid the use of animals in assessing the neutralising potency of antitoxins, the efficacy testing of vaccines will still require their administration to animals to elicit an antibody response. There is concern that species differences in immune response diminish the justification of this use of animals. Research should be devoted to developing human-derived systems in which vaccine efficacy can be tested, thereby avoiding all use of animals in the control of such materials.

References

1 Jeffcoate SL, Corbel MJ, Minor PD, Gaines-Das RE, Schild GS: The control and standardisation of biological medicines. Proc Royal Soc Edinburgh 1993;101B:207-226.

2 Schantz EJ, Johnson EA: Properties and use of botulinum toxin and other microbial neurotoxins in medicine. Microbiol Rev 1992;56:80-99.

3 Pai LH, Pastan I: Immunotoxins and recombinant toxins for cancer treatment, in DeVita VT, Hellman S, Rosenberg SA (eds): Important Advances in Oncology, Philadelphia, JB Lippincott Company, 1994, pp 3-19.

4 Sesardic D, Wong MY, Gaines-Das RE, Corbel MJ: The First International Standard for Antitetanus Immunoglobulin, Human: pharmaceutical evaluation and international collaborative study. Biologicals 1993;21:67-75.

5 Melville-Smith ME, Seagroatt VA, Watkins JT: A comparison of enzyme-linked immunosorbent assay (ELISA) with the toxin neutralization test in mice as a method for the estimation of tetanus antitoxin in human sera. J Biol Stand 1983;11:137-144.

6 Gentili G, Pini C, Collotti C: The determination of the potency of human tetanus immunoglobulins by an enzyme-linked immunosorbent assay. J Biol Stand 1984;12:167-173.

7 Gupta RK, Siber GR: Comparative analysis of tetanus antitoxin titres of sera from immunized mice and guinea pigs determined by toxin neutralization test and enzyme-linked immunosorbent assay. Biologicals 1994;22:215-219.

8 Simonsen O, Bentzon AW, Heron I: ELISA for the routine determination of antitoxic immunity to tetanus. J Biol Stand 1986;14:231-239.

9 Nyerges G, Lutter J: Influence of the method of preservation of erythrocytes on the correlation between tetanus antitoxin values and by seroneutralization tests in mice. J Biol Stand 1980;8:311-316.

10 Hendriksen CFM, van der Gun JW, Nagel J, Kreeftenberg JG. The toxin binding inhibition test as a reliable in vitro alternative to the toxin neutralization test in mice for the estimation of tetanus antitoxin in human sera. J Biol Stand 1988;16: 287-297.

11 Ekong TAN, McLellan K, Sesardic D: Immunological detection of *Clostridium botulinum* toxin type A in therapeutic preparations. J Imm Methods 1995;180:181-191.

12 Sesardic D, Khan V, Corbel MJ: Targeting of specific domains of diphtheria toxin by site-directed antibodies. J Gen Microbiol 1992;138:2197-2203.

13 Sesardic D, Corbel MJ: Testing for neutralising potential of serum antibodies to tetanus and diphtheria toxin. Lancet 1992;340:737-738.

14 Knight PA, Tilleray J, Querminet J: Studies on the correlation on the range of immunoassays for diphtheria antitoxin with the guinea-pig intradermal test. Develop Biol Standard 1986; 64:25-32.

15 Sesardic D, Hoy CS, McKenna A, Corran PH, Feavers IM: Characterization of immunogenicity of diphtheria toxin, in Freer et al. (eds): Bacterial Protein Toxins. Stuttgart, Jena, New York, Gustav Fischer, 1944, (Suppl 24), pp 508-509.

16 Schiavo G, Poulain B, Benfenati F, DasGupta BR, Montecucco C: Novel targets and catalytic activities of bacterial protein toxins. Trends Microbiol 1993;1:170-174.

17 Pappenheimer AM: Diphtheria toxin. Ann Rev Biochem 1977; 46:69-94.

18 Neglich JG, Metherall JE, Russell DW, Eidels L: Expression cloning of a diphtheria toxin receptor: identity with a heparin-binding EGF-like growth factor precursor. Cell 1992;69:1051-1061.

19 Mekada E, Okada Y, Uchida T: Identification of diphtheria toxin receptor and a nonproteinous diphtheria toxin-binding molecule in Vero cell membrane. J Cell Biol 1988;107:511-519.

20 Miyamura K, Nishio S, Ito A, Muata R, Kono R: Micro cell culture method for determination of diphtheria toxin and antitoxin titres using Vero cells. J Biol Stand 1974;2:189-201.

21 Aggerbeck H, Heron I: Improvement of a Vero cell assay to determination of diphtheria antitoxin content in sera. Biologicals 1991;19:71-76.

22 Montecucco C, Schiavo G: MicroReview: Mechanism of action of tetanus and botulinum neurotoxins. Mol Microbiol 1994;13:1-8.

23 Schiavo G, Benfenati F, Poulain B, Rossetto O, Polverino de Laureto P, DasGupta BR, Montecucco C: Tetanus and botulinum B neurotoxin block transmitter release by proteolytic cleavage of synaptobrevin. Nature 1992;359:832-835.

24 Blasi J, Chapman ER, Link E, Binz T, Yamasaki S, DeCamilli P, Sudhof T, Niemann H, Jahn R: Botulinum neurotoxin A selectively cleaves the synaptic protein SNAP-25. Nature 1993; 365:160-163.

25 Shone CC, Quinn CP, Wait R, Hallis B, Fooks SG, Hambleton P: Proteolytic cleavage of synthetic fragments of vesicle-associated membrane protein isoform-2 by botulinum type B neurotoxin. Eur J Biochem 1993; 217:965-971.

26 Lawrence GW, Weller U, Dolly JO: Botulinum A and the light chain of tetanus toxin inhibit distinct stages of Mg-ATP-dependent catecholamine exocytosis from permeabilised chromaffin cells. Eur J Biochem 1994;222:325-333.

Dr. D. Sesardic, Division of Bacteriology, National Institute for Biological Standards and Control, Blanche Lane, South Mimms, Potters Bar, Hertfordshire EN6 3QG, UK

Brown F, Cussler K, Hendriksen C (eds): Replacement, Reduction and Refinement of Animal Experiments in the Development and Control of Biological Products.
Dev Biol Stand. Basel, Karger, 1996, vol 86, pp 319-320

Investigation of the Product Equivalence of Monoclonal Antibodies for Therapeutic and in Vivo Diagnostic Human Use in Case of Change from Ascitic to in Vitro Production

G. Schäffner, M. Haase, S. Giess

Paul-Ehrlich-Institut, Langen, Germany

A number of monoclonal antibodies currently being tested in clinical trials have been developed and produced from the ascitic fluid of mice. Recently, the acceptance of in vivo-produced monoclonal antibodies is decreasing for several reasons. In changing from in vivo to in vitro production, the pharmaceutical manufacturer has to alter the whole production, requiring tremendous additional development and production costs. A favourable regulatory environment is required to encourage manufacturers to change from ascitic production to in vitro cell culture and to avoid delay in granting a marketing authorization for a successful antibody produced by new technology. Changes, amendments or improvements of the manufacturing process, even minor qualitative and quantitative changes in the active substance itself, would require a new application according to the legal regulations of the European Union and of the national drug laws of the Member States.

We propose a procedure to take previously evaluated and development-related data into account by demonstrating the equivalence of two forms of one monoclonal antibody obtained before and after the introduction of a change in the manufacturing process. Criteria are described which could be useful to demonstrate product equivalence of both forms of the antibody by thorough in vitro and pharmacological testing to minimize the need for further clinical evaluation of the new product. The thorough in vitro analysis of parameters such as isotype, subclass, affinity, microheterogeneity, molecular weight, primary and secondary structure, structural integrity of the antibody molecule, glycosylation pattern, specificity, cross-reactivity and biological potency with subsequently performed pharmacological/toxicological evaluation of biodistribution, half-life and safety provide sufficient data for a decision on the need for further clinical trials.

The full report will be published in Biologicals 1995;23:253-259.

Dr. G. Schäffner, Paul-Ehrlich-Institut, Paul-Ehrlich-Str. 51-59, D-63225 Langen

Brown F, Cussler K, Hendriksen C (eds): Replacement, Reduction and Refinement of Animal Experiments in the Development and Control of Biological Products.
Dev Biol Stand. Basel, Karger, 1996, vol 86, p 321

SESSION X

Poster Session

Brown F, Cussler K, Hendriksen C (eds): Replacement, Reduction and Refinement of Animal Experiments in the Devolepment and Control of Biological Products.
Dev Biol Stand. Basel, Karger, 1996, vol 86, pp 323-349

Evaluation and Validation of a Single-Dilution Potency Assay Based Upon Serology of Vaccines Containing Diphtheria Toxoid: Analysis for Consistency in Production and Testing

A.M. Akkermans, C.F.M. Hendriksen, F.R. Marsman, H.J.M. van de Donk

National Institute of Public Health and Environmental Protection, PO Box 1, 3720 BA Bilthoven, The Netherlands

Part of the quality control of the diphtheria component of adsorbed combined vaccines is an in vivo bio-assay of the final product to provide assurance that the toxoid is capable of inducing protective immunity. The use of the multi-dilution assay as an indicator of the potency of a toxoid vaccine has been the subject of criticism for some time (Stainer et al 1986), first because of the large expenditure in laboratory animals and second, because under certain conditions precise knowledge of the potency of toxoid vaccines is of little practical use.

The new procedure relies on previous experience with multi-dilution assays and is based on proof of consistency in production and testing, evidence of a highly significant regression of response against dose and justification of the assumption of linearity and parallelism of the dose-response relationship for the vaccine under study. These principles have been used to provide data on consistency of production and testing of diphtheria toxoid vaccines at the National Institute of Public Health and Environmental Protection (RIVM) in The Netherlands. One group of animals is immunized with one dilution of the vaccine under test, and another with one dilution of the reference preparation. Vaccine dilutions are chosen in such a way that, if the test vaccine induces a significantly higher immune response than the reference preparation, assurance is provided that the test vaccine contains at least the minimum required level of international units (Knight 1989).

The actual performance of a single-dilution assay has been tested in a simulation study, using retrospectively the data of multi-dilution potency assays on 27 consecutive batches of diphtheria toxoid. From the results of the evaluation study it is concluded that, under conditions of Good Manufacturing Practice (GMP) and continuous monitoring of the test system, introduction of a single-dilution assay based on serology in routine potency testing of the diphtheria component of DPT-polio vaccines at the RIVM is a realistic approach. The method is recommended to manufacturers who produce large series of batches annually.

Detection of a Point Mutation in Poliovaccines with a Water Soluble Carbodiimide Reaction of DNA Heteroduplexes

G. Amexis, W.M. Bertling, J. Löwer

Paul-Ehrlich-Institute, 63225 Langen, Germany

The establishment of a specific and sensitive molecular biological method for the detection of defined point mutations which correlate with neurovirulence in a poliovirus population is an approach for the replacement of animal tests. In our model a 200 bp fragment in the region of the point mutation at base 472 in the 5'-untranslated region of poliovirus, Sabin 3 was PCR-amplified and subsequently cloned into appropriate vectors for the production of sufficient amounts of dsDNA. The first step involves the formation of heteroduplexes of wildtype and attenuated Sabin 3 Poliovaccine. DNA fragments were denatured for 10 min at 100°C, and re-annealed overnight. The second step is the modification of the generated heteroduplexes by a water-soluble carbodiimide (CDI). This reagent binds covalently with mismatched bases. CDI-modified heteroduplexes block the subsequent primer extension at the point of modification. In a second primer extension step with labelled primers the site of the mutation can be localized. In our model the mutation at base 472 (T → C) in Sabin 3 was specifically detected. By enhancing the sensitivity of this system, a detection level for point mutations of 0.1% should be achieved.

Determination of Protective Erysipelas Antibodies in Pig and Mouse Sera as Possible Alternatives to the Animal Challenge Models Currently Used for Potency Tests

R. Beckmann, H. Gyra, K. Cussler

Paul-Ehrlich-Institute, Veterinary Department, D-63207 Langen, Germany

Animal challenge tests are widely used to measure the potency of bacterial vaccines. Pigs are infected to demonstrate efficacy of erysipelas vaccines in clinical trials whereas mice are used in batch potency tests of inactivated vaccines according to EP and CFR requirements. These tests cause severe suffering and a large number of animals is needed.

Recently, protective antigen structures of *Erysipelothrix rhusiopathiae* in the range from 64-66 kDa have been characterized. We have developed an ELISA in which an alkaline lysate extract containing these protective antigens of *E. rhusiopathiae* is used to measure protective antibodies. Several batches of different inactivated erysipelas vaccines were examined according to the EP potency test. Simultaneously, additional mice were vaccinated for the serological test and bled after three weeks. The individual antibody titres of these mice were used to estimate the potency of the vaccine batches. The results of the ELISA and of the offical test method were compared.

In addition, three batches of different erysipelas vaccines, which had been shown to be efficacious in the pig challenge test, were used to immunize pigs. Blood samples were taken before vaccination and three to six weeks after. The erysipelas antibody titres were measured by an ELISA, which had been adapted for pig serum. The results of the serological method were compared with the results from the challenge tests in pigs and mice.

Detection of Clostridium Perfringens Toxins Using Cell Cultures

E. Borrmann, F. Schulze

Federal Institute for Health-Related Consumer Protection and Veterinary Medicine, Jena, Germany

Clostridium perfrigens types A, B, C, D and E produce major lethal toxins designated alpha-, beta-, epsilon- and iota-toxin. We investigated the possibility of using cell cultures as an indicator of the presence of the *Clostridium perfrigens* toxins alpha, beta and epsilon. The toxins were prepared from culture filtrates of *Clostridium perfrigens*. The purification of the toxins was carried out by ammonium sulphate precipitation, dialysis and freeze-drying. Epsilon-toxin was activated by 0.1% trypsin. The effectiveness of the activation was determined by comparison of the activated and non-activated samples on cell cultures. Cytotoxic effects of the toxins on cell cultures were compared after three days by microscopic examination, neutral red assay and tetrazolium MTT assay. We investigated seven different cell lines and concluded that the MDCK cell line was suitable for the detection of epsilon-toxin and MA 104 cell line for the beta-toxin. Under our conditions the MTT assay was better suited than the neutral red assay for the analysis of the cytotoxic effects on cell cultures.

Hepatitis B Virus Polymerase Chain Reaction - A Possible Alternative Method to the in Vivo Assay Using Chimpanzees

M. Chudy, M. Weeke-Lüttmann, J. Löwer

Paul-Ehrlich-Institute, Paul-Ehrlich-Str. 51-59, 63225 Langen, Germany

Blood-derived products may harbour the risk of the contamination with blood-borne viruses. In order to assure the safety of such biologics, validation studies have to be performed to show that the purification process can remove or inactivate relevant viruses. Currently, there is no in vitro infectivity assay for human hepatitis B virus (HBV). The most reliable in vivo system uses primates (chimpanzees). The costs to run a study with chimpanzees are very high. Moreover, the availability of these primates is restricted. Therefore, the replacement of animal experiments in the quality control of biological products is important.

As an example it will be demonstrated that the polymerase chain reaction (PCR) is a suitable method for the detection of HBV in an HB vaccine derived from plasma of an HB surface antigen carrier. The results reveal that all tested plasma pools used for the production of the vaccine aims at the elimination of contaminating viruses. The detection of viral nucleic acid with the help of the PCR can replace the in vivo infectivity assay.

Guidelines for the Choice of an Alternative to Proposed Animal Experiments

M. Dawson

Department of Pharmaceutical Sciences, University of Strathclyde, University of Strathclyde, Glasgow G1 1XW, Scotland

When applying for a licence in Great Britain to conduct animal experiments it is a requirement to state that alternatives have been considered. In most other countries this is, if not a legal requirement, at least a recommendation.

However there is no definitive list of alternatives which may conveniently be consulted. Applicants must search the literature and are left not knowing whether their search has been comprehensive and up-to-date. In Britain, the Universities Federation for Animal Welfare (UFAW) is compiling such a list, as a small brochure, which can conveniently be consulted and also contains indications for further reading where more detail may be found, and the state of validation. The list has sections on computers and mathematical modelling, physical methods, bacteria, yeasts and other small organisms, subcellular tests, vertebrate cells, Drosophila, brine shrimps, plants, eye tests, skin tests, fertile hens' eggs, teratogenicity, cephalopods, microgravity and post-marketing surveillance.

The methods are intended to apply to toxicity testing and to pharmacological and physiological experiments. It is pointed out that some experiments may not be replaceable by a non-animal method, but that these might be usable for part replacement or preliminary assessment. It is hoped that this publication will fill an undesirable gap in the literature on alternatives.

Animal Ethics and Welfare – the Australian Way

A. Derks

Arthur Webster Pty Ltd, 2152 Castlereagh Road, Penrith, NSW, 2750, Australia

The use of animals for research in Australia is governed by a code of practice by which the Australian government sets goals of refining, reducing and replacing animals used for research.

The Code of Practice covers a number of key areas. These include:

1. Animal Care and Ethics Committees;
2. Licensing for research and animal supply;
3. Recording of animal usage;
4. Auditing of establishments by animal welfare representatives.

Arthur Webster Pty Ltd must comply with the code to maintain its accreditation and thereby to produce veterinary vaccines. This presentation will demonstrate how Websters, through its Animal Care and Ethics Committee, institutes the requirements of the Code of Practice. The overriding strength of the Code is that it actually does have legal status. All institutions have to re-apply for research licences annually. To be granted licences applicants have to demonstrate that they:

a) are linked to a properly functioning Animal Care and Ethics Committee, the standards being set within the Code;

b) are committed to reducing levels of animals used. This is demonstrable by submission of annual usage figures and scientific presentations outlining progress achieved and expected;

c) comply with the standards of animal welfare outlined in the Code. This is assessed by periodic independent inspections.

A Serological Potency Assay for *Clostridium chauvoei*

P.J. Hauer, M.S. Whitaker, L.A. Henry

National Veterinary Services Laboratories, VS, APHIS, USDA, Ames, IA 50010, USA

An indirect enzyme-linked inmmunosorbent assay (ELISA) has been developed to quantify antibody specific for a highly purified *C. chauvoei* flagellar antigen in pooled rabbit sera. The test compares optical densities of the unknown to a standard at several serum dilutions and uses parallel lines to compute a relative potency. Sera from rabbits vaccinated with *C. tetani, C. sordellii, C. perfrigens* types C and D, *C. septicum, C. haemolyticum*, and *C. novyi* have been evaluated in the assay and show minimal cross-reactivity. Relative potency results from the serological assay for vaccines of varying potencies compared favourably with guinea challenge results.

Identifying and Purifying Protective Immunogens from Cultures of *Clostridium chauvoei*

P.J. Hauer, R.F. Rosenbusch

National Veterinary Services Laboratories, VS, APHIS, USDA, Ames, IA 50010, USA, and the Veterinary Medical Research Institute, Iowa State University, Ames, IA 50011, USA

Proteins from cultures of *C. chauvoei* were separated using preparative isoelectric focusing and evaluated for their ability to induce protective immunity in guinea pigs. Two proteins were identified as protective immunogens. Highly purified flagellar protein induced protective immunity, as did an 88 kilodalton protein with an isoelectric point of 4.96. Guinea pigs immunized with these proteins protected no anti-haemolytic antibodies, yet were protected from a challenge of 100 LD_{50}. Western blots with sera from vaccinated cattle and rabbits demonstrated the immunogenicity of the two proteins in those species.

Development of an Enzyme-Linked Immunosorbent Assay for Potency Testing of Erysipelas Bacterins

L.M. Henderson, K.F. Scheevel, D.M. Walden

National Veterinary Services Laboratories, Veterinary Services, APHIS, USDA, Ames, IA 50010, USA

Clinical signs of swine erysipelas can be controlled by vaccination with live or killed products containing the causative agent, *Erysipelothrix rhusiopathiae*. Potency testing of the killed products (bacterins) requires evaluation of the ability of the serial (numbered lot) to stimulate a protective immune response. Currently, a mouse vaccination-challenge assay is used to evaluate the potency of many erysipelas bacterins. We have recently reported the development of a hybridoma secreting monoclonal antibodies that recognize an essential protein involved in the protective response to challenge-exposure with virulent *E. rhusiopathiae*. We report here the development of an enzyme-linked immunosorbent assay based on the use of the monoclonal antibody, ERHU1-B60-91, that is a suitable alternative to the mouse potency assay currently in use. Specificity and cross-reactivity have been demonstrated. The assay requires the use of a reference bacterin produced in the same manner as the test serial. Some products may be tested without elution; others methods are presented. Bacterins may be assayed before the addition of adjuvant with an in vitro potency assay which would alleviate animal welfare concerns, reduce human exposure to the agent, and result in savings of time and costs.

Characterization of a Monoclonal Antibody for in Vitro Potency Testing of Erysipelas Bacterins

L.M. Henderson[1], P.S. Jenkins[2], K.F. Scheevel[1], D.M. Walden[1]

[1] National Veterinary Services Laboratories, Veterinary Services, APHIS, USDA, Ames, IA 50010, USA
[2] Cell and Hybridoma Facility, Iowa State University, Ames, IA 50011, USA

Erysipelothrix rhusiopathiae, a small Gram-positive rod, is the causative agent of swine erysipelas and causes disease in humans and other host species. Clinical signs of swine erysipelas can be controlled by vaccination. There are a number of killed erysipelas bacterins licensed for use in swine. Many of these products use a mouse vaccination-challenge model for potency testing. A hybridoma secreting monoclonal antibodies specific for a protective immunogen of *E. rhusiopathiae* was constructed to facilitate the development of an in vitro potency assay. Ascites fluid from the hybridoma, ERHU1-B60-91 (patent pending), was shown to provide passive protection from challenge-exposure to a virulent strain of the agent (*E. rhusiopathiae* E1-6P) in swine. The hybridoma recognizes a highly conserved essential protein of 64 to 66 kDa. Immunogens in the 64 to 66 kDa complex have been shown to be involved in a protective immune response in swine, and a serologic response to the protein has been correlated with protection in swine. The hybridoma is suitable for the development of an assay for quantifying the protective antigenic content of erysipelas bacterins as an alternative to the mouse potency assay.

Interlaboratory Validation of in Vitro Serological Assay Systems to Assess the Potency of Tetanus Toxoid in Vaccines for Veterinary Use

C.F.M. Hendriksen[1], J. Woltjes[2], A.M. Akkermans[1], J.W. van der Gun[1], F.R. Marsman[1], M.H. Verschure[2], K. Veldman[2]

[1] National Institute of Public Health and Environmental Protection (RIVM), 3720 BA Bilthoven, The Netherlands
[2] Central Veterinary Institute (CDI), 8200 AB Lelystad, The Netherlands

An inter-laboratory validation study was carried out in seven laboratories to evaluate the suitability of in vitro serological assay systems for the assessment of the potency of tetanus toxoid in single and multicomponent vaccines for veterinary use. Nine commercial vaccines and one experimental tetanus toxoid preparation were selected for immunization purposes according to Method A of the European Pharmacopoeia (1990). Titres of tetanus antitoxin in guinea-pig and rabbit serum samples (pooled and individual) were estimated by indirect ELISA, toxin binding inhibition (ToBI) test, passive HA test and by the prescribed standard toxin neutralization (TN) test in mice. Estimates of the potency by in vitro tests and by the TN test were in good agreement for the various vaccines tested and for antitoxin titres of individual serum samples. On statistical analysis intra- and inter-laboratory variation was acceptable for ELISA and ToBI test, but larger variation was seen in the HA test. It is concluded that ELISA and ToBI are suitable in vitro assay systems for assessing the potency of tetanus toxoid in batches of single and multicomponent vaccines for veterinary use. Rigid standardization of the HA test is essential before this test can be used for the same quality control purpose.

Use of ELISA to Quantify the Antitoxin Content of Commercial Equine Tetanus Antitoxin

D.R. Kolbe

National Veterinary Services Laboratories, VS, APHIS, USDA, Ames, IA 50010, USA

An indirect enzyme-linked immunosorbent assay (ELISA) has been developed to quantify the antitoxin in commercial equine tetanus antitoxin. The serum from horses, hyperimmunized with an antigen prepared from a toxin-producing strain of *Clostridium tetani,* is currently tested in guinea pigs using a toxin neutralization test (TNT) against a standard toxin and antitoxin. The in vivo test measures antitoxin content through effectiveness of protection of guinea pigs injected with diluted mixtures of antitoxin, based on the volume of antitoxin in the final container and the number of toxin units claimed on the container label. The ELISA, which replaces the guinea pig TNT, is designed to measure the antitoxin content of test samples containing 100 to 1500 antitoxin units per ml. The in vitro test requires less time to complete and eliminates pain caused to animals in the guinea pig TNT. Test results correlated very well when commercial equine tetanus antitoxin was assayed for antitoxin content by ELISA and the guinea pig TNT.

Enzyme-Linked Immunosorbent Assay Assessment of Bovine Viral Diarrhoea Virus Antigen in Inactivated Vaccines Using Polyclonal or Monoclonal Antibodies

L.R. Ludemann[1], *J.B. Katz*[2]

USDA/APHIS, National Veterinary Services Laboratories,
[1] Veterinary Biologics Laboratory and
[2] Diagnostic Virology Laboratory P.O. Box 844, Ames, IA 50010, USA

An enzyme-linked immunosorbent assay (ELISA) procedure was developed to assay the cytopathic and non-cytopathic bovine viral diarrhoea (BVD) virus strains used in inactivated vaccines licensed by the United States Department of Agriculture. The assay uses a biotin-labelled, staphylococcal protein A purified polyclonal BVDV antibody (Bad) from a calf hyperimmunized against NADL, Singer, C24v, a field isolate and New York-1 (NY-1) strains. The Bab recognized the following reference strains of BVD virus: NADL, NY-1, C24v, Singer, and a field isolate. Monoclonal antibodies (Mab) directed against gp48 and gp53 of the Singer strain of BVDV could detect only the Singer and the NY-1 strains. None of the Mabs tested could differentiate between cytopathic and non-cytopathic BVD virus strains. In vaccines containing multiple viral and bacterial components, the Bab was specific for the BVDV fraction. Two vaccines not recognized by the Bab differed from the others in the type of adjuvant. The formation of antigen-adjuvant complexes during vaccine production may inhibit the ability of Bab to detect BVDV antigens in an ELISA format. This ELISA procedure enables the detection of BVDV antigens and demonstrates the potential for in vitro testing of inactivated BVDV vaccines in place of the currently required host animal testing.

Replacement of Foetal Bovine Serum with Bovine Colostrum and Serum-Based Medium as a Supplement for in Vitro Culture of Vero and CHO Cells

R.M. Ölander

National Public Health Institute, Helsinki, Finland

Foetal bovine serum (FBS) is widely used as a cell culture supplement for both Vero and CHO cells. From both an economical and an ethical point of view the replacement of FBS with other biological fluids would be preferable. In this study a commercially available mixture (Viable™AC-2, Valio Bio-products Ltd, Turku, Finland) containing an ultrafiltrate fraction of bovine colostrum (UF) (6.7%,) and adult bovine serum (BS) (1%) was tested in the concentrations 1-14% as an FBS substitute in the diphtheria (Vero) and pertussis (CHO) toxin neutralization assays, respectively. Both cell lines were cultivated in 10% FBS + MEM before the actual test took place. According to R. Pakkanen (In Vitro Cell Dev Biol 30A: 295-299, May 1994) FBS can effectively be substituted with 8% UF/BS without any weaning period.

1% Viable™AC-2 supplement in MEM was enough to obtain a good Vero cell growth in the cell control wells on the microtitre plate used in the diphtheria toxin neutralization assay. The supplement had no inhibitory effect on the diphtheria toxin or on the positive serum samples or WHO standard serum compared to 2% FBS supplemented medium.

CHO-K1 cells, used in the pertussis toxin neutralization assay, did not grow in MEM supplemented with 1-8 % of Viabler™ AC-2. In higher concentrations (10-14%) the cell growth was minimal after two days. After four days the growth had reached an acceptable level compared to two days with 10 % FBS supplement, but the clustering effect of the pertussis toxin was still very difficult to confirm. These results indicate that further testing with e.g. proline-supplemented different basal media is needed.

Serological Potency Testing of Vaccines Against Progressive Atrophic Rhinitis

V. Öppling, M. Kusch, P. Rübmann, K. Cussler

Paul-Ehrlich-Institute, Paul-Ehrlich-Str. 51-59, D-63225 Langen, Germany

Atrophic rhinitis in pigs is a commercially important disease worldwide. Recently a protein toxin of *Pasteurella multocida* has been identified as the causative agent inducing the progressive form of the disease. *Pasteurella multocida* toxoid is considered the most important antigen in vaccines used for immunoprophylaxis. At the moment there is no generally accepted method for potency testing of this protective vaccine component. Manufacturers have therefore developed their own, product-specific tests. Several serological methods as well as lethal challenge tests are used. The diversity of these methods does not allow any comparison between potency results. A serological method for potency testing the *Pasteurella multocida* toxoid component is demonstrated, applicable for all vaccines registered in Germany against progressive atrophic rhinitis. After immunizing guinea pigs twice at intervals of three weeks with one fifth pig dose, no distress and only minimal local reactions were observed. Determination of antibodies to *Pasteurella multocida* toxin in a cell culture neutralization assay on embryonic bovine lung cells gives good reproducible results. Two weeks after booster immunization a maximum titre was reached.

Thirty batches belonging to ten different vaccines were examined for their ability to induce antibodies against *Pasteurella multocida* toxin in this system. A lyophilized serum with an average titre of $1{:}2^{3,26}$ was used as reference. With one exception all vaccines induced antibody titres higher than the reference. Batches of vaccine which failed to induce antitoxin in guinea pigs were also unable to stimulate seroconversion in pigs.

Evaluation of the *Leptospira pomona* ELISA and its Correlation with the Hamster Potency Assay

K.W. Ruby

National Veterinary Services Laboratories, USDA/APHIS, Ames, IA 50010, USA

The National Veterinary Services Laboratories have developed an in vitro assay for measuring the relative potencies of leptospiral bacterins containing serovar *pomona*. Development of in vitro assays for testing veterinary biologicals provides advantages over in vivo assays. In vitro assays are more precise and faster when compared with in vivo assays using laboratory animals. In vitro assays also reduce the number of animals required for assessment of biologicals. However, the correlation of in vitro with in vivo data is not always exact. Here we report on the evaluation of the *L. pomona* ELISA and its correlation with the hamster potency assay. For evaluation of the ELISA, all biologics firms marketing leptospiral bacterins in the United States were invited to participate in the study. All participants had favourable comments regarding the performance, sensitivity, and specificity of the ELISA. Correlation studies were conducted using 10 serial two-fold dilutions of a monovalent aluminium hydroxide-adjuvanted *L. pomona* bacterin to vaccinate hamsters according to the Code of Federal Regulations. These same dilutions were also evaluated in the ELISA protocol. Results indicated that the *L. pomona* ELISA correlated with the hamster potency assay.

Development of an in Vitro Assay for Measuring the Relative Potency of Leptospiral Bacterins Containing Serovar *Canicola* and its Correlation to the Hamster Potency Assay

K.W. Ruby[1], D.M. Walden[1], M.J. Wannemuehler[2]

[1] National Veterinary Services Laboratories, USDA/APHIS, Ames, IA 50010, USA
[2] Department of Veterinary Microbiology and Preventive Medicine, Veterinary Medical Research Institute, Iowa State University, Ames, IA 50011, USA

Currently, potency of leptospiral bacterins is evaluated by a hamster assay in which a minimum of 80% of vaccinated hamsters must survive a virulent challenge. We have developed an in vitro assay for measuring the relative potency of bacterins containing *Leptospira interrogans* serovar *canicola* using a monoclonal antibody (Mab), designated as 4DB, prepared against a non-adjuvanted Triton X-100 extract of the National Veterinary Services Laboratories challenge strain of serovar *canicola* (Moulton strain). The assay is an enzyme-linked immunosorbent assay (ELISA) and involves capturing antigen from bacterins with anti-canicola polyclonal rabbit sera, followed by reaction with the 4DB ascites fluid, anti-mouse IgM peroxidase-labelled antibody, and colour substrate (ABTS). Bacterins adjuvanted with aluminium hydroxide or oil require treatment to elute antigen from the adjuvant before use in the ELISA protocol. Non-adjuvanted bacterins and bacterins containing adjuvants other than aluminium hydroxide or oil require no treatment before use in the ELISA protocol. The ELISA appears to correlate with the hamster potency assay. Here, we report on the development of an in vitro assay for measuring the relative potency of bacterins containing serovar canicola and its correlation to the hamster potency assay.

A Non-Lethal Bioassay for Potency Testing of Therapeutic Clostridial Neurotoxins

D. Sesardic, T. Ekong, K. H. McLellan, R. Gaines Das

Division of Bacteriology and Informatics Laboratory, National Institute for Biological Standards and Control, Potters Bar, Hertfordshire EN6 3QG, UK

The type A neurotoxin produced by *Clostridium botulinum is* a potent and irreversible neuromuscular blocking agent which causes paralysis by preventing the release of neurotransmitters from motor neurons. This specific action has resulted in the increasing use of the toxin in the treatment of a number of neuromuscular disorders involving previously untreatable and often painful muscle spasms. At present, the only assay with the specificity and sensitivity necessary to estimate accurately the potency of botulinum toxin in therapeutic preparations is bioassay, in which lethality is used as the end point. This is also the only assay currently accepted by regulatory authorities for this purpose. Hence, refinement of this assay, particularly with respect to the severity of the end point, is urgently required.

We have now developed a non-lethal method for estimating the potency of therapeutic botulinum toxin type A, based on an assessment of the degree of flaccid paralysis that results from injection of mice with toxin into the inguinocrual region. The scoring system developed, although subjective, showed a very good quantitative relationship to the dose (linear between 0.05-0.25 MLD_{50} units/dose). There was excellent agreement between the scores obtained by two independent observers, who were blind to the treatment (Rs = 0.87, n = 448) and between those obtained at 24 and 48 hours by the same observers (Rs = 0.94). There was an excellent correlation between the relative potency estimates of eight therapeutic preparations of toxin, obtained using the conventional lethal and the new non-lethal assays (Rs = 0.98, $P<0.05$). In the non-lethal assay the animals remained in good health and showed minimal signs of distress. Hence, these preliminary studies provide a sound basis for further validation.

Steps Towards the Development of a Replacement in Vitro Test for Quantification of Antibodies to Tetanus Toxin during Potency Testing of Vaccine

J. Southern, H. Clements, L. Spenceley

South African Institute for Medical Research, Serum & Vaccine Department, Johannesburg, South Africa

The potency test for tetanus-containing vaccines requires assay of toxin-neutralizating antibodies elicited by vaccination of guinea-pigs with dilutions of the vaccine. There is no validated in vitro assay for toxin neutralization; hence mice are used. An alternative is the demonstration and quantification of antibodies which will react with purified toxin. Agglutination of red blood cells, sensitized with purified tetanus toxin is a potential method, and this is under evaluation.

Tetanus toxin produced by the dialysis culture used for routine vaccine production was purified and the coupling of this to erythrocytes from different species investigated. These suspensions were standardized and compared using antisera raised during potency tests of vaccines.

Tetanus toxin purified by various methods has been used to sensitize red cells from sheep, geese and turkeys, using glutaraldehyde and lysine. Standardization of agglutination tests continues. Although not yet fully validated, this in vitro test may prove a suitable replacement for the animal test. A similar test may be developed for other toxins.

Is the Mouse Weight Gain Test Outdated as a Model for the Assessment of Pertussis Toxicity?

I.V. Straaten-van der Kappelle, J. van der Gun, A. van der Ark, C. Hendriksen

National Institute of Public Health and Environmental Protection, PO Box 1, 3720 BA Bilthoven, The Netherlands

A collaborative study has been carried out to establish the reliability and relevance of five test systems to assess the toxicity of whole cell pertussis vaccine. Six vaccines, including both «normal» and «abnormal» products with respect to toxicity, were tested. The study was based on three in vivo bioassays; the Mouse Weight Gain (MWG) test, the Leukocytosis Promoting Factor (LPF) test and the Histamine Sensitizing Factor (HSF) test, and two in vitro test systems; the Chinese Hamster Ovary (CHO) clustering test and the Lymulus Amoebocyte Lysate (LAL) endotoxin test. Fourteen laboratories in various countries participated in the study.

In almost all participating laboratories the MWG test was not sensitive enough to assess toxicity of whole cell pertussis vaccines. The LPF, HSF and CHO tests appeared to be more sensitive, but considerable intra-laboratory variation was seen. Significant variation in test results also occurred in the LAL test. It is concluded that, on the issues of optimalization and stringent standardization, HSF, CHO and in particular the LPF test might be better models to assess pertussis toxin (PT) activity than the MWG test. A great disadvantage of the HSF test is the large number of mice needed to test one vaccine and it is therefore not an alternative for the MWG test. An inhibition ELISA has been used to estimate levels of PT. With this test it is possible to measure both detoxified and active PT, and it could therefore be of value for pre-screening purposes, just as the LAL test is for endotoxin activity. We propose to perform the toxicity tests (LPF and CHO) only in case the PT inhibition ELISA exceeds a warning limit.

Importance of Experiments in Poliovirus-Susceptible Transgenic Mice for Evaluating Current Potency Tests of Inactivated Polio Vaccine (IPV)

R. Taffs[1], E. Dragunsky[1], T. Nomura[2], K. Hioki[2], I. Levenbook[1], E. Fitzgerald[1]

[1] FDA, Rockville, MD, USA
[2] CIEA, Kawasaki, Japan

While reduction or elimination of experiments in animals is an essential goal of vaccine test development, the adequacy of in vitro alternatives must be assured. This is problematic for vaccines directed against pathogens having a natural host range restricted to humans. With the creation of transgenic (Tg) mouse strains carrying the human poliovirus receptor, it is now possible to address the acceptability of alternative tests using a fundamentally new approach, resulting in more relevant criteria by which test alternatives can be evaluated. Our investigations of IPV-mediated protection against poliomyelitis in the Tg-21 strain of mice have identified suitable immunization and challenge regimens for IPV testing, showing that a single intramuscular vaccination and an intraperitoneal challenge are sufficient to model disease protection. One outcome of experiments so far performed is a discrepancy between sero-conversion and protected status in a number of mice, indicating that immune system components besides neutralizing serum antibodies may participate in protection against poliovirus challenge in these mice. This necessitates the study of the ability of IPV to generate cellular immunity important to disease resistance, and we suggest that current potency tests of IPV need to be evaluated in the context of disease resistance and not exclusively on the basis of sero-conversion. The protection model in Tg mice is likely to have an important role in making the transition to in vitro IPV tests which fully correlate with protection and therefore are capable of reliably evaluating vaccine efficacy.

Quality Control for Screening-Assays by Means of Control Charts: Polio Vaccine Testing as an Example

I. Vogel

Federal Institute for Serum – and Vaccine Control, Possingergasse 38,
A-1160 Vienna, Austria

It is important for quality control laboratories that all assays are validated. Besides determination variability and reproducibility, the results of the assays should be monitored and should show no substantial variation over a period of time. This monitoring could be done by means of control charts.

The procedure is exemplified by means of polio vaccine testing. This test is carried out on trivalent vaccines in cell cultures ($CCID_{50}$). Based on a sequence of results from such tests (evaluated by Probit-regression) on a standard preparation, control limits are established. Subsequent tests should be within these limits. It is shown that both systematic trends as well as single outliers could immediately be detected. This could be helpful in finding systematic changes in the potency of the standard or within the testing system as early as possible and also in locating handling errors.

Purification, Characterisation and Production of a Monoclonal Antibody to *Cl. Novyi* Type B Alpha Toxin

K. Whitworth

Hoechst UK Ltd, Milton Keynes, United Kingdom

Cl. novyi type B alpha toxin was purified using ammonium sulphate precipitation followed by chromatography. The toxin was shown to be a protein with a molecular weight of approximately 200 kDa by SDS-polyacrylamide gel electrophoresis. The necrotising toxin was lethal in mice and had no haemolytic activity against horse red blood cells as expected.

The production of monoclonal antibodies against the purified toxin was initiated using an in vitro immunisation technique and fusion supernatants were found to react with the alpha toxin. Further cloning will be carried out and assay development will commence with a view to introducing an ELISA to be used as an in-process control test for toxoids of *Cl. novyi* type B.

Comparison of Rocket Immuno-Electrophoresis (RIE) and ELISA for Determination of Tetanus Antibodies in Tetanus Immunoglobulins

A. Zott

Paul-Ehrlich-Institute, Langen, Germany

The tetanus antibody content of immunoglobulins is determined by a toxin neutralisation test in mice as described in the European Pharmacopoeia. To reduce animal trials, 12 years ago a series of tests at this Institute had shown comparable results to the toxin neutralisation test by using counter electrophoresis, radial immunodiffusion and a nephelometric method. Continuing this work, Tetanus-ELISA and rocket immunoelectrophoresis (RIE) were compared to each other by taking the result in the pharmacopoeia test (> 308 IU/ml) as a basis. A series of 25 ELISA repetitions showed a range from 182-415 IU/ml for the preparation.

The tests by RIE were performed ten times with a range from 325-342 IU/ml. The higher variability of results in the ELISA system might be explained by a) ELISA being a method which is more likely to detect (very) small amounts of antibody than testing concentrated solutions and b) the complex procedure being more likely to produce a high variability. In 28% of the ELISA the product would have failed to pass the minimum requirement of 250 IU/ml. In contrast to the ELISA all RIE tests passed the EP-requirements. These results show the superiority of this system as it is simple and reproducible where the preparation can be used in the concentrated form. Additionally, RIE results can easily be documented by copying the plates or by scanning them into a computer system.

A Proposal for the Development of a Standardised Programme for the Calculation of Probit Analysis

A. Zott, M. Schwanig

Paul-Ehrlich-Institut, Langen, Germany

In recent years, a substantial improvement has been achieved in the standardisation of design and implementation of animal assays of toxoid vaccines. A comparative evaluation of more than 200 sets of animal assay results of various manufacturers shows that there are considerable differences in the use of mathematical evaluations between manufacturers. One problem is the use of different algorithms for probit analysis calculations on computer. With real data from a test, such as plan, dose levels, total number of animals and number of survivors, different results are obtained depending on the computer programme applied. These differences, often only minor, are of great significance when the results are on the borderline for linearity and parallelism of the dose-response curve and when the lower confidence limit of potency is close to the release limit for the vaccine.

In conclusion, the following proposals can be made:

1. with the high variability of animal tests, the highest precision of the evaluation method (maximum number of iterations etc.) does not play a decisive role: for the routine analysis of animal test results a standardised computer programme for all vaccine manufacturers and corresponding control institutes would be more important;
2. a standardised programme must allow the use of different levels and numbers for doses of the standard and test preparations. For the presentation of results, a standardised layout limited to the essential elements would improve clarity.

Brown F, Cussler K, Hendriksen C (eds): Replacement, Reduction and Refinement of Animal Experiments in the Development and Control of Biological Products.
Dev Biol Stand. Basel, Karger, 1996, vol 86, p 351

Summary and Recommendations

Brown F, Cussler K, Hendriksen C (eds): Replacement, Reduction and Refinement of Animal Experiments in the Development and Control of Biological Products.
Dev Biol Stand. Basel, Karger, 1996, vol 86, pp 353-358

Summary and Recommendations

H.J.M. van de Donk

RIVM, Bilthoven, The Netherlands

The Replacement, Reduction and Refinement of the use of experimental animals (the three R's of Russel and Burch [1]) was considered at a symposium held at the Paul Ehrlich Institute (PEI) in Langen, Germany on 2-4 November 1994. There were 35 lectures, 26 posters and three discussion sessions. Approximately 150 participants from 25 countries attended.

The use of animals was put into historical perspective, illustrated with the case of diphtheria prophylaxis one century ago, which contributed much to public health, but at the same time boosted the use of animals. However, a re-evaluation of the use of animals is presently underway. Russel and Burch's principle of the three R's, endorsed by the European Union directive 86/609/EEC, was discussed in the light of numbers of animals used and avoidable suffering. The role of the European Centre for the Validation of Alternative Methods (ECVAM), one of the organizing institutes of the symposium, was also explained.

International Activities

The first part of the programme dealt with the activities of a number of international organizations in upholding the three R principle.

The European Pharmacopoeia (EP) is already applying the three R's principle to some extent, especially regarding pyrogen testing and testing of blood products and hormones. Moreover, the EP supports and coordinates validation studies which are necessary for appropriate use of alternative methods. These activities are being carried out in close collaboration with laboratories which have acquired the necessary experience, and with licensing authorities. The EP will examine, together with ECVAM, the possibility of collaborating on specific subjects of common interest, taking into account the mutual programmes in progress and areas of competence.

The World Health Organization (WHO) regularly reviews and adapts its requirements to enable, where possible, a reduction to be made in animal usage. Examples include tests on virus seed lots of measles, mumps and rubella vaccines, replacement of multi-dilution assays by single-dilution assays for components of diphtheria-tetanus-pertussis vaccines, and the possibility of using validated alternative methods for the potency testing of toxoids. The substitution of Vero cells for primary monkey kidney cells as cell substrates for the production of poliomyelitis or other viral vaccines will require fewer monkeys. Also, cell lines are made available by WHO for quality control, thus further avoiding the use of primary cells. In future assuring quality may rely more on the establishment of consistency and less on potency testing, allowing alternative tests quite different from methods currently available.

The Animals and Plant Health Inspection Service (APHIS) of the US Department of Agriculture (USDA) allows, in particular cases, the use of validated in vitro alternatives for potency testing, the use of serological tests and sequential use of animals. APHIS plans to allow euthanasia in some cases instead of lethal end-points.

The European Union legislation for veterinary vaccines still allows repeated testing at national level, which leads to inefficient use of animals, as was stated by representatives of both industry and licensing authorities. Industry would appreciate more flexibility from the authorities towards alternative methods, even though such tests might not measure identical phenomena.

The Japanese National Institute of Health (JNIH) has been able to reduce the use of monkeys for neurovirulence testing of oral poliomyelitis vaccine by one third, has replaced the use of rabbits in testing for pyrogens by in vitro endotoxin testing for some protein products, and has succeeded in using the same guinea-pigs for both general safety and endotoxin testing.

A workshop organized by ECVAM in April 1994, on «Alternatives to Animal Testing in the Quality Control of Immunobiologicals», drafted many recommendations, some of which had already been taken on board by international organizations. A code of practice was also drafted by workshop participants. Conclusions from a symposium held in June 1994 on possible alternatives for the quality control of veterinary vaccines were also reported. The use of international databases and the implementation of international coordination were recommended. Open-minded authorities may add another two R's: Regulatory Revolution. Animal welfare should always be one of the issues considered in debating test methods or evaluating licence applications.

It was proposed by representatives of one licensing authority and an animal health institute to reduce in a stepwise manner and to finally eliminate batch target animal testing for veterinary vaccines depending on whether consistency had been achieved, and on the extent of safety data generated during the development of the product.

The European Federation of Animal Health (FEDESA) considers it justified to accept alternative methods under certain conditions, although the European Union, by directive 92/18/EEC, indicated that the challenge assay is the preferred method for establishing potency of veterinary vaccines. FEDESA has produced a list of cases where the correlation of serological indicators with immune protection has been shown.

New Developments

Several new developments for production, testing and research methodologies, mainly in vitro, were presented and discussed.

New in vitro assays

The neurovirulence test of oral poliomyelitis vaccine (OPV) in monkeys must still be performed on each batch of bulk product, in contrast to measles vaccine, where neurovirulence testing is confined to seed lots. Two alternative tests for neurovirulence of OPV, one using transgenic mice and the other using PCR, were presented but the need for scientific criteria to evaluate and validate these tests was stressed.

An assay to establish the potency of inactivated poliomyelitis vaccine (IPV) in rats based on the test developed by Van Steenis has been developed and statistically evaluated; this test is a good candidate to replace the current test in monkeys.

Antigen determination of hepatitis A vaccine using different assays has been evaluated. Two in vitro tests and one serological assay in animals were compared. Results showed that the in vitro assays have smaller variations and are suitable for characterizing intermediate products and to replace the in vivo assay described in the draft EP monograph.

ELISA systems have been developed to determine glycoprotein and nucleoprotein of rabies vaccines, and these might be considered as replacement candidates for the current challenge assay. Correlation between the newly developed tests and the current test has been demonstrated. However, the systems await further evaluation and validation, especially in the light of the different virus strains used for production.

Data obtained from serology with monoclonal antibody-based ELISA tests in calves previously immunized with vaccines to protect against a number of bovine viruses has proved useful for evaluating relative potencies in the licensing procedure.

Replacement and Reduction of animals in research and development of new vaccines are unlikely. This was illustrated by research developments in the field of measles vaccines, vaccines to protect against immunodeficiency viruses, and new presentation and delivery systems, none of which can be successfully accomplished without the use of animal models. However, one speaker considered that well-defined subunit vaccines will eventually replace whole virus vaccines, resulting in biochemical and immunological quality control testing of these products in place of animal testing. Also, peripheral blood mononuclear cells can be used to assess the pathogenicity of simian immunodeficiency viruses, avoiding testing in large numbers of monkeys. A reverse transcriptase-PCR method had been developed to detect rapidly the Newcastle disease virus in poultry vaccines. This method could be extended to detect other extraneous agents in poultry vaccines or vaccines to be used in swine or cattle.

Establishing in vitro methods for DTP vaccine

Alternative methods for assaying antitoxin activities are also under investigation. In vitro serological assays can be used to replace the toxin neutralization test in animals as described in the US potency test for toxoid vaccines. However, the level of toxin used is critical.

A collaborative assay organized by the EP has demonstrated that a serological method (Vero cell assay) can replace existing lethal and intradermal challenge assays used to determine the potency of diphtheria vaccines. However, for combinations of diphtheria toxoid and other antigens, as in DTP vaccines, it has proven difficult to meet assay validity criteria. It was poposed during the discussion that the reproducibility of these assays might be improved by making a combined DTP reference preparation, by the preparation of an antitoxin reference preparation, and by the establishment of an international facility for production and distribution of reagents such as monoclonal antibodies.

To establish the potency of tetanus toxoid vaccine, a toxin binding inhibition test (ToBI) has been developed, based on serology in immunized mice. This test has been successfully validated in one laboratory.

Potency testing of D and T vaccines could be even further refined to reduce the number of animals used by performing a multi-dilution assay on a single pilot vaccine. The results of this assay would be used to release a batch of nonadsorbed concentrated bulk toxoid for vaccine production. Then the final adsorbed vaccine product would be tested in a simplified potency assay using an undiluted fraction of a human dose in a few guinea-pigs.

Several other possible refinements for potency testing of DTP vaccines were discussed. Passive haemagglutination tests based on covalently binding antigens to turkey red blood cells can be used to quickly establish the levels of anti-D and anti-T antibodies. A serological assay for the determination of the potency of pertussis vaccine correlates well with the challenge assay, but needs to be validated in a multi-laboratory study. A guinea-pig model has been developed to assess in a single animal system the potency of diphtheria, tetanus, acellular pertussis and *Haemophilus influenzae* b (Hib) antigens by serological assays. Potencies of Hib vaccines reflect antibody responses in humans.

Single-dilution assays can further reduce animal usage and can be justified provided dosages of the reference preparation and of the test vaccines are carefully chosen.

The increasing knowledge of the mechanisms underlying the protection offered by vaccines supports the development of in vitro assays to characterize bacterial vaccines. Examples of in vitro assays already available are the CHO-cell clustering and the Limulus amoebocyte lysate assays. However, biologicals will still need to be characterized with respect to their biological activity. Many currently available in vitro methods rely on non-functional endpoints and may give misleading information. Future in vitro tests should be based on a better understanding of the mode of action of the biological, such as binding to a relevant receptor. The already available Vero cell test for the determination of diphtheria toxoid and the emerging possibility for the use of the intracellular target for tetanus toxin are intriguing examples.

Monoclonal antibodies

With respect to the manufacture of monoclonal antibodies, a procedure has been proposed to allow manufacturers to switch from ascites to culture technology without repeating full clinical investigation. This would entail thorough in vitro testing and subsequent pharmacological and toxicological evaluation.

The general safety test

Retrospective evaluation of abnormal toxicity (general safety) testing of both veterinary and human vaccines revealed some specific effects due to the pertussis component and to differences in animals rather than any inherent defect in batches. Hence it can be concluded that this test may be of limited value once consistency has been established.

It was suggested that FEDESA and EP jointly draft a study protocol to evaluate the combination of potency and safety testing for veterinary vaccines to avoid separate double dose safety testing.

Workshops

Several workshops were held for in-depth treatment of general subjects.

Workshop on harmonization

In an open discussion between a panel and the audience on harmonization of animal tests, the urgency to reduce duplication caused by differences between the various national regulations for human and veterinary vaccines was stressed. Representatives from the United States Pharmacopoeia (USP), EP, JNIH, WHO, the US Food and Drug Administration (FDA), USDA, FEDESA, PEI, Istituto Superiore di Sanita and Connaught Laboratories were included in the panel.

Brief presentations were given on the impact of the lack of harmonization on the safety tests for veterinary vaccines, quality control of DTP and OPV, abnormal toxicity tests, and pyrogen tests. In addition, the impact of harmonization on the developing world as well its legal implications were mentioned.

It was agreed that overwhelming evidence is available that the current United States, WHO, European and Japanese regulations for these products have for many years ensured that only safe and potent products are released on these markets. It was also confirmed that WHO and other requirements include paragraphs which allow the acceptance of alternative methods by the licensing authorities in different countries when they give equivalent assurance of the safety and potency of the products. Duplication of animal tests due to different levels of acceptance of alternative methods should be avoided.

The panel members and the audience agreed that a mutual acceptance of the current animal tests required by the various authorities should be possible in principle as a first step towards global harmonization of regulations.

Finally, it was unanimously recommended to urge those international groups already involved in harmonization, such as International Conference on Harmonization (ICH), WHO, the Pharmacopoeia Discussion Group of the EP, the Japanese Pharmacopoeia, USP, and the relevant licensing authorities to give highest priority to harmonization of tests using animals.

Workshop on statistics and validation

Validation studies have to address both reliability and relevance of the test method. Before a validation study can be launched, the test must be proven fit for use. Problems likely to occur during the validation study should be identified prior to the study.

To be able to use the method for a long period of time, the reagents critical to the execution of the test need to be identified. In case of a change in the origin of such a reagent, revalidation is needed. It was stressed that only one change at a time should be evaluated. If several modifications are to be evaluated, this should be done in a sequential manner.

Concepts such as validity, accuracy, precision, sensitivity, and specificity were explained. The need to standardize the data by a database system was emphasized, as were the concepts of intra-and inter-laboratory variability, the statistical evaluation and the presentation of the study. Guidelines on validation are issued by the Veterinary Medicines Directorate of the United Kingdom and are also included in the USP.

Workshop on serological methods and cell cultures

A number of topics were discussed, including validation of the Vero cell test for the estimation of potency of batches of diphtheria toxoid, the reading of this test system, the relevance of current potency tests for toxoids as models for immunogenicity in humans, the single-dilution assay, and the safety test. With regard to the potency test for diphtheria, test data were presented showing a good correlation between the in vitro serological test and the in vivo test. A large scale collaborative study on potency testing of diphtheria toxoids initiated by the FDA is in preparation. A homologous reference diphtheria antiserum, produced in guinea-pigs, will be used. Institutes and laboratories are invited to participate.

The use of the mouse model as an alternative to the guinea-pig model for the assessment of the potency of diphtheria toxoid was also discussed. It was recommended to re-evaluate the data of the EP collaborative study on diphtheria toxoid to analyze the effect of different mouse strains on the estimation of potency. Furthermore, it was recommended to specify the requirements for potency tests of diphtheria toxoid obtained in mice and in guinea-pigs. The Expanded Programme on Immunization (EPI) was invited to include a single-dilution test in its monographs on toxoid vaccines. Effects on the outcome of the potency test were discussed with respect to the number of dilutions used, making vaccine dilutions in saline, and the mouse strains used.

Conclusions and Recommendations

Besides the recommendations summarized above which resulted from the workshops, it was recommended that human data be used as far as possible to underpin the relevance of current potency scientifically. The general safety test in animals was discussed in the light of its discriminating power, and it was recommended that the relevance of this test be re-evaluated. Finally, it was recommended that more insight be gained in the mechanisms of immunogenicity, as it was believed that understanding these mechanisms will ultimately result in expediting the introduction of alternative methods.

This symposium demonstrated that sound science and animal welfare can go together well!

Reference

1 Russell WMS, Burch RL. The principles of humane experimental technique. London: Methuen, 1959.

List of Participants

ADAM M.M., National Institute of Hygiene, 1966 Gyalt út 2-6, H-1966 Budapest, Hungary.

AGGERBECK H., Statens Seruminstitut, Artillerivej 5, DK-2300 Copenhagen S, Denmark.

ANDERS C., Fort Dodge Laboratories, Finisklin Industrial, Sligo Est., Ireland.

ANGERMANN A., Biotest Pharma GmbH, Landsteinerstr. 5, D-63225 Dreieichenhain, Germany.

AUERBACH J., Behringwerke AG, P.O. Box 1140, D-35001 Marburg, Germany.

BALLS M., ECVAM, IRC Environment Institute, I-21020 Ispra VA, Italy.

BARON D., Pasteur Merieux Serums & Vaccins, Parc Industriel d'Incarville, B.P. 101, F-27101 Val-de-Reuil Cedex, France.

BEVAN R.E., Cyanamid Webster PTY LDT, P.O. Box 234, 2153 Baulkham Hills NSW, Australia

BEVILACQUA J., Connaught Labs. Ldt., 1755 Steeles Ave West, Willowdale Ont. M2R 3T4, Canada.

BLUMRICH M., Behringwerke AG, P.O. Box 1140, D-35001 Marburg, Germany.

BORRMANN E., Federal Institute for Veterinary Medicine, Naumburger Str. 96a, D-07743 Jena, Germany.

BROCK B.D., Lederle-Praxis, 401 N Middletown Road, Pearl River, New York 10965, USA.

BRÖCKEL M., Behringwerke AG, P.O. Box 1140, D-35001 Marburg, Germany.

BROWN F., IABS, Plum Island Animal Disease Center, P.O. Box 848, Greenport, 11944 New York, USA.

BRUCKNER L., Institute of Virology and Immunoprophylaxis, CH-3147 Mittelhäusern, Switzerland.

BUJANOWSKI-WEBER J., Fresenius AG, Am Haag 7, D-82166 Gräfelfing, Germany.

BURKHARDT F., Med. Chem. Labor Thun, Postfach 126, CH-3608 Thun, Switzerland.

CARUSO A., Lederle-Praxis, 401 N Middletown Road, Pearl River, New York 10965, USA.

CASADAMONT M., Pasteur Merieux S & V, B.P. 101, F-27100 Val-de-Reuil, France.

CASTLE P., European Pharmacopoeia, B.P. 907, F-67029 Strasbourg, France.

CHINO F., National Institute of Health, Gakuen 4-7-1 Musashimurayama 208, Tokyo, Japan.

CHUDY M., Paul-Ehrlich-Institut, Paul-Ehrlich-Str. 51-59, D-63225 Langen, Germany.

CICHUTEK K., Paul-Ehrlich-Institut, Paul-Ehrlich-Str. 51-59, D-63225 Langen, Germany.

CIDDA C., Alfa Biotech S.p.A, Via Castagnetta 7, I-00040 Pomezia (Roma), Italy.

CONNOR N., Evans Medical Ldt., Gaskill Road, Speke, Liverpool Merseyside L24 9GR, United Kingdom.

CUSSLER K., Paul-Ehrlich-Institut, Paul-Ehrlich-Str. 51-59, D-63225 Langen, Germany.

DABBAH R., U.S. Pharmacopoeia, Rockville, Maryland 20852, USA.

DE LEEUW W.A., Veterinary Health Insp., P.O. Box 5406, NL-2280 HK Rüswük, The Netherlands.

DE REE H., Solvay Duphar B.V., P.O. Box 900, NL-1390 DA Weesp, The Netherlands.

DEMPSTER R.P., Mallinckrodt Veterinary NZ Ldt., Private Bag 908, Upper Hutt, New Zealand.

DERKS A.A., Cyanamid Webster PTY LDT, P.O. Box 234, 2153 Baulkham Hills NSW, Australia.

DUCHENE M., SmithKline Beecham Biologicals S.A., 89, rue de l'Institut, B-1330 Rixensart, Belgium.

DULAR U., Bureau of Biologics Health Protection Branch, RM 16 B HPB-BLDG Tunney's Pasture, Ottawa Ontario K1A OL2, Canada.

ENGELMANN H., Sächsisches Serumwerk Dresden, Zirkusstrasse, D-01069 Dresden, Germany.

FABRE I., Agence du Médicament, Direction des Laboratoires et des Contrôles, 14, rue Ecole de Pharmacie, F-34000 Montpellier, France.

FITZGERALD E.A., FDA Center for Biologics, Rockville, Maryland 20852, USA.

FOLKERS C., FEDESA, v.H. Hubarlaan 5, NL 1217 LJ Hilversum, The Netherlands.

GENTY E., Rhône Merieux, 29, Avenue Tony Garnier, F-69007 Lyon, France.

GERDIL C., Pasteur Merieux, 1541 Avenue M. Merieux, F-69280 Marcy L'Etoile, France.

GODDARD R.D., Central Veterinary Laboratory, Woodham Lane, New Haw, Addlestone Surrey KT15 3NB, United Kingdom.

GOLDENTHAL E.A., Connaught Laboratories, 1755 Steeles Ave W., Willowdale, 416 Ontario L0G 1M0, Canada.

GOMMER A.M., RIVM, P.O. Box 1, NL-3720 BA Bilthoven, The Nertherlands.

GOODMAN, S.A., U.S. Department of Agriculture, 6505 Belcrest Road, Hyattsville, MD 20782, USA.

GRÖTSCH W., Labor L+S GmbH, Mangelsfeld 4, D-97708 Bad Bocklet, Germany.

GRUBER F.P., FFVFF-Altex, Biberlinstr. 5, CH-8032 Zürich, Switzerland.

GRUNE-WOLFF B., Zebet, Diedersdorfer Weg 1, D-12277 Berlin, Germany.

GUPTA R.K., Massachusetts Public Health, Biologic Laboratories, 305 South Street, 02130 Boston MA, USA.

HAMPTON T., Evans Medical Ldt., Gaskill Road, Speke, Liverpool Merseyside L24 9GR, United Kingdom.

HANSEN A., National Board of Health, Frederikssundsvej 378, DK-2700 Bronshoj, Denmark.

HANSEN G.A., Statens Seruminstitut, Artillerivej 5, DK-2300 Copenhagen S, Denmark.

HANSPER M., Forschungszentrum Jülich GmbH, PT BEO 21, D-52425 Jülich, Germany.

HANSSON U.T., Swedish Fund for Research without Animal Experiments, Edelcrantzvägen 1, S-12654 Hägersten, Sweden.

HARTMANN H., Bundesamt für Veterinärwesen, Schwarzenburgstr. 161, CH-3097 Bern-Liebefeld, Switzerland.

HARTUNG T., University of Konstanz, Biochemical Pharmacology, D-78434 Konstanz, Germany.

HAUER J.P., National Veterinary Services Lab., P.O. Box 844 Ames, Iowa 50010, USA 515.

HENDRIKSEN C., RIVM, P.O. Box 1, NL-3720 BA Bilthoven, The Nertherlands.

HERDEN P., Procter & Gamble Pharmaceuticals, Dr.-Otto-Röhm-Str. 2-4, D-64331 Weiterstadt, Germany.

HLINAK A., Institut für Virologie Fachbereich Veterinärmedizin Freie Universität Berlin, Luisenstrasse 56, D-10117 Berlin, Germany.

HOFBAUER C., Luitpold Pharma GmbH, Zielstattstr. 9, D-81379 München, Germany.

HOLST J., Dept. of Vaccine NIPH, Geitmyrsveien 75, N-0462 Oslo, Norway.

HORAUD F., IABS, Institut Pasteur, 25, rue du Docteur Roux, F-75724 Paris Cedex 15, France.

HOUGHTON S., Hoechst Animal Health, Walton Manor, Walton Milton Keynes MK7 7AJ, Great Britain.

HUNGERER K.D., Behringwerke AG, P.O. Box 1140, D-35001 Marburg, Germany.

JADHAV S.S., Serum Institute of India Ldt., 212/2, Hadapsar, Pune M.S. 411028, India.

JENNINGS M., Research Animals Department RSPCA, Causeway, Horsham RH12 1 HG, United Kingdom.

JUNGBÄCK C., Paul-Ehrlich-Institut, Paul-Ehrlich-Str. 51-59, D-63225 Langen, Germany.

KLOCKMANN U., Behringwerke AG, Emil-von-Berhing 76, D-35001 Marburg, Germany.

KNIGHT P., Biologicals QA, Bld 114, Wellcome Research Labs, Langley Court, Beckenham, Kent BR3 3BS, United Kingdom.

KOVACS F., Boehringer Ingelheim Vet., Bingerstrasse 173, D-55216 Ingelheim/Rhein, Germany.

KRASSELT M.,RIVM, P.O. Box 1, NL-3720 BA Bilthoven, The Nertherlands.

KREEFTENBERG J.G., Connaught Laboratories, 1755 Steeles Avenue West, Willowdale, Ontario M2R 3T4, Canada.

KURTH R., Paul-Ehrlich-Institut, Paul-Ehrlich-Str. 51-59, D-63225 Langen, Germany

LAMB D., Connaught Laboratories, P.O. Box 187, Swiftwater Pennsylvania, USA.

LEFRANÇOIS S., Agence du Médicament, Direction des Laboratoires et des Contrôles, 14, rue Ecole de Pharmacie, F-34000 Montpellier, France.

LUCKEN R., Veterinary Medicines Directorate, Woodham Lane, New Haw, Addlestone, Surrey KT15 3NB, United Kingdom.

LUND A., Central Veterinary Laboratory, P.O.B. 8156 Dep., N-0033 Oslo, Norway.

MARBEHANT P., SmithKline Beecham Animal Health, 1, Rue Laid Burniat, B-1348 Louvain-La-Neuve, Belgium.

MASSON P., EVIC-CEBA, 48, rue Jean Duvert, F-33290 Blanquefort, France.

METZ B., BMFT, D-53170 Bonn, Germany.

MILSTIEN J.B., World Health Organization, 20 Ave Appia, CH-1211 Geneva 27, Switzerland.

MINOR P., NIBSC, Blanche Lane, South Mimms, Potters Bar Hertfordshire EN6 3QG, United Kingdom.

MOGENSEN H.J., Statens Seruminstitut, Artillerivej 5, DK-2300 Copenhagen S, Denmark.

MONCEF S.M., SmithKline Beecham Bio, 89, rue de l'Institut, B-1330 Rixensart, Belgium.

MONTAGNON B., Pasteur Mérieux Sérums & Vaccins, 1541, Avenue Marcel-Mérieux, F-69280 Marcy L'Etoile, France.

MOOS M., Paul-Ehrlich-Institut, Paul-Ehrlich-Str. 51-59, D-63225 Langen, Germany.

NEUBERT A., Impfstoffwerk Dessau-Tornau GmbH, Postfach 214, D-06855 Rosslau, Germany.

NIEDRIG M., Robert Koch-Institut, Nordufer 20, D-13353 Berlin, Germany.

NOSTITZ D., Regierungspräsidium Dessau, Postfach 87, D-06839 Dessau, Germany.

ÖLANDER R.M., National Public Health Inst., Mannerheimintie 166, SF-00300 Helsinki, Finland.

OSTERHAUS A.D.M.E., Erasmus University, P.O. Box 1738, 3000 DR Rotterdam, The Netherlands.

PICK H., Riemser Tierarzneimittel GmbH, An der Wiek 07, D-17498 Riemserort, Germany.

PROCHAZKOVA S., State Inst. for Drug Control, Srobárova 48, P.O.B. 87, 10041 Prague 10, Czech Republic.

REBER G., Behringwerke AG, P.O. Box 1140, D-35001 Marburg, Germany.

REDHEAD K., NIBSC, Blanche Lane, South Mimms, Potters Bar, Hertfordshire EN6 3QG, United Kingdom.

REED N.E., Central Veterinary Laboratory, Woodham Lane, New Haw, Addlestone, Surrey KT15 3NB, United Kingdom.

RELYVELD E.H., Institut Pasteur, 6, rue du Sergent Maginot, F-75016 Paris, France.

ROBERTS B., Mallinckrodt Veterinary, Breakspear Road, South Harefield, Uxbridge, United Kingdom.

RODENBACH C., Bayer AG, Animal Health, Monheim, Geb. 6210, D-51368 Leverkusen, Germany.

ROMMEL E., SmithKline Beecham Biologicals, rue de l'Institut 89, B-1330 Rixensart, Belgium

ROOIJAKKERS J.M.E., Erasmus University, Dr. Molewaterplein 50, NL-3000 DR Rotterdam, The Netherlands.

ROTH F., Institut für Tropentierhygiene, Kellnerweg 6, D-37077 Göttingen, Germany.

SABOURAUD A., Pasteur Mérieux S.V., 1541 Avenue M. Mérieux, F-69280 Marcy L'Etoile, France.

SANTIROCCO N., Alfa Biotech S.p.A, Via Castagnetta 7, I-00040 Pomezia (Roma), Italy.

SCHÄFFNER G., Paul-Ehrlich-Institut, P.O. Box 1740, D-63207 Langen, Germany.

SCHLITT A., Stiftung zur Einschränkung von Tierversuchen, Kaiser Str. 60, D-55116 Mainz, Germany.

SCHNEIDER B., Med. Hochschule Hannover, Konstanty-Gutschow-Str. 8, D-30625 Hannover, Germany.

SCHOBER-BENDIXEN S., Immuno AG, Uferstr. 15, A-2304 Orth a.d. Donau, Austria.

SCHOLTHOLT J., Bundesverband der Pharmazeutischen Industrie e.V., Karlstr. 2, D-60329 Frankfurt/Main, Germany.

SCHUHMACHER C., Vetoquinol Biotechnologie, rue du Chêne St. Anne, B.P 189, F-70200 Lure, France.

SCHÜTT I.D., Bundesministerium für Gesundheit, Am Probsthof 78A, D-53121 Bonn, Germany.

SCHWANIG M., Paul-Ehrlich-Institut, Paul-Ehrlich-Str. 51-59, D-63225 Langen, Germany.

SESARDIC D., NIBSC, Blanche Lane, South Mimms, Potters Bar, Hertfordshire EN6 3QG, United Kingdom.

SMIDOVA V., Sevac Ldt., Korunni 108, 1010103 Prague, Czech Republic.

SOUTHERN J.A., SAIMR Serum & Vaccine, P.O. Box 28999, Sandringham 2131, South Africa.

SPIESER J.M., European Pharmacopoeia, B.P. 907, F-61029 Strasbourg Cedex 1, France.

STAINER W.D., Stainer Associates, Richmond Hill, 109 Regent St. Ontario L4C 9P3, Canada.

STRAATEN V.D. KAPPELLE I., RIVM, P.O. Box 1, NL-3720 BA Bilthoven, The Nertherlands.

STRAUGHAN D.W., Frame, 29 Kennedy Road, Shrewsbury, Shropshire SY3 1 AB, United Kingdom.

STÜNKEL K., Pitman-Moore GmbH, Im Langen Felde 3-5, D-30938 Burgwedel, Germany.

TAFFS R.E., FDA Center for Biologics Evaluation and Research, Building B, Room 126, 5516 Nicholson Lane, Kensington, Maryland 20895 USA.

THIELE S., Solvay Veterinär GmbH, Karstrasse 70, D-41068 Mönchengladbach, Germany.

TOLLIS M., Instituto Superiore di Sanita', V. le Regina Elena 299, I-00161 Roma, Italy.

VAN DE DONK H., RIVM, P.O. Box 1, NL-3720 BA Bilthoven, The Nertherlands.

VAN DER ARK A., RIVM, P.O. Box 1, NL-3720 BA Bilthoven, The Netherlands.

VAN DER GUN J., RIVM, P.O. Box 1, NL-3720 BA Bilthoven, The Nertherlands.

VAN DER KAMP M., NCA, Yalelaan 17, De Uithof, NL-3584 CL Utrecht, The Netherlands.

VANHOOREN G., Institute for Hygiene and Epidemiology, Juliette Wytsmanstraat 14, B-1050 Brussels, Belgium.

VANNIER P., CNEVA, B.P. 53, F-22440 Ploufragan, France.

VEENSTRA S., SmithKline Beecham, rue de l'Institut 89, B-1330 Rixensart, Belgium .

VERBOVEN A., NV Upjohn SA, Rijksweg 12, B-2870 Puurs, Belgium.

VINCENT-FALQUET J.C., Pasteur Merieux S.V., 1541 Avenue M. Merieux, F-69280 Marcy L'Etoile, France.

VITKOVA E., State Inst. for Drug Control, Srobárova 48, P.O.B. 87, 10041 Prague 10, Czech Republic.

VOGEL I., Serumprüfungsinstitut, Possingergasse 38, A-1160 Wien, Austria.

WEBER C.W.J., J.C.W. Weber & Co Inc., Wymbolwood Beach RRI, Wyevale Ontario L0L 2T0, Canada.

WEBSTER C.J., FEDESA SmithKline Beecham, Place de l'Université 16, B-1348 Louvain -La-Neuve, Belgium.

WERNER G., Berlin-Chemie AG, Glienicker Weg 125, D-12489 Berlin, Germany.

WHITWORTH K., Hoechst Animal Health, Walton Manor, Walton, Milton Keynes MK7 7AJ, United Kingdom.

WINSNES R., The Norwegian Medicines Control Authority, Sven Oftedals vei 6, N-0950 Oslo, Norway.

WOOD D., NIBSC, Blanche Lane, South, Mimms, Potters Bar Hertfordshire EN6 3QG, United Kingdom.

XUEREF C., Pasteur Merieux Sérums & Vaccins, 1541 Avenue M. Merieux, F-69280 Marcy L'Etoile, France.

ZÄNKER S ., FEDESA, rue Defacaz, B-1050 Brussels, Belgium.

ZIGTERMAN G., Intervet International, P.O. Box 31, NL-5830 AA Boxmeer, The Netherlands.

ZÖBISCH H., TAD Pharmazeutisches Werk GmbH, Postfach 720, D-27472 Cuxhaven, Germany.

ZOTT A., Paul-Ehrlich-Institut, Paul-Ehrlich-Str. 51-59, D-63225 Langen, Germany.

Index of Authors

Aggerbeck H.: 297
Akkermans A.M.: 199, 323, 335
Amexis G.: 325
Anderson R.: 283

Balls M.: 11
Beckmann R.: 326
Benscoter K.: 283
Bertling W.M.: 325
Bevilacqua J.M.: 121
Borrmann E.: 327
Brantschen S.: 129
Brechtbühl K.: 175
Bruckner L.: 175
Burkhardt F.: 129

Castle P.: 21
Ceccini D.: 283
Chino F.: 53
Chiu S.W.: 121
Chudy M.: 328
Cichutek K.: 167
Clements H.: 343
Cordioli P.: 147
Cussler K.: 325, 339

Dawson M.: 329
Derks A.: 330
Di Pasquale I.: 147
Di Trani L.: 147
Dittmar M.T.: 167
Dragunsky E.: 345

Ekong T.: 342

Finkel-Jimenez B.: 129
Finkler H.: 157
Fitzgerald E.: 345
Fultz P.: 167

Gaines Das R.: 342
Giess S.: 319
Glück R.: 129
Gommer A.M.: 217
Goodman S.A.: 41
Grachev V.:31
Griffin P.: 283
Griffiths E.: 31
Groen J.: 137
Gupta R.K.: 207, 283
Gyra H.: 324

Haase M.: 319
Hartung T.: 85
Hauer P.J.: 330, 331
Henderson L.M.: 332, 333
Hendriksen C.F.M.: 3, 199, 271, 323, 335, 344
Henry L.A.: 331
Heron I.: 297
Hioki K.: 345
Hofmann M.A.: 175
Huet M.: 225

Jadhav S.S.: 245
Jenkins P.S.: 334
Jungbäck C.: 157

Katz J.B.: 337
Knight P.A.: 185
Kolbe D.R.: 336
Kreeftenberg J.G.: 121
Kurth R.: 167
Kusch M.: 339

Lery L.: 225
Levenbook I.: 345
Löwer J.: 325, 328
Lucken R.N.: 67, 97
Ludemann L.R.: 337

Marsman F.R.: 323, 335
McLellan K.H.: 342
Milstien J.: 31
Minor P.D.: 113
Montanez-Ortiz L.: 283

Nomura T.: 345

Oelander R.M.: 337, 338
Osterhaus A.: 137

Padilla A.: 31

Redhead K.: 263
Relyveld E.H.: 225
Roberts B.: 97
Rooijakkers E.: 137
Rosenbusch R.F.: 332
Rost B.: 283
Rübmann P.: 339
Ruby K.W.: 339, 340

Sauer A.: 85
Schäffner G.: 319
Schneevel K.F.: 332, 333
Schulze F.: 327
Schwanig M.: 349
Sesardic D.: 311, 342
Shinde Y.P.: 245
Siber G.R.: 207, 283
Southern J.: 343
Sparkes J.D.: 121
Spenceley L.: 343
Stainer D.W.: 303
Staüber N.: 175
Straughan D.W.: 11, 65

Taffs R.: 345
Tollis M.: 147

Uittenbogaard J.: 137

Van de Donk H.: 199, 271, 323
Van der Ark A.: 271, 344
Van der Gun J.: 199, 335, 344
Van der Kamp M.D.O.: 73
Van Herwijnen J.: 137
Van Straaten-van de Kappelle I.: 271, 344
Verschueren C.: 49
Verschure M.H.: 335
Vignolo E.: 147
Vogel I.: 346

Wagener S.: 167
Walden D.M.: 333, 334, 341
Wannemuehler M. J.: 341
Webster C.J.: 103
Weeke-Lüttmann M.: 328
Wendel A.: 85
Whitaker M. S.: 331
Whitworth K.: 347
Woltjes J.: 335
Wood D.J.: 79

Xu J.: 283

Young L.: 121

Zänker S.: 49
Zott A.: 346, 349

New insights into the genetic basis of antibiotic resistance and pathogenicity

Genetics of Streptococci, Enterococci and Lactococci

Editors: Joseph J. Ferretti, Michael S. Gilmore, Todd R. Klaenhammer, Fred Brown

This comprehensive volume, based on an international conference organized by the American Society of Microbiology and the International Association of Biological Standardization, brings together the most up-to-date information concerning the genetics and molecular biology of streptococci, enterococci and lactococci. These Gram-positive bacteria have attracted increasing interest following recent outbreaks of toxic systemic disease involving streptococci, outbreaks of panresistant enterococci, and advances in the design and engineering of novel antimicrobials based on lactococcal products. The framework of discussion is established by the keynote lecture on the origins of streptococcal genetics, presented by Maclyn McCarty, one of the co-discoverers of DNA as the transforming principle. The following 92 papers, grouped into 5 lecture sessions, discuss recent findings and developments regarding gene transfer and antibiotic resistance, the genetics of pathogenic streptococci, oral streptococci, lactococci and the genetics of bacteriocin production. Drawing together a large amount of new data, this book will be essential reading for microbiologists, geneticists, biotechnologists, dental researchers and epidemiologists.

Selected Contents

Keynote Lecture
McCarty, M.: Origins of Streptococcal Genetics

Session I: Gene Transfer and Antibiotic Resistance
Clewell, D.B. et al.: The Conjugative Transposon Tn*916* of *Enterococcus faecalis:* Structural Analysis and Some Key Factors Involved in Movement
Wirth, R. et al.: Evolution of the *Enterococcus faecalis* Sex Pheromone System
Wirth, R.; Marcinek, H.: In vivo Gene Transfer by *Enterococcus faecalis*
Clermont, D. et al.: Old and New (Tn*3708*) Mobile Chromosomal Elements in Streptococci and Enterococci
Sahm, D.F.; Gilmore, M.S.: High-Level Gentamicin Resistance Among Enterococci
Rice, L.B.; Murray, B.E.: β-Lactamase Producing Enterococci

Session II: Pathogenic Streptococci
Gibson, C. et al.: Regulation of Host Cell Recognition in *Streptococcus pyogenes*
Podbielski, A. et al.: Molecular Characterization of the M Type 49 Group A Streptococcal (GAS) *virR* Gene
Hollingshead, S.K.; Bessen, D.E.: Evolution of the *emm* Gene Family: Virulence Gene Clusters in Group A Streptococci
Malke, H. et al.: The Streptokinase Gene: Allelic Variation, Genomic Environment, and Expression Control

Session III: Oral Streptococci
Kuramitsu, H. et al.: Analysis of Glucan Synthesis by *Streptococcus mutans*
Jacques, N.A.: Extracellular Sucrose Metabolism by *Streptococcus salivarius*
Burne, R.A. et al.: Regulation of Fructan Degradation by *Streptococcus mutans*
Zuccon, F.M.: Physical Map of *Streptococcus mutans* GS-5 Chromosome

Session IV: Lactococci
Davidson, B.F. et al.: Genomic Organization of Lactococci
Godon, J.-J. et al.: Molecular Analysis of the *Lactococcus lactis* Sex Factor
Israelsen, H. et al.: Environmentally Regulated Promoters in Lactococci
Miereau, I. et al.: Genetics of Peptide Degradation in Lactococcus: Gene Cloning, Construction and Analysis of Peptidase Negative Mutants
Leenhouts, K.J.: Integration Strategies and Vectors

Session V: Bacteriocins
De Vos, W.M. et al.: Genetics of the Nisin Operon and the Sucrose-Nisin Conjugative Transposon Tn*5276*
Simpson, W.J. et al.: A Lantibiotic Gene Family Widely Distributed in *Streptococcus salivarius* and *Streptococcus pyogenes*
Nes, F. et al.: Genetics of Non-Lantibiotics Bacteriocins

Developments in Biological Standardization, Vol. 85
Edited by the International Association of Biological Standardization
ISSN 0301–5149

Fields of Interest: Immunology; Microbiology; Genetics, Pharmacology, Biotechnology, Dental Medicine, Epidemiology

Genetics of Streptococci, Enterococci and Lactococci
4th International Conference Organized by the American Society for Microbiology and the International Association of Biological Standardization, Santa Fee, N.M., May 1994
Editors: Ferretti, J.J.; Gilmore, M.S. (Oklahoma City, Okla.); Klaenhammer, T.R. (Raleigh, N.C.); Brown, F. (Greenport, N.Y.)
XIV + 666 p., 201 fig., 48 tab., soft cover, 1995
CHF 420.– / DEM 503.– / USD 365.25
Prices subject to change
DEM price for Germany, USD price for USA only
ISBN 3–8055–6207–1

KI 96226

KARGER

Symposia Series in Immunobiological Standardization

Vol. 1: Talloires 1965 Rabies
Vol. 2: Munich 1965 Neurovirulence of viral vaccines
Vol. 3: Stockholm 1965 Biotechnical developments in bacterial vaccine production
Vol. 4: Royaumont 1965 Immunological methods of biological standardization
Vol. 5: London 1966 Laboratory animals
Vol. 6: Utrecht 1966 Adjuvants of immunity
Vol. 7: Marburg/L. 1967 Assay of combined antigens
Vol. 8: Lyon 1967 Foot-and-Mouth Disease: Variants and immunity
Vol. 9: Paris 1967 Pseudotuberculosis
Vol. 10: London 1967 Biological assay methods of vaccine and sera
22nd Symp., London 1968 Standardization of immunofluorescence. Ed. Blackwell
Vol. 11: London 1968 Rubella vaccines
Vol. 12: Tunis 1968 Brucellosis. Standardization and control of vaccines and reagents
Vol. 13: Utrecht 1969 Pertussis vaccine
Vol. 14: London 1969 Standardization of interferon and interferon inducers
Vol. 15: Berne 1969 Enterobacterial vaccines
Vol. 16: Versailles 1970 Anti-lymphocytic serum
29th Symp., Brighton 1969 Hazards of handling simians. Ed. Lab. Animals Ltd. (4)
Vol. 17: Frankfurt 1970 BCG vaccine
Vol. 18: Copenhagen 1972 HL-A reagents
Vol. 19: Utrecht 1972 Smallpox vaccines
Vol. 20: London 1972 Influenza
Vol. 21: Lyon 1972 Rabies II
Vol. 22: Monaco 1973 The use of vaccines

Developments in Biological Standardization

Vol. 23: Madrid 1973 Sterilization and sterility testing of biological substances
Vol. 24: San Francisco 1973 Preservatives in biological products
Vol. 25: Lyon 1973 Requirements for poultry virus vaccines
Vol. 26: Budapest 1973 Selected veterinary vaccines
Vol. 27: Budapest 1973 Purity of human plasma proteins
Vol. 28: London 1974 Immunity to infections of the respiratory system in man and animals
46th Symp., London 1974 Breeding of simians and their uses in developmental biology. Ed. Lab. Animals Ltd. (6)
Vol. 29: Geneva 1974 (WHO/IABS) Standardization and control of allergens administered to man
Vol. 30: Milan 1974 Viral hepatitis
Vol. 31: Rabat 1975 Brucellosis (II)
Vol. 32: Paris 1975 (OIE/IABS) Colstridial sera and vaccines for veterinary use
Vol. 33: Douglas 1975 Vaccination by the non-parenteral route
Vol. 34: Budapest 1976 Pyrogenicity, innocuity and toxicity test systems for biological products
Vol. 35: Lyon 1976 Foot-and-mouth disease vaccines
Vol. 36: Lyon Washington D.C. 1976 Freeze-drying of biological products
Vol. 37: Geneva 1976 Standardization of cell substrates for the production of virus vaccines

Vol. 38: London 1977 Biological preparations in the treatment of cancer
Vol. 39: Geneva 1977 Influenza immunization (II)
Vol. 40: Marburg 1977 Standardization of rabies vaccines (III) (WHO/IABS)
Vol. 41: Guadeloupe 1978 Vaccinations in the developing countries
Vol. 42: Paris 1978 Second General Meeting of European Society of Animal Cell Technology
Vol. 43: Brussels 1978 Immunization: Benefit versus risk factors
Vol. 44: Villars 1979 Test methods for the quality control of plasma proteins
Vol. 45: San Antonio 1979 The Standardization of animals to improve biomedical research, production and control
Vol. 46: Oxford 1979 Third General Meeting of European Society of Animal Cell Technology
Vol. 47: Bilthoven 1980 Reassessment of inactivated poliomyelitis vaccine
Vol. 48: Geneva 1980 Standardization of albumin, plasma substitutes and plasmapheresis
Vol. 49: Leetown 1981 Fish biologics: Serodiagnostics and vaccines
Vol. 50: Heidelberg 1981 Joint ESACT-IABS Meeting on the use of heteroploid and other cell substrates for the production of biologicals
Vol. 51: Lyon 1981 Immunization of adult birds with inactivated oil adjuvant vaccines
Vol. 52: Lyon 1981 Herpes virus of man and animal: Standardization of immunological procedures
Vol. 53: Dublin 1982 Enteric infections in man and animals: Standardization of immunological procedures
Vol. 54: Athens 1982 Viral hepatitis: Standardization in immunoprophylaxis of infections by hepatitis viruses
Vol. 55: Copenhagen 1982 Fifth General Meeting of European Society of Animal Cell Technology
Vol. 56: Algeria 1983 IIIrd International Symposium on Brucellosis
Vol. 57: Paris 1983 Monoclonal antibodies: Standardization of their characterization and use
Vol. 58: Budapest 1983 BCG vaccines and tuberculins
Vol. 59: Geneva 1983 Standardization and control of biologicals produced by recombinant DNA technology
Vol. 60: Gardone Riviera 1983 Production and exploitation of existing and new animal cell substrates
Vol. 61: Geneva 1984 Proceedings of the Fourth International Symposium on pertussis
Vol. 62: Stockholm 1985 Diagnostics and vaccines for parasitic diseases
Vol. 63: San Francisco 1985 Use and standardization of chemically defined antigens
Vol. 64: London 1985 Reduction of animal usage in the development and control of biological products
Vol. 65: Amsterdam 1985 Use and standardization of combined vaccines
Vol. 66: Baden 1985 Advances in animal cell technology: cell engineering, evaluation and exploitation
Vol. 67: Melbourne 1986 Standardization in blood fractionation including coagulation factors
Vol. 68: Geneva 1986 Cells – Products Safety
Vol. 69: London 1987 Cytokines: Laboratory and clinical evaluation
Vol. 70: Arlington, 1988 Continuous cell lines as substrates for biologicals
Vol. 71: Utrecht, 1989 Monoclonal antibodies for therapy, prevention and *in vivo* diagnosis of human disease
Vol. 72: Annecy, 1989 Progress in animal retroviruses
Vol. 73: Shizuoka, 1990 Pertussis: Evaluation and research on acellular pertussis vaccines
Vol. 74: Bethesda, MD, 1990 Biological product freeze-drying and formulation
Vol. 75: London, 1990 Virological aspects of the safety of biological products
Vol. 76: Bethesda, MD, 1991 Continuous cell lines – Current issues
Vol. 77: Annecy, 1991 Standardization of the immunopharmacology of natural and synthetic immunomodulators
Vol. 78: Bethesda, MD, 1991 Poliovirus attenuation: Molecular mechanisms and practical aspects
Vol. 79: Ploufragan, 1992 The first steps towards an international harmonization of veterinary biologicals: 1993 and free circulation of vaccines within the E.E.C.
Vol. 80: Heidelberg, 1992 Transmissible spongiform encephalopathies-Impact on animal and human health.
Vol. 81: Cannes, 1992 Virological safety aspects of plasma derivatives.
Vol. 82: Albany, NY, 1993 Recombinant vectors in vaccine development.
Vol. 83: Annecy, 1993 Genetic stability and recombinant product consistency.
Vol. 84: Langen, 1993 Non-target effects of live vaccines.
Vol. 85: Santa Fee, NM, 1994 Genetics of streptococci, enterococci and lactococci

INTERNATIONAL ASSOCIATION OF BIOLOGICAL STANDARDIZATION

WHAT IS THE I.A.B.S. ?

The International Association of Biological Standardization (IABS) was founded in Lyons in 1955 by a group of independent experts who identified an urgent need for an improvement in the quality and comparability of the data being exchanged between scientists working in the research, development, production, standardization and regulation of biological products. To this end, IABS aims to bring together state controllers, manufacturers and research workers from academic institutions and public health organisations.

The IABS is a branch of the International Union of Microbiological Societies and has about 500 members in 50 countries. Every four years, its members elect the Council by postal ballot and its representatives, in their turn, elect the officers.

The duty of the Council is to organise meetings on subjects of contemporary interest in biological products being used in human and veterinary medicine and to promote the editorial activity of the IABS by publishing: i) the proceedings of the international meetings in the symposium series *Developments in Biological Standardization* published by S. Karger, ii) *Biologicals,* a major international journal published quarterly by Academic Press (London), and iii) the *Newsletter* which reports on current IABS activities.

The Council also supports the objectives of the major international public health programmes and organisations, and particularly the World Health Organization.

In 1994, the Council decided to establish an IABS Task Force on Vaccines to provide a focus for our activities in the field of vaccination and immunisation. The role of the Task Force is to foster communication between vaccine manufacturers, scientific research workers and regulatory authorities concerning the quality control, standardization and efficacy of vaccines and vaccine combinations and the exchange of information on the development of new improved vaccines. A major objective of the Task Force will be the organisation of international meetings and the promotion of high quality publications on vaccines in our journal *Biologicals* and in the *Developments in Biological Standardization.*

The Task Force has a membership of 30 leading international experts on vaccination and immunisation working together to identify the major topics in the field of vaccines that need international debate and cooperation.

INTERNATIONAL ASSOCIATION OF BIOLOGICAL STANDARDIZATION

WHY NOT BECOME A MEMBER OF THE IABS?

As a member of the IABS, it is possible to contribute directly to the activity of our Association and to support the research, development, standardization and regulation of biological products. By your membership you would:

- receive our quarterly *Newsletter* which gives information on the activities of the IABS and highlights international news of current interest;
- subscribe to our journal *Biologicals*, published quarterly by Academic Press (London), at a reduced rate;
- take advantage of the discount on booksellers' prices for the volumes in the symposium series *Developments in Biological Standardization*, published by Karger;
- participate fully in our international meetings, with a reduction in the registration fees.

If you are not already a member of our Association and wish to join us, please complete this form and send it to our permanent office:

BIOSTANDARDS
P.O.B. 456
78, av. de la Roseraie
CH-1211 Geneva 4
Switzerland
Fax: (41) 22 702 93 55

or hand it to the Secretariat, if attending a Meeting.

- -

Surname...First name(s)...

Scientific titles...Professional position..

Complete mailing address..

....

Field of expertise...

Tel.. Fax..

Annual fee: Sfr. 170.-, which includes membership, subscription to *Biologicals* and the quarterly *Newsletter* of the IABS;
or Sfr. 90.-, without the subscription to *Biologicals*.

Method of payment:

❍ I have given instructions to my bank to make a credit transfer to the Swiss Bank Corporation, P.O.B. CH-1211 Geneva 11, Switzerland, a/c Biostandards n° 131,652.0.
❍ I enclose my cheque for Sfr................................ made payable to Biostandards
❍ I authorise you to charge Sfr................... to my card: VISA / MASTERCARD / AM. EXPRESS

N°.. Expiry date........................

Signature...

ACHEVÉ D'IMPRIMER SUR LES PRESSES DE L'IMPRIMERIE MÉDECINE ET HYGIÈNE À GENÈVE (SUISSE) FÉVRIER 1996